Beer: A Quality Perspective

A Volume of the Handbook of Alcoholic Beverages Series

Edited by
Charles W. Bamforth

Series editors:
Inge Russell, Charles W. Bamforth and Graham Stewart

AMSTERDAM • BOSTON • HEIDELBERG • LONDON • NEW YORK • OXFORD
PARIS • SAN DIEGO • SAN FRANCISCO • SINGAPORE • SYDNEY • TOKYO
Academic Press is an imprint of Elsevier

Academic Press is an imprint of Elsevier
30 Corporate Drive, Suite 400, Burlington, MA 01803, USA
525 B Street, Suite 1900, San Diego, California 92101-4495, USA
32 Jamestown Road, London NW1 7BY, UK

Copyright © 2009 Elsevier Inc. All rights reserved

No part of this publication may be reproduced, stored in a retrieval system or transmitted in any form or by any means electronic, mechanical, photocopying, recording or otherwise without the prior written permission of the publisher

Permissions may be sought directly from Elsevier's Science & Technology Rights Department in Oxford, UK: phone (+44) (0) 1865 843830; fax (+44) (0) 1865 853333; email: permissions@elsevier.com. Alternatively you can submit your request online by visiting the Elsevier web site at http://elsevier.com/locate/permissions, and selecting Obtaining permission to use Elsevier material

Notice
No responsibility is assumed by the publisher for any injury and/or damage to persons or property as a matter of products liability, negligence or otherwise, or from any use or operation of any methods, products, instructions or ideas contained in the material herein. Because of rapid advances in the medical sciences, in particular, independent verification of diagnoses and drug dosages should be made

British Library Cataloguing in Publication Data
A catalogue record for this book is available from the British Library

Library of Congress Cataloguing in Publication Data
A catalogue record for this book is available from the Library of Congress

ISBN   978-0-12-669201-3

For information on all Academic Press publications
visit our website at www.elsevierdirect.com

Typeset by Charon Tec Ltd., A Macmillan Company. (www.macmillansolutions.com)

Printed and bound in the USA
Transferred to Digital Printing, 2013

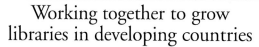

Contents

Series editors	*vii*
List of contributors	*ix*
Preface	*xi*

1 Beer foam: achieving a suitable head — 1
D. Evan Evans and Charles W. Bamforth

Introduction	1
Beer foam physics	3
Foam measurement	8
Beer components that influence foam quality	12
Proteins	13
Non-starch polysaccharides	21
Hop acids	22
Cations	25
Lipids	25
Other foam negative/positive beer constituents	26
Manipulating the brewing process to optimize foam quality	28
Summary	48
Acknowledgments	48
References	48

2 Beer flavor — 61
Paul Hughes

Introduction	61
The flavor unit	61
Brewing raw materials and beer flavor	62
Impact of beer production processes	72
In-pack flavor changes	78
Taints and off-flavors	80
Holistic flavor perception	81
Summary	82
References	82

3 The flavor instability of beer — 85
Charles W. Bamforth and Aldo Lentini

- Factors impacting the shelf life of beer — 88
- The impact of temperature — 95
- The chemistry of flavor change in beer — 96
- An evaluation of processes from barley to beer in the context of flavor instability — 100
- References — 105

4 Colloidal stability of beer — 111
Kenneth A. Leiper and Michaela Miedl

- Summary — 111
- Biological stability — 112
- Importance of whole process to ensure stability — 112
- Chill haze — 114
- Protein in beer — 114
- Beer polypeptides and their functions — 116
- Identifying polypeptides by size — 117
- Identifying polypeptides by amino acid composition — 125
- Identifying polypeptides by hydrophobicity — 130
- Summary of polypeptides — 132
- Polyphenols — 134
- Methods of detecting polyphenols — 135
- Polyphenols in beer — 135
- Haze-forming reactions between polypeptides and polyphenols — 138
- Stabilization treatments — 140
- Combined stabilization system (CSS) — 150
- Other treatments — 150
- Other non-biological hazes — 151
- Testing the effectiveness of beer stabilization — 152
- Haze identification — 154
- References — 154

5 Microbiological stability of beer — 163
Anne E. Hill

- Overview of microbial spoilage — 163
- Outline of the brewing process — 164
- Raw materials — 165
- Wort — 169
- Fermentation — 169
- Storage and finishing — 170
- Packaging and packaged beer — 170
- Dispense — 171
- Detection — 171
- Improving microbiological stability — 178
- Quality control — 180
- References — 181

6 Beer gushing	**185**
Leif-Alexander Garbe, Paul Schwarz and Alexander Ehmer	
Introduction	185
Terminology	186
Physical background of gushing	187
Primary gushing	188
Secondary gushing	201
Summary	204
References	205
7 Beer color	**213**
Thomas H. Shellhammer	
Color perception	213
Measuring color	214
Standard methods for measuring beer color	218
Origins of beer color	221
References	226
8 Beer and health	**229**
Charles W. Bamforth	
Atherosclerosis	232
Hypertension and stroke	235
The digestive system	235
The reproductive system	237
Brain and cognitive function	237
Kidney and urinary tract	239
Age	239
Cancer	240
Allergy	241
References	242
Appendix Practicalities of achieving quality	**255**
Charles W. Bamforth	
Definitions of quality	255
Responsibility for quality	256
Quality systems	256
Quality assurance versus quality control	257
Specifications	258
The cost of quality	258
Statistical process control	259
Process capability	261
Control charts	261
Standard methods of analysis	262
Setting specifications and monitoring performance	270
Hazard analysis critical control points (HACCP)	271
Index	279

Series editors

Inge Russell PhD, DSc, FIBD, FIBiol is the editor-in-chief of the *Journal of the Institute of Brewing*, a visiting professor at Heriot-Watt University, Edinburgh, Scotland, an adjunct professor in the Department of Biochemical Engineering, University of Western Ontario, Canada, and director of the Alltech PhD Program. She has over 32 years experience in the fermentation industry as a research scientist and later directing the Research and Development for Labatt Brewing Company, Canada. She is the author of numerous books and papers as well as co-editor of *Critical Reviews in Biotechnology*.

Charles W. Bamforth PhD, DSc, FIBD, FIBiol, FIAFoST is currently the chair of the Department of Food Science and Technology and the Anheuser-Busch Endowed Professor of Malting and Brewing Sciences at the University of California, Davis. He has held positions as director of research for Brewing Research International and as research and quality assurance manager for Bass Breweries in the UK. He is the author of numerous books and papers as well as being the editor-in-chief of the *Journal of the American Society of Brewing Chemists*.

Graham G. Stewart PhD, DSc, FIBD, FIBiol is emeritus professor of brewing at Heriot-Watt University, Edinburgh, Scotland. Until 2007, he was professor and director of the International Centre for Brewing and Distilling (ICBD) at Heriot-Watt University. He previously held the position of technical director responsible for research and development, quality assurance and technical training at the Labatt Brewing Company, Canada. He is the author of numerous books and papers as well as being co-editor of *Critical Reviews in Biotechnology*.

List of contributors

Charles W. Bamforth
Department of Food Science and Technology, University of California, Davis, CA 95616-8598
Tel.: +1 530-752-9476
E-mail: cwbamforth@ucdavis.edu

Alexander Ehmer
Technical University Berlin, Research Institute for Chemical Technical Analyses, Seestrasse 13, 13353 Berlin
Tel.: +49 30 314-27562

D. Evan Evans
Tasmanian Institutite of Agricultural Research University of Tasmania
Private Bag 54 Hobart Tas 7001 Australia
Tel.: +61-(0)3-6226-2638; +61-(0)3-6226-2642
E-mail: eevans@postoffice.utas.edu.au

Leif-Alexander Garbe
Technical University Berlin, Research Institute for Chemical Technical Analyses, Seestrasse 13, 13353 Berlin
Tel.: +49 30 314-27562
E-mail: Leif-a.garbe@gmx.de; Leif-A.Garbe@tu-berlin.de

Anne E. Hill
International Centre for Brewing and Distilling, School of Life Sciences, John Muir Building, Heriot Watt University, Riccarton Campus, Edinburgh, Midlothian, EH14 4AS
E-mail: A.Hill@hw.ac.uk

Paul Hughes
International Centre for Brewing and Distilling School of Life Sciences, Heriot-Watt University, Edinburgh, EH14 4AS, UK
Tel: +44 131 451 3183
E-mail: P.S.Hughes@hw.ac.uk

Kenneth A. Leiper
International Centre for Brewing and Distilling School of Life Sciences, Heriot-Watt University, Edinburgh, EH14 4AS, UK
Tel.: +44 131 451 3183
E-mail: kaleiper@yahoo.com

Aldo Lentini
Foster's Australia,
National Supply Chain Operations, Beverages: Capability and Improvement,
4-6 Southampton Crescent, Abbotsford, Victoria, Australia 3067
Tel.: (+61 3) 9420 6506
E-mail: LentiniA@fostersgroup.com

Michaela Miedl
The International Centre for Brewing and Distilling (ICBD), School of Life Sciences, Heriot-Watt University, Riccarton EH14 4AS, Edinburgh, Scotland, UK
Tel.: +44-131-451-3467
E-mail: mm60@hw.ac.uk; michaela.miedl@coorsbrewers.com

Paul Schwarz
Barley and Malt Quality, Department of Plant Sciences, North Dakota State University, Fargo, ND 58105
Tel.: 701 231-7732
E-mail: Paul.Schwarz@ndsu.edu

Thomas H. Shellhammer
Department of Food Science and Technology, Oregon State University, 100 Wiegand Hall, Corvallis, OR 97331-6602
Tel.: 541.737.9308
E-mail: Tom.shellhammer@oregonstate.edu

Preface

Recent years have seen the emergence of several excellent books addressing the science and technology of brewing. We felt it important not to produce yet another volume of that type. Instead, we have approached matters somewhat differently. We have focused on the quality of beer and the impact of raw materials and processing on each of the ways by which beer is perceived by the consumer. The ultimate requirement for beer, as for any foodstuff, is its acceptability in the eyes, mouth and mind of the drinker.

Such an approach has once before been taken, in *Essays in Brewing Science* by Michael Lewis and myself (Springer, 2006). The difference is that, in the present work, the diverse authors were asked to "dig deep" into the richness and depth of research that has been devoted to the various quality attributes. On this basis, the aim has been to provide a thorough and extensively referenced treatment of all facets of beer appearance, flavor, stability and wholesomeness.

Upon pouring a beer into the glass, the drinker will make judgments on the acceptability of the product, based solely on what his or her eyes are seeing. We start the journey through the diverse quality criteria with Evan Evans and I considering the factors that determine foam stability. Inadequate foam performance leads drinkers to conclude that the beer is of inferior quality.

Of course, foaming can be done to excess, to the extreme that the contents of a bottle or can may spontaneously spew out when the container is opened. The factors that trigger this unacceptable defect are addressed at length by Leif-Alexander Garbe, Paul Schwarz and Alexander Ehmer.

Having poured the beer successfully into the glass, with the desired amount of foam, the consumer is likely next to scrutinize the clarity of the liquid. Does it have the brilliance demanded – or, in a beer that is necessarily cloudy (such as a hefeweissen), is the extent of turbidity what is expected? Kenneth Leiper and Michaela Miedl cover in detail the impact that raw materials, process and product handling have on haziness.

All beers, of course, have their specific color, whether it is the blackest stout or the palest lager. As for all other facets of product character, the color is impacted by raw materials and the manner by which they are dealt with in the production of beer. Thomas Shellhammer goes deep in his appreciation of this remarkably complex topic.

At last, having critically judged the product for every manifestation of its appearance, the consumer is prepared to raise the glass of beer to be smelled and tasted. Does it taste right? Does it delight? Is it moreish? A myriad of chemical species combine to determine the rich complexity of beer aroma and taste and Paul Hughes walks us through them and how their levels in beer are achieved consistently.

Beer flavor, though, is not static. Most beers are never better than when first brewed and packaged and the longer they are stored before consumption, the less and less appealing they become. Aldo Lentini and I explain the diverse changes that can take place and how they can be minimized.

Whilst most of these flavor changes are due to chemical reactions, any microbial contamination of beer will also lead to taste deterioration, as well as other quality defects, such as turbidity. Microbial contamination during the malting and brewing processes also has serious negative consequences. Anne Hill leads us through the breadth of microbiological concerns.

And so the consumer has hopefully been delighted with a beer that has looked good and tasted good. The question is: Might it have actually *done* them some good? Increasingly the evidence is … yes … and I close out the volume by digging deep into the rich pool of evidence for beer's worthy role as a legitimate component of the diet.

The authors are internationally acclaimed experts in their fields, with representation from great schools in Germany, the United Kingdom, Australia and the United States. They have been brought together to deliver what we feel to be the most detailed and extensively referenced volume about beer quality yet written. The aim has also been to make it highly readable and enjoyable. As such, then, we expect it to prove an indispensable reference tome for researchers in the field and for brewers everywhere needing a one-stop resource for all facets of beer quality. But more, we expect it to appeal to anyone curious about the myriad of factors which make beer the outstanding beverage that it is.

Professor Charlie Bamforth PhD, DSc, FIBD, FIBiol, FIAFoST
California, USA

1

Beer foam: achieving a suitable head

D. Evan Evans and Charles W. Bamforth

Introduction

The aesthetic of drinking beer is to an extent subliminal. The presentation of the beer in the glass in terms of its foam head, clarity/brilliance and color conjure Pavlovian anticipation for the perceptive drinker. There is no disputing the logic that "a beer drinker drinks as much with his or her eyes as with their mouth" (Bamforth et al., 1989). Other chapters in this book will offer insights into flavor, flavor stability, colloidal stability, color and the wholesomeness of beer.

That foam is perhaps one of the most appealing beer qualities is perhaps not surprising since the foam acts as an efficient gas exchange surface pitching aromas towards the drinkers olfactory sensors (Delvaux et al., 1995). As such, it provides a drinker's first tantalizing entrée as to the quality of the beer's flavor, freshness, refreshingness and wholesomeness. Foam is also tactile to the lips and impacts mouthfeel through its stability and its structure (bubble size). In part, this experience is modulated by the degassing of the beer in the mouth, which in turn is a function of beer carbonation/nitrogenation (Todd et al., 1996).

Some may counter, "most beer is drunk from a bottle." Well, yes it is, and beer bottles are often attractive and convenient in their own right, although the drinker is largely missing out on the flavor cues extolled above. Bottles being glass are not opaque, particularly with the current fashion for clear and green varieties, so that they allow the drinker to clearly view foam formed as a result of the drinking action. In Belgium in particular, with branding of glassware, glass shape and material (Delvaux et al., 1995), the glass has become an art form that almost supplants the bottle. Lastly, in attempting to compare beer with wine, in at least more up market settings, it would certainly be considered to be passé or uncouth to drink wine from the bottle!

No brewer can afford to have their carefully crafted company or brand image downgraded by poor customer experiences. Given that the budget line for most brewers in the traditional beer countries is being expanded primarily by the provision of high margin premium beers, drinkers are perhaps more likely to pour the contents of the bottle in to a glass to savior the benefits outlined above. If not from direct experience, "alpha-beer drinkers" who do pour their beer into a glass or customers who have purchased their beer dispensed off tap (draught) may prejudice their colleagues' brand perception based on perceived excellent or poor experiences. Thus insurance of an excellent experience for these influential drinkers and occasions is paramount in maintaining brand appeal.

But what are the features of good foam quality? Typically, this is defined by a combination of its stability, quantity, lacing (glass adhesion or cling), whiteness, "creaminess" (small homo-disperse bubbles) and strength. Here "beauty" is definitely in the eye of the beholder as consumers discriminate between beers based on their foam characteristics. These choices have been found to diverge between genders, race or even region (Bamforth, 2000a; Smythe et al., 2002).

Beer drinker 1 (a Belgian?)		
"Lively beer with good head"	"Flat beer"	"Great lacing, top beer!"
Beer drinker 2 (a Londoner?)		
"Ripped off, foam is not beer!"	"That is more like it, a full measure of beer"	"A dirty glass"
Beer drinker 3 (a Lady?)		
"Will this foam stick to my lip and wreck my make-up?"	"Not very lively and appealing beer"	"This glass has not been properly cleaned before filling"

Figure 1.1
The perceptions of three different drinkers of beer foam quality.

Bamforth (2000a) concluded that "there is a divide between consumers who like to see stable (but not excessive) head or foam but a clean glass at the end of drinking and those who favour a lacing pattern on the glass." More recently, it was demonstrated that men generally rate foam lacing higher than do women (Roza et al., 2006). In addition, a female colleague pointed out the "obvious." Some women tend to be adverse to foam because the thought of its adherence to their lips is unsettling, in that it could spoil their carefully applied make-up. Overall, these quandaries can perhaps be simply outlined pictorially as shown in Figure 1.1.

It is the intention of this chapter to comprehensively appraise the science underpinning beer foam quality. The discussion will begin by considering the basics of foam physics and how these principles can be applied to the task of measuring foam quality then extend to the basic biochemical constituents such as protein species and hop acids that combined determine foam quality. This holistic understanding is aimed at providing brewers with the best possible knowledge so that they may manipulate the raw materials, process and options for delivery to consistently produce the optimal foam quality that their targeted customers demand.

Beer foam physics

Elementary to an understanding of beer foam quality are the principles of beer foam physics that underlie the interaction of the various beer components, dispense and finally customer presentation, be it in a glass or some other container. The principles concerned have been comprehensively described by Dr Albert Prins' group of Wageningen (Prins and Marle, 1999; Prins, 1988; Ronteltap, 1989; Ronteltap et al., 1991), and by others such as Walstra (1989), Fisher et al. (1999) and Bamforth (2004a). These authors have simplified the somewhat complex physics involved into the following fundamental but interrelated events:

- Bubble formation and size
- Drainage
- Creaming (bubble rise or beading)
- Coalescence
- Disproportionation

Bubble formation and size

Despite beer being supersaturated with carbon dioxide, bubbles will not form spontaneously unless nucleation occurs (Figure 1.2), promoted by a particle, fiber or scratch in the glass (Prins and Marle, 1999) or the dispense mode, be that tap (Carroll, 1979) or bottle (Skands et al., 1999). These nucleation sites should ideally be small to create smaller bubbles that create foam that is most appealing to the drinker (Bamforth, 2004a). A desirable attribute of nitrogenated beers, due to the lower partial pressure of nitrogen gas compared to CO_2, is the production of much smaller bubbles (Carroll, 1979; Fisher et al.,

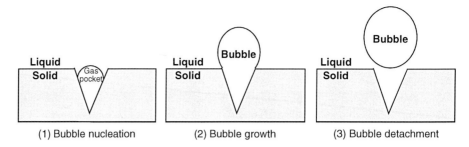

Figure 1.2
Schematic representation of the bubble formation sequence (after Ronteltap et al., 1991).

1999). Such principles are applied in the use of nucleated glassware such as the "headkeeper" style (Parish, 1997) or as a partial function of widgets (Brown, 1997; Browne, 1996) which will be discussed later in the chapter. Finally, the control of dispense angle and low dynamic surface tension leads to smaller bubbles with a homeodisperse size distribution, which results in the desirable "creamy" foam characteristic (Ronteltap et al., 1991).

The factors governing the size of bubble that is generated in nucleation are described in the equation (1.1)

$$\text{Bubble radius} = \left[\frac{3R_m\gamma}{2\rho g}\right]^{1/3} \quad (1.1)$$

where
R_m = radius of nucleation site (m)
γ = surface tension (mN m^{-1})
ρ = relative density of the beer (kg m^{-3})
g = acceleration due to gravity (9.8 m s^{-2})

The radius of the nucleation site is very significant, but surface tension and specific gravity (relative density) are less important.

Drainage

Upon its formation, foam is usually termed to be "wet." The excess beer in the foam rapidly drains by gravity to produce "dry," well drained foam in which more subtle effects of drainage can be discerned (Figure 1.3). In the dry foam, continued beer drainage by gravity and "Plateau border suction" weakens the bubble film eventually leading to bubble collapse (Ronteltap et al., 1991). Ronteltap et al. (1991) concludes that the counteracting forces to drainage are viscosity of the beer, capillary effects and the beer's surface viscosity. The influence of beer viscosity is consistent with beer foam being observed to be more stable at lower temperatures. This explanation was more recently simplified to conclude that surface viscosity as opposed to bulk viscosity was most important

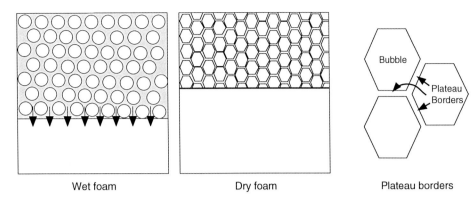

Figure 1.3
Schematic representation of drainage. In dry foam the bubbles take on a honeycombe-like polyhedric structure showing well defined plateau borders between bubbles (after Prins and Marle, 1999).

(Bamforth, 2004a). In practice this interpretation, appears to be confirmed by the observation that non-starch polysaccharides that increase bulk viscosity, such as β-glucans, have a negligible influence on beer foam stability (Lusk et al., 2001a,b).

Liquid drainage from foams is addressed by the formula:

$$Q = \frac{2\rho g q \delta}{3\eta} \quad (1.2)$$

where
Q = flow rate (m³ s⁻¹)
η = viscosity of film liquid (Pa s)
ρ = relative density of the beer
q = length of Plateau border (m)
g = acceleration due to gravity
δ = thickness of film (m)

Creaming

Creaming or "beading" is defined as the appealing spectacle in beer of bubble recruitment into the foam (Figure 1.4) which should ideally be sustained for the duration of consumption (Bamforth, 2004a). Combined, the nucleation activity, the surface tension, beer density and CO_2 content determine the level of creaming. Accepting the central role of a nucleation site, it was found that CO_2 content was the most influential variable because the typical ranges for surface tension and density were usually not large enough to make a discernable difference in creaming (Lynch and Bamforth, 2002).

Creaming can be explained by the model

$$a_n^0 = 3.11C + 0.0962\gamma - 218\rho + 216 \quad (1.3)$$

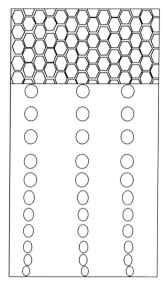

Figure 1.4
Schematic representation of creaming.

where
a_n^0 = initial nucleation activity
γ = surface tension
ρ = density
C = carbon dioxide content (vol. CO_2/vol. beer)

Coalescence

Coalescence in foams is defined as the merger between two bubbles (Figure 1.5) caused by the rupture of the film between the bubbles to produce a larger, less stable and less appealing bubble (Ronteltap et al., 1991). With unspoiled or high quality beer this mechanism is of limited importance (Ronteltap, 1989). However, if highly hydrophobic material such as lipid, oily snacks (i.e. crisps), lipstick, cleaning agent or dirty glasses contact the beer this effect can be catastrophic to beer foam stability. Commonly known as the "hydrophobic particle mechanism" (i.e. impact of lipids) or "particle spreading mechanism" (i.e. impact of detergents), such small disruptive particles, when positioned in the bubble film, rapidly initiate coalescence (Ronteltap et al., 1991).

Disproportionation

Disproportionation, also known as Ostwald ripening, is defined as the bubble fusion or foam coarsening process resulting from inter-bubble gas diffusion (Ronteltap et al., 1991; Bamforth, 2004a). By this process, gas from smaller bubbles with higher Laplace pressure diffuse into larger bubbles with lower Laplace pressure (Figure 1.6). Thus, smaller bubbles disappear

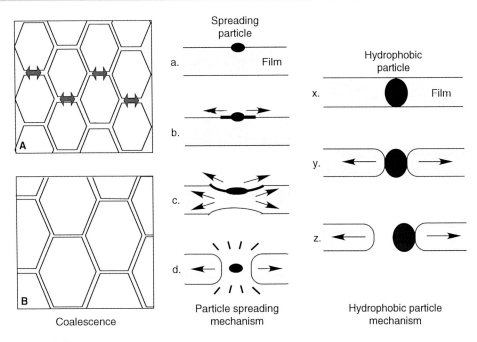

Figure 1.5
Schematic representation of coalescence (A, B), the particle spreading mechanism (a, b, c, d), and hydrophobic particle mechanism (x, y, z) (after Ronteltap et al., 1991).

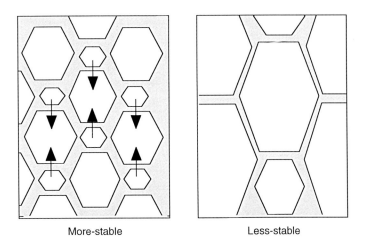

Figure 1.6
Schematic representation of bubble disproportionation.

and larger bubbles become even larger, resulting in "bladdery" bubbles that are less attractive (Bamforth, 1999). As gas diffusion is an important factor in disproportionation, it is easy to again see the substantial benefit of nitrogen gas for foam stability because of its relatively lower aqueous solubility compared to CO_2 (Carroll, 1979; Mitani et al., 2002; Bamforth, 2004a). Thus the combination of gas content and bubble film thickness which is in turn related to drainage rates are the most important factors in limiting disproportionation (Bamforth, 2004a).

Disproportionation has been modeled according to the De Vries equation:

$$r_t^2 = r_0^2 - \frac{4RTDS\gamma t}{P\theta} \quad (1.4)$$

where
r_t = the bubble radius at time t
r_0 = bubble radius at the start
R = the gas constant (8.3 J K^{-1} mol^{-1})
T = absolute temperature (°K)
D = the gas diffusion coefficient (m^2 s^{-1})
S = the solubility of the gas (mol m^{-3} Pa^{-1})
γ = the surface tension
t = time (s)
P = pressure
θ = the film thickness between bubbles

The enormous benefit that low levels of nitrogen gas have on foam stability is explained on account of its vastly lower solubility than carbon dioxide, meaning that it is less able to dissolve in the liquid interface between bubbles and pass from one bubble to the next. Film thickness is also important, and this will be impacted primarily by drainage rates (see earlier) but also by any surface active materials that enter into the bubble wall and interact to achieve a framework capable of maintaining film integrity.

Foam measurement

Analytical assessment of beer foam is complicated by the diverse range of visual characters that determine foam quality. Perhaps the lack of a universally agreed assessment method is due to the necessity to measure the different components of foam quality that include stability, quantity, lacing, whiteness, "creaminess" and strength. In addition, consumer dispense needs to be divided broadly into pouring from bottle or can packaging, or from a dispense tap, each having differing dynamics of foam generation. Similarly, off tap foam generation is a function of the wide variety of delivery devices and orifices used (Carroll, 1979). This makes it nigh impossible for any single analytical assessment apparatus to provide a comprehensive evaluation of foam quality. The ultimate is determined by an expert assessment-panel but these are expensive, unwieldy and subject to the biases based on gender, race and socio economic background (Bamforth, 2000a; Smythe et al., 2002; Roza et al., 2006). In this it has been argued that "beer foam stability is arguably the most important, for if the foam is not stable, the other characteristics are of little consequence" (Evans et al., 2003). Equally it was suggested (Bamforth, 1999) that lacing is the most significant parameter, as it will not occur without the prerequisites of foam formation and head retention and it is the most recognizable manifestation of foam quality as perceived across a crowded bar. However more recent research (Bamforth, 2000a; Roza et al., 2006) generated results that suggest that not everybody appreciates lacing. The most commonly used methods for commercial and research foam quality evaluation are NIBEM, Sigma head value (SHV), Rudin head retention (Rudin) or Ross and Clark that measure foam stability (Table 1.1, Figure 1.7).

Table 1.1
A summary of the attributes of a selection of foam analysis tests

Foam quality assessment method	Beer Status[a]	Foam formation	Foam measurement	Primary measure	Citation
Expert panel	Gassed	Pouring or tap	The eyes	Combined	Various evaluation panels
Blom	Degassed	CO_2 sparging	Drainage	Stability	Blom (1937)
Ross and Clark	Degassed	Pouring	Drainage	Stability	Ross and Clark (1939)
Rudin	Degassed	CO_2 sparging	Drainage	Stability	Rudin (1957)
Micro Scale	Degassed	Homogenizer	Drainage	Stability	St John-Coghlan et al. (1992), Hung et al. (2005)
Sigma head value	Gassed	Pouring	Drainage	Stability	ASBC (1992)
LG automatic	Gassed	CO_2 sparging	Drainage	Stability	Rasmussen (1981)
NIBEM or NIBEMT	Gassed	CO_2 flashing	Foam collapse[b]	Stability	Klopper (1973)
Constant pour test	Gassed	Pouring	Foam collapse[c]	Stability	Constant (1992)
Cylinder pour test	Gassed	Pouring	Foam collapse rate	Stability	Vundla and Torline (2007)
FCT	Gassed	Pouring	Foam collapse[c]	Stability	Yasui et al. (1998), Ferreira et al. (2005)
Shake test	Degassed	Shaking	Foam collapse	Stability	Kapp & Bamforth (2002)
Lacing-Cling	Degassed	CO_2 sparging	Amount of lace[d]	Lacing	Jackson and Bamforth (1982)

[a]Gassed = carbonated or nitrogenated, Degassed = gas removed from beer before measurement.
[b]Foam collapse measured by electrode.
[c]Foam collapse measured by video or image analysis.
[d]Amount of lacing collected measured by spectrophotometric absorbance.

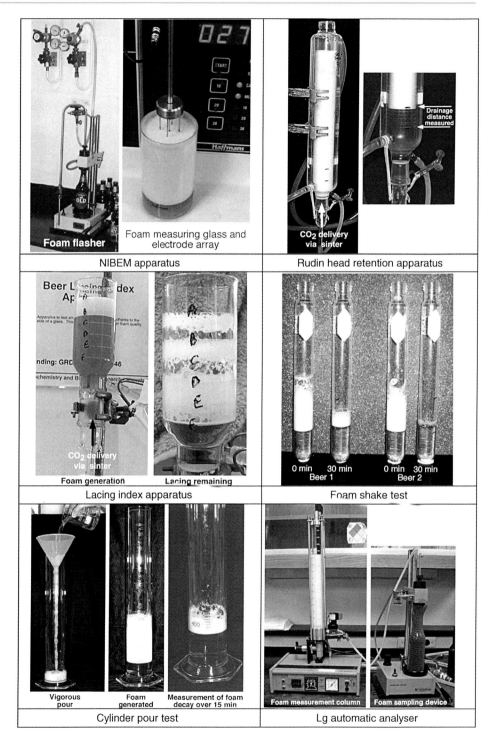

Figure 1.7
A selection of foam analysis procedures.

As almost all common evaluation procedures essentially measure foam stability (Table 1.1), it is notable that the proportion of foam produced is excessive compared to the dispense provided for or by the consumer (Figure 1.7). In general, the procedures that are reliant on foam generation by "natural" pouring techniques tend to be inconsistent, while "artificial" foam generation by gassing through porous frits, "flashing" or by other devices produces foams that are not typical of consumer dispense (Constant, 1992). This is particularly so with respect to the amount of foam generated that it is much too wet or contaminated with beer compared to genuine foam (Leeson et al., 1990). Considering a natural pour from a bottle, it has been observed that normal variation in the diameter and scoring of the bottle opening substantially reduced the reproducibility of a robotic pour test (Skands et al., 1999). Further, it was observed that the pivot point in pouring a bottle was important for the quantity and quality of the foam generated (Yasui et al., 1998). Further complications ensue with emulating the tilt of the glass and control of the rate of beer dispense which is customary for normal beer dispense in trade or by the consumer.

The exception, in terms of the generation of a somewhat typical amount of foam is the lacing index method (Jackson and Bamforth, 1982; Figure 1.7). This test specifically measures the amount of lacing adhering to the glass. This test generates a relatively small amount of foam (17 mm), but measures the amount of lacing spectrophotomically which provides a clue as to why generally large amounts of foam are produced in foam quality tests (Figure 1.7). That is producing a consistent foam is not easy, after which, measuring relatively small changes in the amount of foam, as with typical consumer dispense, is difficult.

It was optimistically suggested that "perhaps the ideal foam assessment method will evaluate beer foam by utilizing digital cameras combined with image analysis software employing algorithms that accurately match consumer assessment" (Evans and Sheehan, 2002). The early signs that such an analysis procedure would be developed were encouraging as gauged by proposals by Rasmussen (1981), Haugsted and Erdal (1991), Hallgren et al. (1991), Constant (1992), Mulroney et al. (1997), Yasui et al. (1998), Skands et al. (1999), Anger et al. (2002). Given the substantial developments and reductions in cost of computing power, digital photographic equipment, and image analysis in the last 15 years, it is disappointing that a relatively time efficient, simple to operate, not unduly expensive, device with unrestricted availability has not eventuated for wide spread adoption across the brewing industry. However, the cost of developing the algorithms to analyze the foam is perhaps the constraint. In part this was overcome by Ferreira et al. (2005) who utilized commercially available Kodak Scientific Imaging Systems 1D software (Rochester, NY) for finding bubbles on electrophoresis gels, blots, etc.

Pragmatically, the NIBEM, and to a less extent now Rudin, SHV or the Ross and Clark procedures are the most frequently used methods for QC/QA purposes in the brewing industry (Bamforth, 1999) which suggests that they must at least have some value in identifying batches that are likely to cause problems in the trade. As such Evans and Sheehan (2002) highlighted the need for comparison between these traditional foam assessment methods, visual assessment and any new device developed. In this endeavor, the simple model foam standard proposed by Vundla and Torline (2007) may better enable comparisons between foam analysis procedures and between laboratories.

Surprisingly, comparisons between foam evaluation produces are relatively rare in the literature. Yasui et al. (1998) developed a digital image analysis device that calculates a "foam collapse time" (FCT) value as a measure of foam quality for beer poured from a bottle. Regression analysis showed that the FCT value was significantly correlated with visual assessment of overall foam quality ($r = 0.638$) and stability ($r = 0.486$) but not with the Rudin, NIBEM or SHV tests. The Rudin, NIBEM and SHV tests were significantly correlated with each other ($r > 0.769$). Similarly, Ferreira et al. (2005) also confirmed that the SHV test was not correlated with their version of the FCT test. More recently, Lewis and Lewis (2003) concluded that the Constant method of foam analysis and a pour test (similar to Ross and Clark) were valid measures of foam stability while the SHV test was not. In contrast, Hughes (1997) compared a number of foam analysis procedures including NIBEM, Rudin, Ross and Clark and lacing index with the visual assessment ranking by a trained panel of judges and concluded that "perceived foam quality on the whole matches the ranking" given by the foam analysis procedures tested. Most recently, it has been concluded that both the Lg-automatic foam tester and the NIBEM-TPH ($r = 0.89$, variant measuring at consistent temperature, atmospheric pressure and humidity) are consistent and good predictors of visually judged foam quality (Roza et al., 2006; Hung et al., 2005). In all, these reassurances of valid foam analysis are comforting since our understanding of the impact of beer components and process conditions on foam stability are largely based on these evaluation procedures.

Beer components that influence foam quality

Much research activity has been devoted to determining the key beer components that influence beer foam quality. In the past many of these investigations have applied reductionist scientific principles to identify these key foam determinants, in the hope of identifying one component to manipulate to optimize foam quality. Combined, these studies have established that foam quality, essentially its stability, is promoted by interactions between proteins (>5 kDa) contributed by malt and hop acids (Asano and Hashimoto, 1980; Bamforth, 1985). The proteins identified have included protein Z, LTP1 and various hordein-derived species. In addition to these widely reported species, proteomic techniques with mass spectrometric evaluation were recently applied to detect a range of other proteins species in twice-foamed foam (2x foam, Hao et al., 2006), although these analyses did not ascertain if these proteins were significant beer foam components. Various divalent metal cations have also been shown to effectively cross-link hop iso-α-acids to strengthen the bubble film. Components such as polyphenols and non-starch polysaccharides (i.e. β-glucan and arabinoxylan) have been attributed potential minor roles in promoting foam quality. Other components such as lipids, basic amino acids and high levels of ethanol have been shown to destabilize beer foam. The attributes and contributions of these "yin and yang" components have recently been comprehensively reviewed by Evans and Sheehan (2002). On the whole it would appear that the combined contributions of the promoting components such as proteins are incremental but the destabilizing effects of lipids can be disastrous to foam quality.

To achieve these evaluations, physiochemical measurements of individual components have been made to determine each constituent's foam promoting ability. These assessments include hydrophobicity (Slack and Bamforth, 1983; Onishi and Proudlove, 1994; Yokoi et al., 1994), fluorescence recovery after photo bleaching (Clark, 1991; Hughes and Wilde, 1997), surface dilational rheology (Douma et al., 1997; Hughes and Wilde, 1997), and surface-viscometric activity (Maeda et al., 1991; Yokoi et al., 1989). Of these characteristics it is now considered that protein hydrophobicity is probably of most importance (Bamforth, 1999). However, Evans and Sheehan (2002) concluded, "these physical considerations are only, at best, a useful guide to the foam promoting behaviour of beer components." This is because the number of potential foam active components are certainly numerous and their interactions are complex. Therefore the practical value of each individual foam component can only be resolved in experiments where their impact on foam quality is evaluated in the context of the whole beer system. In the last decade there have been a number of investigations that have adopted this more holistic approach (Evans et al., 1999c; Evans et al., 2003; van Nierop et al., 2004; Kapp and Bamforth, 2002; Brey et al., 2002; Lewis and Lewis, 2003; Bamforth and Kanauchi, 2003; Bamforth, 2004a) to determine the interactions and interplay between components that produce optimal foam quality.

In the next section, our focus will be to summarize the attributes and importance of the various foam promoting and inhibiting components. For all these components to influence beer foam quality, they must be resilient and able to survive the relatively challenging malting and brewing process conditions which confronts them with heat and enzymatic denaturation/modification, substantial changes in pH, stabilization treatments (haze, pasteurization) and the potential to be utilized or adsorbed by the fermenting yeast.

Proteins

Lipid transfer protein (LTP)

There are two members of the lipid transfer protein family expressed in barley, LTP1 and LTP2. LTP2 is expressed in the aleurone layer during the early stages of grain development (Kalla et al., 1994) and small amounts have been purified from malt (Jones and Marinac, 1995). However, LTP2 has yet to be found in beer in significant quantities or associated with foam quality. LTP1, formerly known as PAPI, is expressed primarily in the aleurone layer late in grain development and in the early stages of germination (Mundy and Rogers, 1986; Skiver et al., 1992). It is a ~9.7 kDa hydrophobic protein that is concentrated in beer foam (Sorensen et al., 1993; Lusk et al., 1995; Lusk et al., 1999) and ELISA has estimated that it constitutes up to 1% of total malt protein (Evans and Hejgaard, 1999).

During the kettle boil, LTP1 is irreversibly denatured, which substantially reduces its immuno reactivity to the barley LTP1 antibodies (Bech et al., 1995a,b; Lusk et al., 2001a,b) but improves its foam promoting properties (Sorenson et al., 1993; Bech et al., 1995a,b; Lusk et al., 1995). Lusk et al. (2001a,b) compared the level of un-denatured LTP1 ("native") with denatured LTP1 ("foam") by two quantitative ELISAs specific to each LTP1 type due to the use of specific monoclonal antibodies. These authors found that the level of LTP1 decreased in the cooled wort to ~25% of the peak level observed during mashing (Figure 1.8). Comparison of Figures 1.14 and 1.15 (shown later in this chapter) indicate that LTP1 is reduced to between ~5 and ~20% of the level in pre-boil wort as a result of kettle boiling (depending on the severity of the boil) using the ELISA developed by Evans and Hejgaard (1999) that used polyclonal antibodies raised against barley LTP1. Thus despite heat denaturation and Maillard glycation (Perrocheau et al., 2006), the level of LTP1 measured by this ELISA appears to provide a useful estimate of beer LTP1 content as the reduction in LTP1 level due to boiling is similar to that found by Lusk et al. (2001a,b). Perhaps this is a result of the novel format of this ELISA that uses

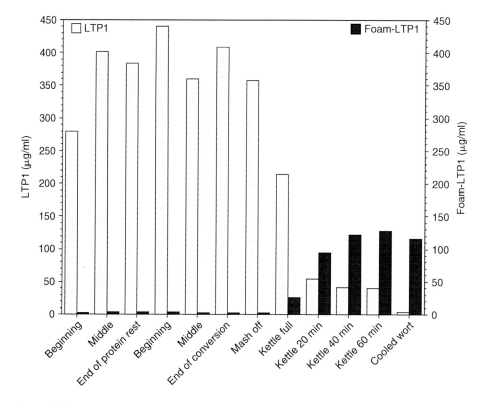

Figure 1.8
The level of "native" LTP1 (open bars) and "denatured" foam-LTP (black bars) in the brewhouse measured by two ELISAs specific to each LTP1 type (extracted from Lusk et al., 2001a,b).

a combination of sheep and rabbit antibodies so as to avoid the requirement to conjugate a reporting agent (i.e. horse radish peroxidase) to the secondary antibody. This presumably reduces potential steric hindrance in the binding two antibodies to LTP1 in the ELISA. Despite the uncertainty as to how completely the Evans and Hejgaard (1999) LTP1 ELISA measures denatured LTP1, van Nierop et al. (2004) found this measure to be very useful in determining final beer foam quality as is explained below.

LTP1 would appear to have different modes of action in relation to beer foam quality. Firstly in isolation, beer LTP1 has excellent foam generation but poor foam stabilizing properties. The foam stabilizing properties are substantially enhanced when it is combined with an isolated LMW hordein/glutelin or HMW foam fraction that contains protein Z (Sorensen et al., 1993). This efficacy with other proteins to provide foam stability was also observed by Douma et al. (1997). Bech et al. (1995a,b) found that increasing the beer LTP1 content resulted in improvements in foam stability when foam performance was assessed by a digital analysis system that was the fore-runner to the Lg automatic tester. Similarly Lusk et al. (1995) also concluded that LTP1 was an important determinant of foam stability as judged by the Constant foam analysis test. Conversely, small scale (600–800 ml) and pilot (50 l) trials that primarily used the Rudin foam analysis procedure found an ambiguous relationship between LTP1 and foam stability with different trials showing positive, negative or no correlation (Evans et al., 1999c, 2003). In part the explanation for these observations might be that the Rudin analysis measures foam stability and takes no account of the amount of foam formed. Alternatively, the second mode of action of LTP1, as a lipid binding protein may explain the discrepancies outlined above. Van Nierop et al. (2004) observed that increased denaturation of LTP1 by boiling could in fact be detrimental to foam stability due to the reduced ability of denatured LTP1 to bind foam destabilizing lipids. Where the level of lipids in beer was low or the level of LTP1 was high there was little impact on beer foam stability, however where the level in beer of LTP1 was low, and the level of lipids high, substantial reductions in foam stability were observed (Table 1.2).

Table 1.2
Comparison of typical wort boiling and beer foam characteristics of three different breweries, including the relative free fatty acid (FFA) levels typically found in their beers (table from van Nierop et al. (2004))

Brewery	A	B	C
Location	Coastal	Inland	Coastal
Altitude	~200 m	~1600 m	~150 m
Boil temperature	~102°C	~96°C	~102°C
LTP1 (μg/ml)	2–3	17–35	2–3
FFA (mg/ml)	2.84	1.12	0.70
Foam	Low	High	High

Carlsberg Research Laboratories were granted a patent in 1999 for the manipulation of LTP1 to improve foam (Bech et al., 1999). Essentially the patent claim covers all beverage preparation processes that increase the level of LTP1 to improve foam quality using LTP1 as an additive or the use of recombinant technology to increase the levels of LTP1 in the beverage. Evans and Sheehan (2002) concluded that "the range in variation for the level of LTP1 found in current barley varieties and their wild relatives is considered to be prior art, so the approach of accessing the naturally occurring variation for LTP1 in currently available genetic material was not covered by the patent." Similarly, brewer optimization of wort boiling practices (van Nierop et al., 2004) would also not be expected to be covered by the patent.

Hordeins

Hordeins, the major storage proteins of barley, consist of a complex polymorphic mixture with major grouping B, C, D and γ designated on the basis of the molecular size by electrophoresis (Kreis and Shewry, 1992). By their nature, being prolamines, they are insoluble in aqueous solutions (Osbourne, 1924) and require proteolytic hydrolysis to be rendered water-soluble. They are also relatively rich in proline (and glutamine) which predisposes at least some hordein fragments to the undesirable formation of chill haze (Asano et al., 1982). As for LTP1, hordein fragments have been found to partition and concentrate in beer foam in techniques such as foam collection, foaming towers or the simple sequential re-foaming (Sorensen et al., 1993; Sheehan and Skerritt, 1997; Evans et al., 2003; Hao et al., 2006). However, care needs to be exercised with such techniques that artificially over or exhaustively foam beer as some foam proteins such as protein Z can be rendered insoluble after one foaming (Lusk et al., 1999).

The search for foam promoting hordein has been somewhat of a holy grail for foam researchers. However, the diversity of hordein groups and their interaction with proteases to enable solubilization produces a myriad of potentially desirable foam promoting species, along with the undesirable haze promoting types. Some investigators have claimed some success in identifying hordein species that are associated with or are claimed to improve foam quality. Sheehan and Skerritt (1997) identified a 23 kDa hordein band, with hordein specific monoclonal antibodies, whose concentration was enriched in foam by a factor of 2.7 (foam-beer ratio), suggesting a protein that was more surface active with the potential to promote foam stability. A 17 kDa protein has also been found to be concentrated in beer foam (Vaag et al., 1999). Subsequent protein sequencing classified this as a new type of hordein – epsilon-1-hordein. This protein was also found to have some sequence homology with the 23 kDa protein identified above. In beer foam analysis experiments using a prototype for the Lg automatic tester, it was found that foam half life was significantly correlated with the combined protein content of the 17 kDa protein and LTP1 ($r^2 = 0.69$) for beers that were moderately carbonated (CO_2 5.1–5.2 g/l) but not when the beers were highly carbonated (CO_2 5.4 g/l). The Carlsberg group has also patented the manipulation of the 17 kDa protein for improving beer foam stability (Vaag et al., 2000). The IFRN group (Norwich, UK) and their collaborators identified, with monoclonal antibodies, a range of hordein-derived polypeptides (3–33 kDa) that were characterized in terms of their hydrophobic character (Kauffman et al., 1994; Mills et al., 1998; Onishi et al., 1999; Hughes et al., 1999). Evans et al. (2003) also developed and produced a number of antibodies (monoclonal and polyclonal) that recognized putative foam promoting hordeins. However, a persistent difficulty of all these studies (except, Vaag et al., 1999) has been translating the antibody identification of putative foam promoting hordein moieties into ELISA's that can be readily used for improvement of foam or diagnosis of beer foam problems.

More recently, a series of experiments investigating the influence on beer foam stability of various model beer foams by Bamforth's group has been enlightening as to the putative role of hordeins. Firstly it was found that the foam stabilizing power of denatured barley albumin fractions was increased "rather more" than that for hordeins while limited proteolysis reduced albumin but enhanced hordein foam stability (Kapp and Bamforth, 2002). However, a later study cautioned that although both hydrolyzed barley hordein and albumin are independently capable of stabilizing foam in model solutions, this capability does not appear to be additive in mixed solutions (Bamforth and Milani, 2004). Bamforth (2004a) was ultimately able to reconcile these results by finding that "hydrolysed hordein appears to selectively enter beer foams at the expense of the more foam-stabilising albuminous polypeptides."

This hypothesis was tested by the inclusion of a proline specific endoproteinase that targets haze promoting hordein (Lopez and Edens, 2005), into brews (Evans et al., 2008). It was expected that the proline endoproteinase treatment would increase foam stability, for at least some malt samples. Table 1.3 shows that the addition of proline endoproteinase to small scale beers made from malt from six varieties had a variable impact on foam quality. With the lacing index test, most varieties were unchanged (varieties A, B, C, E) while two (varieties D, F) showed a small decrease with the inclusion of proline endoproteinase. With the Rudin test the addition of proline endoproteinase can cause either an increase in the foam stability (varieties B, D), no change (variety A) or a small decrease (varieties C, E, F). These observations suggest that there are both foam positive hordeins and foam negative hordeins as was suggested by Evans et al. (2003). The effect of proline endoproteinase is thus determined by the relative proportions of these hordeins in beer that is probably determined by the hordein species composition of the variety being used. Given the heritability of hordein species, it may be possible for barley breeders to select malting varieties that primarily contain foam positive hordeins once the hordein species are characterized.

Protein Z

Protein Z or the "40 kDa" beer protein was the first specific protein species to be proposed to promote foam stability (Kaersgaard and Hejgaard, 1979). Protein Z

Table 1.3
The influence of proline endoproteinase (Brewers Clarex, DSM. Seclin, France) on foam quality for beers made with a small scale brewing procedure (~800 ml, Stewart et al., 1998) (extracted from Evans et al., 2008).

Malt	KI (%)	Proline endoproteinase	Rudin (sec)	Lacing Index
Variety A	40.6	No	100.3	3.61
Variety A		Yes	99.0	3.35
Variety B	40.3	No	95.3	3.61
Variety B		Yes	105.0	3.23
Variety C	40.1	No	132.3	4.01
Variety C		Yes	128.7	4.08
Variety D	45.4	No	106.3	3.90
Variety D		Yes	115.7	2.99
Variety E	43.7	No	100.0	3.20
Variety E		Yes	95.3	3.56
Variety F	46.5	No	118.7	3.97
Variety F		Yes	115.3	3.30
LSD (P < 0.05)	nd		1.3	0.39

is an albumin type protein that is contributed to beer from malt where it has been estimated by ELISA to constitute up to 2% of total protein (Evans and Hejgaard, 1999). Protein Z is actually the sum of the levels of its two isoforms, protein Z4 and protein Z7 that are expressed from two separate but highly related gene families. In barley and malt, protein Z4 is almost always the dominant isoform, accounting for approximately 80% of all protein Z (Evans and Hejgaard, 1999). It has been estimated that protein Z accounts for 10–25% of the non-dialyzable protein in beer making it the most abundant protein species (Kaersgaard and Hejgaard, 1979; Hejgaard and Kaersgaard, 1983). Although it has been found to possess the highest surface viscosity and elasticity properties of all beer proteins (Dale and Young, 1987; Yokoi et al., 1989; Maeda et al., 1991), it has not been shown to be preferentially enriched in foam as has been observed for LTP1 (Sorensen et al., 1993; Leiper et al., 2003). In contrast, Douma et al. (1997) concluded that the interactions between protein Z and other proteins such as LTP1 was very important for foam stability.

The importance of protein Z to beer foam stability has been somewhat contentious. Small scale (600–800 ml) and pilot scale (50 l) brewing experiments generally, but not always, show that the level of protein Z4 is positively correlated with Rudin foam stability (Evans et al., 1999c, 1999, 2003;

Gibson et al., 1996). Similarly trials in which protein Z was at least partially removed from beer by immuno-affinity column before foam testing have shown either minor (Hollemans and Tronies, 1989; Vaag et al., 1999) or substantial (Vaag et al., 1999) decreases in foam stability. This apparent contradiction appears to be resolved by the observation that the contribution of protein Z to foam stability is a function of malt modification (Evans et al., 1999a,b,c; Vaag et al., 1999). That is where malt is relatively under-modified (i.e. KI < 40), the contribution of protein Z is relatively minor, however where malt is more highly modified (i.e. KI > 40) the contribution is more substantial. This could perhaps be, in part, explained by Bamforth's (2004a) prediction that hydrolyzed hordein will exclude albumins such as protein Z from beer.

Lipid binding proteins

Although lipids introduced into beer have an initial deleterious impact on foam stability, the foam stability can recover either fully or partially if the beer is allowed to rest for 24 h (Roberts et al., 1978). The key to this observation is lipid-binding proteins that bind lipid to reduce foam destabilization or even restore foam stability in beer containing unfavorable levels of lipid (Clark et al., 1994; Morris and Hough, 1987; Onishi et al., 1995; Cooper et al., 2002; van Nierop et al., 2004). These proteins originate from malt or wheat and barley adjunct. Mostly these proteins have not been characterized further than being partially purified as beer or malt protein fractions that demonstrate this highly desirable property. The most likely candidates are LTP1 which is well known to bind lipids (Douleiz et al., 1999) or the puroindoline/hordoindoline proteins of wheat (Wilde et al., 1993) and barley (Beecher et al., 2001), respectively. However, Evans and Sheehan (2002) demonstrated that neither puroindoline or hordoindoline survive the brewing process in any meaningful quality. This is not to say that puroindoline or hordoindoline could not be added into the beer sometime after boiling (post fermentation) to provide some foam protection (Wilde et al., 1993; Clark et al., 1994; Dickie et al., 2001). Conversely, van Nierop et al. (2004) concluded that LTP1 does have a role in scavenging of beer lipids to render them harmless to beer foam. Further, this investigation found that the more LTP1 is denatured, the lesser is its ability to bind lipid and protect foam stability.

Protein measurement

Two strategies can be followed for the measurement of beer proteins to assure or improve the foam quality of beer. Firstly, specific protein species such as protein Z4, LTP1 can be targeted using ELISA. The application of these ELISA assessments forms an important part of the understanding of which beer proteins are foam promoting as has been discussed above. ELISAs have also been developed in attempts to measure foam proteins in general such as the "foam protein" ELISA (Ishisbashi et al., 1996, 1997) or the 2x foam ELISA (Evans et al., 1999, 2003). Interestingly, it was later found that the 2x foam and presumably the "foam protein" ELISAs were essentially measuring protein Z4 as the two assays were highly correlated ($r = 0.99$) (Evans et al., 2003). This result was surprising because the protein used for immunization for the 2x foam assay was primarily in the range 7–15 kDa whereas protein Z4 is a 40 kDa protein. Recently, a mass spectrophometric study found protein Z4 fragments in the 7–17 kDa fraction (Hao et al., 2006) that would now appear to explain this discrepancy.

Alternatively, various methods for measuring total protein can be applied to measure foam proteins. Recently, Siebert and Lynn (2005) critically evaluated the potential of a comprehensive selection of these techniques for the estimation of beer protein. Time honored procedures such as Kjeldahl, spectrophotometric determination at 280 nm and the bicinchoninic method (BCA), although reliable in the right application, were not found to be useful due to interference from other substances such as amino acids (Kjeldahl) that were not foam promoting. Therefore assays which select for protein characteristics such as molecular weight >5 kDa and/or hydrophobicity are likely to be most useful. Such an assay is the Bradford Coomassie blue binding (CBB) assay (Bradford, 1976) that has long been known to be correlated with beer foam stability (Dale and Young, 1987; Siebert and Knudsen, 1989). The basis for this correlation is that the CBB assay effectively only measures beer proteins >5 kDa (Lewis et al., 1980; Hii and Herwig, 1982; Compton and Jones, 1985). However, protein amino acid composition, particularly the content of arginine, has a substantial influence on the results of the CBB assay (Compton and Jones, 1985). The implications are that CBB underestimates the level of hordein in beer (Siebert and Lynn, 2005). A potentially useful modification to the CBB test to better measure foam promoting protein has been to separate beer proteins by phenyl Sepharose hydrophobic interaction chromatography and then measure the relative levels of "hydrophilic" and "hydrophobic" proteins with CBB (Bamforth et al., 1993; Bamforth, 1995).

Two other potential protein binding dyes have also recently been evaluated. The fluorescence of 1-anilino-8-naphthalenesulfonate increases when it interacts with hydrophobic proteins. Unfortunately, the high level of background fluorescence in beer and the relatively complex mixture of the proteins in beer negates this fluorochrome having any additional benefits over direct measurement of total protein with CBB (Bamforth et al., 2001). The pyrogallol red-molybdate (PRM) method has been demonstrated not to be interfered with by low molecular weight proteins or amino acids indicating its suitability for use in measuring total protein in beer (Williams et al., 1995). Although it is assumed that amino acid composition will also influence the binding preferences of this dye, this discrimination is potentially valuable for foam protein determination. PRM dye binding is positively biased towards purified LTP1 and the 2x foam fractions compared with CBB which reacts poorly with native LTP1, although somewhat better with LTP1 once it has been denatured (extracted from beer) (Evans and Sheehan, 2002). However, preliminary studies that evaluated 35 beers found that the PRM and CBB assays were highly correlated ($r = 0.95$).

In practice, this leaves brewing chemists with a limited number of options for quality control and brewing process optimization of foam proteins. With the CBB and PRM tests protein denaturation, precipitation, and other modifications

Table 1.4
A summary of the protein composition of 24 Australian and international beers (extracted from Evans et al., 2008), using the ELISAs described by Evans and Hejgaard (1999)

Statistic	CBB protein (μg/ml)	PRM protein (μg/ml)	Protein Z4 (μg/ml)	Protein Z7 (μg/ml)	LTP1 (μg/ml)
Average	125	355	31.1	3.1	2.6
Minimum	73	139	10.6	0.5	0.5
Maximum	192	505	60.0	10.6	5.2
Std. deviation	25	112	13.4	2.1	1.2

during the brewing process select out other proteins that might bind these dyes suggesting that these tests are most useful for quality assurance after the beer is brewed. ELISA's such as those developed for protein Z4, protein Z7 and LTP1 (Evans and Hejgaard, 1999) also are useful for quality control and process optimization. It should however be noted that like the LTP1 ELISA, the protein Z4, and protein Z7 ELISA's were developed using native barley proteins. Thus the denaturation (reduction of disulfide bonds) and Maillard glycation observed as a result of brewing for LTP1 (Perrocheau et al., 2006), could potentially interfere with the ELISA measurement of these proteins in beer. Despite these caveats, the ELISA measurement of protein Z4, and protein Z7 in beer appears to be reasonable as the levels in beer are highly correlated with those from the malt used to produce the beer, despite different wort boiling regime (see Table 1.8). For reference, a summary of typical beer compositions of specific proteins measured by ELISA and protein in general as measured by the CBB and PRM tests are presented in Table 1.4. The ELISAs, due to their inherently greater measurement selectivity, also have wider application in the brewing process chain including the evaluation of malt batches, variety comparison and selection during barley breeding.

Non-starch polysaccharides

The influence on foam stability of high molecular weight, malt-derived, non-starch polysaccharides, such as β-glucan and arbinoxylan or even oligosaccharides via an increase in beer viscosity, is contentious. These components increase beer bulk viscosity, to reduce the drainage of liquid from foam, thus improving its stability (Stowell, 1985; Archibald et al., 1988; Lusk et al., 1995; Evans et al., 1999c; Evans and Sheehan, 2002; Lewis and Lewis, 2003). This conclusion is attractive as it is in line with the basics of beer foam physics (drainage) as outlined by Ronteltap et al. (1991). Further it has been suggested that small quantities of arabinoxylan can effectively both increase bulk viscosity and cross-link with proteins to stabilize foam (Sarker et al., 1998). Interestingly, viscosity was also found to be significantly and positively correlated with lacing index (Evans and Sheehan, 2002) indicating that adhesion of foam to the side of the glass is increased with increasing viscosity. However, other investigations have concluded that β-glucan does not influence foam stability as measured with the Constant foam test (Lusk et al., 2001a,b). Hug and Pfeninger (1980) also found that the inclusion of a β-glucanase preparation during mashing, reduced viscosity but did impact on foam stability as measured by the SHV test. Recently, it has been concluded that malt oligosaccharides do not have a major influence on foam stability (Ferreira et al., 2005). Bamforth (1999) concludes these disagreements are mostly due to the type of foam analysis procedure used to assess stability. Thus, methods such as Rudin, which are more influenced by the drainage factor, are more liable to over rate the importance of viscosity.

Regardless, the improvement of foam stability by increasing viscosity is a course that few brewers can afford to take due to the concomitant problems associated with reduced rates of lautering, rates of beer filtration and hazes and precipitates. Given the widespread use of exogenous cell-wall-degrading enzyme preparations by brewers, it is suggested that beer viscosity is likely to be a minor factor in practice.

Hop acids

Hop acids not only contribute to beer bitterness but are also essential partners with proteins to achieve foam stability. To impart bitterness the hop acids require isomerization traditionally achieved during wort boiling or by catalytic isomerization of hop extracts (Figure 1.9). A further modification that may also be undertaken is to hydrogenate the hop acid, primarily to achieve protection against the formation of light struck flavors. This final modified form is often commonly referred to by its initial trademark name, "tetra hop."

The foam promoting properties of hop acids have long been known and that hop resins, in particular isohumulone, promote foam stability (for review of

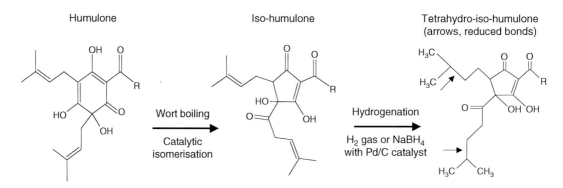

Figure 1.9
Structure and structural modification of humulone.

early work see, Bamforth, 1985). The interaction between hops and proteins in the bubble wall has been attributed to the ionized form of iso-α-acids cross-linking foam proteins (Asano and Hashimoto, 1976). It has been observed that isohumulone was more foam promoting than isocohumulone (Figure 1.10, Diffor et al., 1978). Ono et al. (1983) observed that isohumulones were foam-concentrated to a greater extent than their less hydrophobic counterparts iso-cohumulones. Combined with the finding that more hydrophobic proteins are foam promoting (Slack and Bamforth, 1983; Onishi and Proudlove, 1994; Yokoi et al., 1994), this supports Roberts' (1976) conclusion that the interaction between hops was based on either hydrogen bonding or hydrophobic interactions. As such, the more hydrophobic hydrogenated iso-α-acid hop products would be expected to be more foam promoting. This is indeed the case as shown in Figure 1.10. More recently Lusk et al., (2001a,b) found that tetrahydroisoalpha acids bind with LTP1 in a molar ratio of 23:1. On the basis of protein structural analysis the binding of hop acids was concluded to be the result of both ionic (interactions with basic amino acids) and hydrophobic amino acids. Other natural but minor hop components such as adprehumulone, dihydrohumulone and xanthohumol have been suggested to confer increased benefits over isohumulone for promoting foam stability and lacing (Smith et al., 1998; Wilson et al., 1999).

The impact of hop addition is governed by the hop form and the method by which foam quality is measured (Figure 1.11). Worts were made from two malt samples (variety A and B) and fermented by the small scale brewing procedure (~800 ml, 20 BU hop extract added, Stewart et al., 1998). As the hopping level increases, both foam stability (Rudin) and lacing increases. As with Figure 1.10, tetra-hop had a substantially greater impact on both foam stability

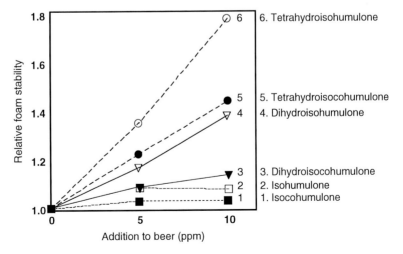

Figure 1.10
Comparison of relative foam stability (a version of the cylinder foam collapse test) of different forms of iso-humulone (from Smith et al., 1998).

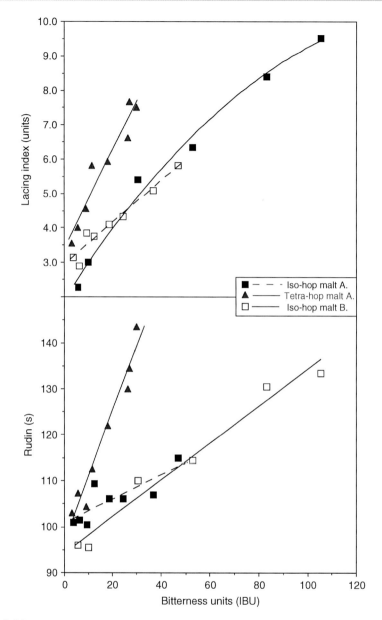

Figure 1.11
Comparison of the influence of hopping rate on foam quality (lacing index and Rudin) using iso-hop and tetra-hop extracts small scale brewing procedure (~800 ml, Stewart et al., 1998) (extracted from Evans et al., 2008).

and lacing than did iso-hop. Interestingly, the magnitude of the increase across the range of hop concentrations is substantially greater for lacing (>300%) than for the stability tests (Rudin ~150%). This observation demonstrates that hop iso-α-acids are of particular importance in determining the ability of beer foams to lace.

Cations

Metal cations are well known to promote beer foam stability and, in the extreme, gushing (Rudin, 1957; Rudin, 1958; Rudin and Hudson, 1958; Archibald et al., 1988). The affinity of isohumulone for cations, deduced by a spectrophometric method, was observed to occur in three affinity groupings: (1) strongest affinity: Mn^{2+}, Al^{3+}, Ni^{2+} and Sn^{2+}, (2) weaker affinity: Mg^{2+}, Zn^{2+}, Ca^{2+} and Ba^{2+}, and (3) no affinity: Li^+, Na^+, K^+ and Rb^+ (Hughes and Simpson, 1995). While cations such as Co^{2+} and Ni^{2+} (extremely toxic) or $Fe^{2+/3+}$ (reduced flavor/colloidal stability) can promote foam stability, their inclusion in beer is very undesirable. More recently, Roza et al. (2006) confirmed that the inclusion of non-toxic metal cations in beer had a significant foam stability benefit, such as the inclusion of as little as 2 ppm of Zn^{2+}. Simpson and Hughes (1994) proposed that multivalent cations bolster foam stability based on reversible cross-linking by hop-derived acids and proteins via "water molecules on the primary amino groups of polypeptides replacing the carbonyl and enolate groups of the iso-a-acids, with binding of the ion-diapole type."

Lipids

Lipids (along with certain detergents) are considered to be perhaps the most foam damaging beer components (Roberts et al., 1978). Certainly the removal of lipids by use of a BSA (a lipid binding protein) affinity column was shown to increase beer foam stability in 7 out of 43 beers evaluated (Dickie et al., 2001). As outlined in the "foam physics" section, small hydrophobic particles such as lipids interfere with the foam stabilizing interaction of protein hydrophobic regions and hop bitter acids and cations (Slack and Bamforth, 1983; Simpson and Hughes, 1994). In other words, lipids in beer promote the coalescence of bubbles that result in rapid beer foam collapse.

Lipids are a rather diverse group of naturally occurring substances. Letters (1992) outlined the most important classes in brewing that include fatty acids (i.e. linoleic acid), glycerides (i.e. triacylglyceride), phospholipids (i.e. phosphatidylcholine), glycolipids (monogalactosidyl-diacylglyceride) and sterols/sterol esters (i.e. ergosterol). It is widely considered that most beer lipid originates from malt and to a lesser extent hops (Blum, 1969; Anness and Reed, 1985; Letters, 1992). Under some conditions yeast can also produce lipids that tend to be of lower molecular weight and include a high proportion of the C_6 or C_8 fatty acids (Narziss et al., 1993). Of all the lipid extracted from malt, hops or yeast, generally only a very small proportion makes it into the finished beer which has been estimated to contain <4 mg/l fatty acid with 1–2 mg/l being typical (Anness and Reed, 1985). The methods used to determine these lipid contents that typically include freeze drying, solvent extraction (typically chloroform and methanol) followed by fatty acid methyl ester analysis by gas chromatography (Anness and Reed, 1985; Jackson, 1981) have ensured that the levels of lipids in beer are not routinely measured. Even the use of solid phase extraction (Wilde et al., 2003) does not appear to improve the efficiency of extraction greatly. As an alternative, Dickie et al. (2001) proposed the use of a BSA affinity column, followed by retesting of foam stability to test beer for lipid damaging effects. Thus the lipid-binding column identifies beers where the lipid damaging effect is important.

A number of investigations have sought to determine which lipids in beer were most foam damaging. Letters (1992) found that the addition of 5 mg/l (fatty acid equivalent) of glycolipid was twice as damaging to foam stability as was phospholipid or neutral lipid. More recently it was found that C_6 to C_{10} fatty acids had little effect on foam stability (Wilde et al., 2003), while fatty acids with longer chain lengths (i.e. C_{16}, C_{18}) were found to be increasingly more foam destabilizing. Bamforth and Jackson (1983) however found that shorter chain fatty acids were significantly inhibitory towards foam lacing. In addition, the degree of fatty acid desaturation with double bonds (i.e. $C_{18:1}$, $C_{18:2}$) increases a fatty acids' foam destabilizing capacity (Wilde et al., 2003; Segawa et al., 2002). In short, long chain, saturated fatty acids are more hydrophobic, hence their more negative foam impact. Recently, Kobayashi et al. (2002) concluded that the most foam destroying fatty acids were the di- and tri-hydroxyoctadecanoic acids (fatty acid hydroperoxides) derived via the action of lipoxygenase on linolenic acid ($C_{18:2}$). A similar conclusion that fatty acid hydroperoxides reduced foam stability was previously made by Yabuuchi and Yamashita (1979). This conclusion recently received substantial validation in that beer made from lipoxygenase-1-less malt had a significantly better foam stability than the beer made from the lipoxygenase-containing control (Hirota et al., 2005, 2006).

Anness and Reed (1985) concluded that the brewing process is very efficient at removing lipids with either the spent grains, hot break, or with the yeast as the role of lipids for yeast nutrition is vital (Thurston et al., 1982). More turbid worts, typically associated with higher levels of lipid result in improved fermentation vigour (Bamforth, 2002) and eventual improved foam stability (Kühbeck et al., 2006a,b,c). As such, it would be expected that most glycerides, glycolipids, phospholipids, sterols and fatty acids would be removed from the beer to support yeast metabolism under normal circumstances. However, the very foam destabilizing di- and tri-hydroxyoctadecanoic acids (Kobayashi et al., 2002) would probably not be utilized by yeast, so remain to potentially destabilize the beer foam. It should also be remembered that, although lipids introduced into beer have an initial deleterious impact on foam stability, the foam stability can recover either fully or partially if the beer is allowed to rest for 24 h (Roberts et al., 1978). This underlines the importance of the level of lipid binding proteins in the beer, their state and their level relative to lipids (van Nierop et al., 2004).

There is no evidence that the essential oils from hops have any impact on foam stability at the levels found even in the "hoppiest" products.

Other foam negative/positive beer constituents

There are a small number of other beer constituents that appear on the whole to be foam negative but have in some investigations been suggested to also be foam promoting. In general, the influence of these components is considered to be minor. The first of the dual personality beer constituents is ethanol. It would be expected that ethanol influences foam stability as it substantially lowers both static and dynamic surface tension (Fisher et al., 1999). Brierley et al. (1996) found with a micro-conductivity test that high levels of ethanol reduced foam stability while low concentrations have been found to improve foam stability as judged by the Rudin test. Lewis and Lewis (2003) observed a relatively weak positive association between ethanol and foam stability as measured by the Constant method. Certainly, one readily observes that following the application of high concentrations of ethanol to foam, it collapses immediately (in the lacing index test), which supports Bamforth's (1999) contention that ethanol can act in an analogous fashion to lipids – disrupting interactions between proteins and iso-α-acids.

The relationship between beer foam quality and polyphenols has been rarely studied. Bamforth (1999) suggested that polyphenols could act in a way analogous with iso-α-acid, cross-linking proteins in the bubble wall. Indeed, Lewis and Lewis (2003) found a strong association between the levels of polyphenols and foam stability. In addition, in model systems catechin can promote the foaming of lactoglobulin (Sarker et al., 1995). Conversely, polyphenols participate in the precipitation of proteins, particularly haze active proteins, during boiling but also during other phases of the brewing process (Asano et al., 1984; Lewis and Serbia, 1984; Yasui et al., 1998). Such a negative role by not only precipitating out haze active but foam promoting proteins has been observed by Evans et al., (1999c). Certainly, the small scale brewing trials conducted with Caminant (ant-28, proanthrocyanidin-free; Jende-Strid, 1997) or hull-less malt samples produced beers with lower polyphenol contents, higher protein content and Rudin foam stability (Evans and Sheehan, 2002). However, brewing experiments using other proanthrocyanidin-free lines showed little difference in foam stability from their parent controls (Back et al., 1999; Delcour et al., 1987; Fukuda et al., 1999; Jende-Strid, 1997; O'Donnell, 1987; von Wettstein et al., 1980). This suggests that in practice polyphenols are of minor importance for foam stability.

The pH of beer has often been found to be a factor in foam stability. The multiple regression models of Melm et al. (1995), showed improved prediction of foam stability when pH was included as a variable. The amphipathic nature of both proteins and hop acids mean their surface characteristics and charge will change with pH. Certainly fermentation and the carbonation of beer with CO_2 will lower the pH. Lower pH values would be expected to result in greater dissociation of hop acids and protein charge to aid migration into foam and their interaction. Increasing concentrations of CO_2 will also of course push the equilibria of bubble nucleation towards formation, a positive contribution towards creaming and bubble recruitment that will improve foam stability (Lynch and Bamforth, 2002). Basic amino acids (arginine > lysine > histidine) interfere with the protein – iso-α-acid interaction to inhibit lacing (Furukubo et al., 1993, Honno et al., 1997). This finding was proposed to explain the influence of several brewing factors including pH in mashing, wort, wort aeration and malt ratio on lacing. In model systems Bamforth and Kanauchi (2003) found higher foam stabilities as the pH was raised through the range 3.8–4.6.

Manipulating the brewing process to optimize foam quality

The preceding discussion and evaluation has summarized the current understanding of what components in beer are important and could potentially be manipulated to optimize foam quality. The emphasis in the statement is on the word "optimize." In the beginning of this chapter, customer expectations for foam quality were placed at the fore because they are the brewers' objective for brand placement. This can be a complicated assignment because customer preferences differ not only on nationality but also on gender and regional lines (Bamforth, 2000a; Smythe et al., 2002; Roza et al., 2006). Perhaps this is a tall order in this current age of brewery consolidation and global beer brands.

In reductionist science terms, one is often left with the impression from the literature that the objective is to maximize (or minimize) the parameter in question, particularly with foam quality. This somewhat black and white view is rather limiting. For instance if the objective was to produce highly stable and foam-able beer, the blinkered choice would be to use *Fusarium* infected malt. This would produce very foam-able, gushing beer that would perhaps initially be exciting and entertaining but would soon be viewed as rather pointless as half the package would reside on the ceiling in extreme cases, not to mention the undesirable health side effects of the attendant mycotoxins. So generally rather than wanting either of the extremes, that being more or less foam, some subtlety is required by the brewer to meet their customers' desires. This optimization of beer foam quality will ideally require that a certain amount of foam to be produced, of a certain stability, with an appealing color (degree of whiteness), degree of creaminess, which during the course of consumption will leave the desired amount of lacing so as to best satisfy the consumer. Fortunately, there are a number of opportunities for the brewer to manipulate foam quality by the selection of raw materials, modifying the brewing process, along with palliative choices for using foam enhancing additives and devices. These potential options will be outlined in the sections that follow.

Raw material selection

Malt selection

A brewers "knee-jerk" reaction with any perceived deficiency in beer quality or brewing efficiency is to first pass the blame on to the quality of the malt and the maltster. While this conclusion may have elements of truth, the fault is more in the specification requested by the brewer rather than any deficiencies in the maltster's ability to practice their trade.

The first option is to use malt with the appropriate levels of foam promoting proteins. Although Roberts et al. (1978) showed that beer can be diluted to around 30% concentration before rapid loss in foam stability (Rudin), it is clear from many other studies that changes in the amount and composition of beer foam proteins will alter the foam stability of beer (for a comprehensive review see Evans and Sheehan, 2002). On the whole, these changes tend to be incremental. As barley/malt protein content rises, so does the level of protein

Z4 and protein Z7 (Evans et al., 1999a,b,c) because the expression of protein Z is under control similar to that of other seed storage proteins such as hordeins (Giese and Hejgaard, 1984). However, as in most varieties, protein Z4 is the dominant isoform, 80% on average, the level of protein Z in malt is essentially governed by the level of protein Z4 (Evans and Hejgaard, 1999).

The levels of protein Z4 and Z7 in barley and malt are determined by genotype. The levels of protein Z4 can be divided into three categories, high (>1000 μg/g malt), intermediate (200–1000 μg/g malt) and low (<200 μg/g malt) (Table 1.5). Protein Z7 can also be divided into two categories, high (>150 μg/g malt) and low (<150 μg/g malt). Genetic mapping showed that one location on the short arm of chromosome 4H controlled the level of protein Z4 while a single location at the distal end of chromosome 5H controlled the level of Z7 (Evans et al., 1999a,b,c). These locations are consistent with the locations of the structural genes for protein Z4 and protein Z7 (Hejgaard, 1984;

Table 1.5
Summary of protein Z4 and Z7 categories for malt from a selection of varieties surveyed in 1998, 1999 and 2000 (Evans, unpublished data using ELISA described by Evans and Hejgaard, 1999)

High protein Z4 (>1000 μg/g malt)		High protein Z7 (>150 μg/g malt)	
Alexis[E]	Grimmett[A]	Arapiles[A]	Moravian III[U]
Arapiles[A]	Harrington[C]	Barque[A]	Morex[U]
Barque[A]	Lindwall[A]	Bowman[U]	Parwan[A]
Bonanza[U]	Picola[A]	Chebec[A]	Pirkka[E]
Bowman	Sloop[A]	Franklin[A]	Sloop[A]
Caminant[E]	Tallon[A]	Grimmett[A]	Stirling[A]
Chariot[E]	Unicorn[J]	Harrington[C]	Unicorn[J]
Gairdner[A]		Karl[U]	
Intermediate protein Z4 (200–1000 μg/g malt)		**Low protein Z7 (<150 μg/g malt)**	
Chebec[A]	Schooner[A]	Alexis	Gairdner[A]
Fitzgerald[A]	Stirling[A]	Barque[A]	Lindwall[A]
Parwan[A]		Bonanza[U]	Picola[A]
		Caminant[E]	Schooner[A]
Low protein Z4 (<200 μg/g malt)		Chariot[E]	Tallon[A]
Karl[U]	Morex[U]	Fitzgerald[A]	
Moravian III[U]	Pirkka[E]		

Origin of variety: [A] Australia, [E] Europe, [U] USA, [C] Canada, [J] Japan

Evans et al., 1995). Thus the levels of protein Z4 and protein Z7 are both determined by the expression of a small number of major genes with at least two alleles. Interestingly, the level of protein Z4 appears to be coordinated with the β-amylase thermostability type (Evans, unpublished data), Sd2L, Sd2H and Sd1 (Eglinton et al., 1998; Kihara et al., 1998; Paris and Eglinton 2002; Polakova et al., 2003; Ovesna et al., 2006). So far, intermediate levels of protein Z4 are associated with the Sd2L type while high levels of protein Z4 are associated with the Sd2H and Sd1 types. The exception is some descendents of Morex (Sd1, USA) and Pirkka (Sd1, Finland) that have low levels of protein Z4. The preservation of these associations is interesting as the protein Z4 gene is located on the short arm while the β-amylase gene is located on the long arm of chromosome 4H (Powling et al., 1981; Nielsen et al., 1983). Given the relatively long genetic difference between these loci, more genetic recombination would perhaps have been expected. In practice these observed associations are useful because most commonly used, high diastatic power varieties are generally Sd1 (or Sd2H), so that high levels of protein Z4 are generally being delivered to breweries. As such, selection by brewers, maltsters and malting barley breeders is relatively simple once the characteristics of a variety are known.

The ability of maltsters and brewers to select the level of LTP1 in malt is unfortunately not so straight forward. To date no simple association between barley genotype or protein content has been observed that predicts the level of LTP1. Table 1.6 shows that for a mixture of Australian, European, Canadian and Japanese lines grown in South Australia in 1997, the level of LTP1 between these relatively genetically diverse varieties is very similar. The one association that has been observed is that barley grown in more humid or wet environments appears to have substantially higher levels of LTP1. Table 1.7 clearly shows that the level of LTP1 grown in the drier Australian environment has substantially lower levels of LTP1 than barley grown in more humid Mississippi, USA environment. This observation is perhaps not surprising because there is a growing consensus that one of the primary biological functions of LTP1 is as a plant defence protein (Douliez et al., 1999). It follows that in environments where that barley is challenged by pathogens or insects, the level of LTP1 would most likely be higher. While the selection of malt with high LTP1 is theoretically possible to increase foam stability, this course of action may not be viable. Recent research has suggested that conditions that stimulate the accumulation of plant defence proteins, not only increase the level of LTP1 but also the level of lipoxygenase (Wackerbauer et al., 2005). As outlined earlier, lipoxygenase is responsible for the production of the highly foam damaging fatty acid hydroperoxides, so that the exercise may be self defeating or worse if too much of the lipid binding ability of the LTP1 is lost due to boiling (see following section for discussion).

The understanding of the level of foam promoting hordeins in barley or malt is not as well defined as for protein Z and LTP1. This is despite a number of studies identifying foam promoting hordeins proteins in beer, generally using immunological procedures (i.e. ELISA) that should be applicable to evaluation of malt or barley (Mills et al., 1998; Kauffman et al., 1994; Sheehan and Skerritt, 1997; Onishi et al., 1999; Vaag et al., 1999; Evans et al., 2003). Of these

Table 1.6
Average malt varietal content of LTP1 for a selection of barley varieties grown in Australia 1997 (Evans, unpublished data using ELISA described by Evans and Hejgaard, 1999)

Variety	LTP1 (µg/g)	σ	n
Alexis[E]	313	24	2
Arapiles[A]	421	65	18
Chariot[E]	518	130	2
Chebec[A]	437	48	7
Fitzgerald[A]	457	167	7
Franklin[A]	369	75	27
Gairdner[A]	465	146	9
Grimmett[A]	376	74	13
Harrington[C]	468	112	15
Lindwall[A]	286	43	5
Manley[C]	415	128	2
Parwan[A]	297	58	5
Picola[A]	460	7	3
Schooner[A]	477	92	30
Sloop[A]	621	189	9
Stirling[A]	513	227	14
Tallon[A]	383	84	10
Unicorn[J]	541	229	2

Origin of variety: [A] Australia, [E] Europe, [C] Canada, [J] Japan

studies only Vaag et al. (1999) was able to show variation in the level of the 17 kDa hordein between samples of malt although no indication was given if this variation was genetic or environmental. Evans et al. (1999, 2003) reported a promising antibody to a foam protein fraction, 2x foam, that almost certainly reacted with B and C hordeins in immunoblots. However, the quantitative ELISA developed with the 2x foam antibody was later found to essentially be measuring the level of protein Z4. It is suggested that an antibody to a similar foam promoting protein fraction and the ELISA developed using it (Ishibashi et al., 1996, 1997), were also essentially measuring protein Z4. Overall, perhaps the difficulty in translating antibodies that recognize foam promoting hordeins in beer to evaluation of their levels and composition in malt is related to the requirement for the water insoluble hordeins in barley to first be hydrolyzed into water-soluble fragments by proteases.

Table 1.7
Comparison of the level of LTP1 in malt made from barley grown in Australia 1997 with that from malt grown in Mississippi in 1996. (Evans, unpublished data using ELISA described by Evans and Hejgaard, 1999)

Statistic	LTP1 (µg/g)	
	Australia	Mississippi, USA[a]
Mean	434	1082
σ	140	132
minimum	230	796
maximum	898	1315
n	169	39

Malt provided by Dr Berne Jones, USDA, Madison WI.
[a]Lines include breeders lines, Morex, Robust, Stander and Excel.

For brewers wishing to adjust the foam characteristics of their beers, the most pertinent question is, do the levels and compositions of foam promoting proteins in malt influence the eventual level of foam protein that ends up in beer? Firstly, the malting and brewing process, especially those parts in which significant heat is applied (i.e. kilning, wort boiling) reduce the level of foam proteins including protein Z and LTP1 (Ishibashi et al., 1996, 1997; Kakui et al., 1999; Evans and Hejgaard, 1999; van Nierop et al., 2004). In terms of malt, the degree of kilning or roasting was particularly important in the level of foam protein that was subsequently extractable (Ishibashi et al., 1997). Consequently, the more highly colored the malt, the less foam active protein will be available for extraction into the beer. Vaag et al. (1999) found for two malt and beer (pilot brew) sets that higher levels of LTP1, 17 kDa foam promoting hordein and protein Z in malt resulted in higher levels of these proteins in beer. Table 1.8 shows an extensive comparison of the influence of malt foam protein levels on beer. For protein Z4 and protein Z7 the correlation was strong ($r > 0.80$) with the level of malt and beer protein Z4 being significantly correlated with the level of CBB total protein. However, the level of malt LTP1 was not significantly correlated with levels in beer. The LTP1 association will be discussed further in the impact of the brewing process section that follows.

Malt modification

The degree of malt modification would be expected to have both positive and negative influences on foam stability, dependent on how foam stability is measured as discussed previously (non-starch polysaccharides section). Non-starch polysaccharides such as β-glucan and arabinoxylan are broken down during malt modification, a desirable outcome for brewing solid–liquid separation processes. Therefore, regardless of whether such components have a foam promoting effect or not, increasing their level is not a practical way of increasing foam stability.

Table 1.8
Correlation of beer protein content and composition with the level of malt protein Z4, protein Z7 and LTP1 for pilot and small scale brews (data extracted from Evans et al., 1999c)[a]

	Beer (µg/ml)			
	Protein[b]	Protein Z4	Protein Z7	LTP1
Beer (µg/ml)				
Protein Z4	**0.57****	–	–	–
Protein Z7	−0.12	–	–	–
LTP1	0.36	0.28	–	–
Malt (µg/g)				
Protein Z4	**0.62****	**0.85****	−0.38	–
Protein Z7	−0.21	−0.39	**0.81****	−0.32
LTP1	−0.06	−0.18	−0.37	−0.38

[a] $n = 25$, 9 pilot brews and 16 small scale brews.
[b] Protein, measured by the Bradford coomassie blue binding assay.

For malt proteins, Bamforth (1985) concludes that optimal proteolysis is required for the production of yeast assimilable nitrogen, to render starch accessible to enzymes but on the other hand to ensure that enough protein survives to stabilize the foam. However, proteins such as protein Z and LTP1 are characterized by their resilience to proteases. Certainly it is clear that protein Z4 levels do not decrease with increasing modification, but rather increase slightly (Evans et al., 1999a,b,c). A similar conclusion can also be made for LTP1 (Evans, unpublished data). This leaves the hordeins, which in the main require some proteolytic action to become water soluble in the first place which suggests that there is some, yet to be defined, optimal level of malt modification for foam stability. The proline endoproteinase experiment (Table 1.3) indicates that there are foam positive and negative hordeins but the characterization of the foam value of individual species has yet to be investigated. In terms of Rudin foam stability, Evans et al. (1999a,b,c) concluded on average, that a 1% increase in Kolbach Index was responsible for a 1 sec decrease in Rudin foam stability. As a significant portion of the decrease in foam stability is due to β-glucan and arabinoxylan break down, it is suggested that the impact of foam protein modification is perhaps relatively small at normal levels of modification.

Malt lipoxygenase

The recent observation that fatty acid hydroperoxides were the most foam destabilizing of lipids (Kobayashi et al., 2002) and the development of viable lipoxygenase-1-less malting varieties (Hirota et al., 2005, 2006) has focused attention in the impact of lipoxygenase, not only on stability of flavor but also foam. In conventional varieties, lower levels of malt lipoxygenase are beneficial for foam stability (Kobayashi et al., 1994). Lipoxygenase can perhaps be reduced by increased severity of kilning, due to its relatively low thermostability (Schwarz and Pyler, 1984; Yang and Schwarz, 1995), but such practices will have ramifications for other thermolabile but desirable malt enzymes (i.e. β-amylase, β-glucanase). It has also been found that the removal of acrospires can improve the foam stability of the subsequent beer (Tada et al., 2004; Nishida et al., 2005). The improvement of foam stability was also in part attributed to the reduction in the levels of foam inhibiting basic amino acids in the beer (Tada et al., 2004). In addition, the level of trans-2-nonenal was also reduced in the beer produced from grist where the acrospire fraction had been removed (Nishida et al., 2005). As trans-2-nonenal is an end point of the pathway initiated by lipoxygenase, these results also suggest an influence by this enzyme. Overall, the selection of malts with low or null levels of lipoxygenase appears an attractive option to increase foam stability in addition to beer flavor.

Adjunct selection

It is brewers dogma that if certain adjuncts, particularly wheat are added to beer, increased foam stability and quality result (Birtwistle et al., 1962; Leach, 1968; Bateson and Leach, 1969; Stowell, 1985; Kakui et al., 1999). This desirable effect from wheat was found to vary in magnitude between variety and wheat source (Bateson and Leach, 1969; Kakui et al., 1999). Perhaps as Bamforth (1985) suggests, this can be attributed to the provision of "a greater quantity of less extensively degraded polypeptide to wort than malt." It is certainly true that wheat production is less restricted in the upper limits of grain protein content where malting barley is generally limited to between 9.5 and 12% total protein. Wheat is also a rich source of arabinoxylan which encouraged Stowell (1985) to conclude that the wheat effect was at least in part due to increased wort viscosity, rather an undesirable virtue for modern brewing practice. The use of certain wheat flours has also been noted to decrease the size of foam bubbles (Kakui et al., 1999). These investigators used a foam-active protein ELISA to track these proteins in wheat and barley. Interestingly, 30% more of the wheat foam active proteins were lost during brewing as compared to those from malted barley. As wheat is also a rich source of puroindoline (Douliez et al., 1999), it is possible that these lipid binding proteins are reducing wort lipid loads even though they do not survive into the beer (Evans and Sheehan, 2002). Other cereals, most notably rice, are recognized for their lack of contribution of protein to beer, hence the use of this adjunct would be expected to reduce foam stability. Overall, it is possible that judicious selection of adjunct can be a viable means to optimizing foam stability and quality.

Hop selection

Hop acids in combination with beer proteins perform a pivotal role in determining the stability and quality of beer foam. As indicated previously, the foam is also the conduit for conveying aroma and the first presentation of bitterness to the consumer. In this, the composition of hops is obviously of prime importance *vis a vis* the relative proportion of the aroma compounds in the essential oils relative to the bittering acids. The relative proportions of isohumulone to coisohumulone, whether "natural" or chemically modified has an important influence on foam stability and lacing (Diffor et al., 1978, Figures 1.10 and 1.11). Hence hop products with higher levels of isohumulone compared to isocohumulone will have somewhat more stable foams while hydrogenated iso-α-acids will tend to increase stability and lacing rather more. Similarly hop products with increased levels of minor hop constituents such as adprehumulone, dihydrohumulone and xanthohumol may also provide options to modify foam quality and stability (Smith et al., 1998; Wilson et al., 1999). However, this option awaits the development of commercially viable products with these characteristics.

It is widely accepted in the brewing industry that foam stability is substantially improved by the inclusion of hydrogenated iso-α-acid hop products (Figure 1.10, Smith et al., 1998). This improvement extends to the amount of lacing as demonstrated in Figures 1.11 and 1.12. Exclusive use of tetra hop in the lacing index test produces a lace that is highly adherent to the glass, so much so that the foam adheres from the top point of foaming and then at the subsequent "sip" points in the test (Figure 1.12(a)). This provides a "curtain-like" lacing effect on the glass. Even a relatively small proportion of tetra hop with respect to isomerized hop is sufficient to provide this adherence although there the curtain effect is less pronounced and associated with a lower lacing index, thus providing distinct lacing rings down the side of the glass (Figure 1.12(b)). Finally, when only isomerized hop is used there is a distinct "slippage" of the foam adherence from the sipping point (Figure 1.12(c)). The central role of hopping

 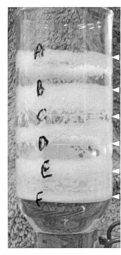 

(a) "Tetra hopped beer (11 BU "tetra") Lacing index = 5.66

(b) Iso-"tetra" hopped beer, (23 BU, ~5 BU "tetra") Lacing index = 4.65

(c) Iso-hopped beer (41 BU) Lacing index = 2.49

Figure 1.12
Pictures of the lacing index test glass at the completion of the test using commercial beers hopped in a variety of ways. The arrows indicate the "sip" marks on the glass.

as a determinant of lacing was also observed by Archibald et al. (1988). As such, hop selection provides a useful means for brewers to tailor their beer brands to their customer's preferences. In this, brewers in their beverage development and market targeting need to keep in mind that male drinkers typically appreciate lacing more than do female drinkers (Roza et al., 2006).

Brewers considering the use of hydrogenated iso-α-acid hop products need to be aware of some of the potential disadvantages. Firstly, if the beer is somewhat over-foamed as in the Cylinder pour foam test, predominantly tetra hopped beers will eventually degrade to produce a foam that is reminiscent of "whipped egg-white, icebergs" and lace that is lumpy and powdery in consistency (Figure 1.13(a)), compared to the more consistent though less stable foam from a predominantly iso-hopped beer (Figure 1.13(b)). There is

(a) Cylinder pour test of tetra hopped beer showing: "whipped egg-white, ice-bergs" of powdery lace and foam

(b) Cylinder pour test of an iso-hopped beer: at the beginning and completion of test

Figure 1.13
Comparison of the impact of hop form in the Cylinder pour test.

also some debate over the flavor impact of use of modified hop products such as tetra hop in terms of bitterness and bitterness quality (Hughes, 2000; Weiss et al., 2002; Cvengroschová et al., 2003). Some experienced beer tasters have even commented that hydrogenated hop products can impart a slight odor of "vulcanized rubber" on the beer (Anon.). Thus, unless the brewer requires the light stabilizing effects of hydrogenated hops, a small addition of hydrogenated iso-α-acid hop to the overall hop load may be a useful tool in optimizing foam quality.

The brewing process

In the Brewhouse

In terms of foam stability, the brewhouse processes of mashing and boiling are a compromise between solublization of proteins and the requirement that they are not unnecessarily lost due to proteolytic action or thermal denaturation. One must also not forget that foam quality is just one of the beer qualities the brewer is seeking to optimize in a commercially efficient process. Of the foam promoting proteins, protein Z and LTP1 are both tolerant of high temperatures and resistant to proteolysis (Hejgaard, 1977; Jones et al., 1995; Kaersgaard and Hejgaard, 1979; Perrocheau et al., 2006) which enables their survival and resilience to the brewing process. The situation with hordeins is more of a balance, as the insoluble barley hordein needs to be made water soluble, but proteolysis must not proceed too far. The extent of proteolysis is essentially determined by the malting process where Jones (2005) estimates that two thirds of barley proteolysis occurs, the other third occurs during mashing.

Grist to water ratio has been shown to be critical to protein extraction in mashing. With the practice of high gravity brewing, it is clear that extraction of hydrophobic, thus desirable foam promoting proteins, is substantially reduced (Cooper et al., 1998a,b; Stewart et al., 2006). Lower mashing in temperatures, particularly $< 55\,°C$ (Palmer, 2006; Evans et al., 2005; Jones, 2005) would be expected to promote increased proteolysis and potential loss of foam promoting proteins. Increased proteolysis would also presumably result in the over production of FAN with its attendant proportions of basic amino acids, not efficiently utilized by yeast, resulting in foam inhibition by these amino acids (Furukubo et al., 1993). That proteolysis occurs in mashing has been questioned (Lewis and Serbia, 1984). A low temperature rest during mashing would also be favorable to lipoxygenase and the consequent production of foam damaging fatty acid hydroperoxides (Schwartz and Pyler, 1984; Kobayashi et al., 1994; Yang and Schwartz, 1995; Kuroda et al., 2002).

Conversely, higher mashing in temperatures would inhibit protease and lipoxygenase activity to perhaps improve foam stability. Certainly, Ishabashi et al. (1997) observed that mashing in at $71\,°C$ increased the level of foam promoting protein in wort. This was confirmed by Sheehan and Skerritt (1997), who suggested that higher mashing temperatures lead to greater survival of hordein-derived polypeptides resulting in increase foam stability. Narziss et al. (1982a,b) observed that higher mash temperatures (>70 °C) promoted the formation of glycoproteins with improved foam promoting ability. Haukeli et al. (1993) also reported an increase in foam potential by increasing mashing-in temperature from 50 °C to 60 °C. In addition, the level of 40 kDa protein (protein Z) and subsequent beer foam stability was observed to be higher in malt that had been wet rather than dry milled (Kano and Kamimura, 1993). Overall, the consensus of opinion appears that mashing in at temperatures 65 °C or greater will have some benefits in increasing foam stability. However, mashing in at temperatures over 65 °C results in reduced wort fermentability due to the thermal inactivation of some starch hydrolyzing enzymes including β-amylase (Evans et al., 2005).

In the brewhouse the critical point for foam stability is wort boiling, not surprisingly the part of the process that has the most extreme temperature conditions. Wort boiling of course has a number of key functions including hop acid extraction and isomeration, termination of remaining malt enzymatic activity, wort concentration, color/flavor enhancement, protein denaturation, and precipitation of haze active proteins and lipids in the trub. Heat denaturation of LTP1 has also been shown to be required to bring out the foam promoting characteristics of LTP1 (Sorensen et al., 1993; Bech et al., 1995a,b; Lusk et al., 1995; Jegou et al., 2001). Boiling is also the most potent stage in the brewing process for the formation of Maillard products (i.e. melanoidins) due to the high concentration of peptides, amino acids and sugars in an aqueous environment with high heat. Protein–carbohydrate Maillard products have long been associated with improving the foam promoting characteristics of beer proteins (Roberts, 1975; Jackson and Wainwright, 1978; Lusk et al., 1995; Hughes, 1997). Not surprisingly, Curioni et al. (1995) observed that glycosylated protein Z had improved foam stability characteristics. Increased glycation and reduction of the structure conserving the four di-sulfide bridges within LTP1 has been shown to contribute to its better adsorption to air–water interfaces (Jegou et al., 2001). It was also found that denaturation of LTP1 was substantially improved at temperatures <100 °C when a reducing agent was included (Perrocheau et al., 2006). The improved foam promoting capability may be explained by the observation that glycosylation improves the flexibility of proteins which enhances their foam promoting ability (Townsend and Nakai, 1983; Le Mestre et al., 1990).

It is to be expected that there is a limit as to the extent of protein denaturation and wort boiling that is desirable. Narziss (1993) observed that increasing the counter-pressure on the wort in an external boiler system could create mean boiling temperatures between 103 and 110 °C, with accelerated protein precipitation, DMS stripping, hop acid extraction and isomerization and, reduced boiling time by 30–40%. However, the higher boiling temperatures were observed to reduce beer foam stability, flavor balance and body relative to temperatures below 103 °C. Therefore, boiling temperatures above 103 °C were not recommended. Van Nierop et al. (2004) considered that wort boiling systems where there was high ΔT-high temperature differential at the wort boiler surface could substantially reduce the level of foam promoting proteins. A more recent evaluation of wort boiling practices provides some possible options for reducing this effect with improved design of the wort heater component of the kettle (Andrews and Axcell, 2003). Further investigation in this area would be expected to assist in optimizing the level and conformation of foam promoting proteins in beer.

Extensive investigation of the influence of wort boiling on the levels of foam promoting proteins has proved enlightening (van Nierop et al., 2002, 2004). This investigation was a result of the location of a company's facilities variously either near sea level or at ~1600 m. At lower altitude wort boiling temperatures were found to be ~102°C while at higher altitudes, 96°C. The lower wort boiling temperature at high altitudes resulted in much higher levels of ELISA measurable foam protein (see Table 1.2). The subsequent investigation confirmed that most foam positive protein and in particular LTP1 were lost during boiling (Figure 1.14).

A later investigation focusing on wort boiling found that the plots for foam protein loss were divergent (Figure 1.15). The level of LTP1 decreases rapidly within the first 15 min while the level of protein Z4 and CBB measurable protein is reduced more gradually and consistently during the boil. Overall, the level of reduction in LTP1 is 97% compared to 47% for protein Z4 and 29% for CBB protein. As the LTP1 antibody was raised to native barley LTP1, van Nierop et al. (2004) concluded that once LTP1 was denatured and glycosylated

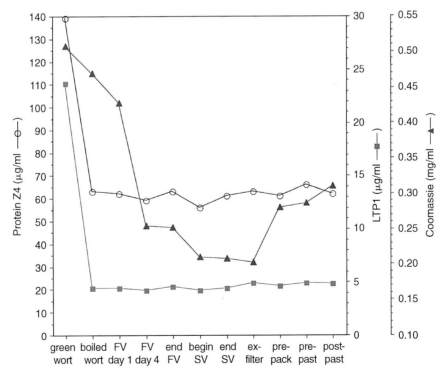

Figure 1.14
The effect of brewing process on the levels of protein Z4 and LTP1 (extracted from van Nierop et al., 2002).

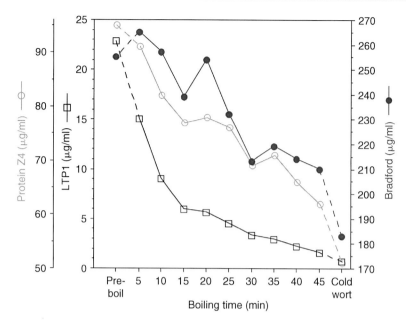

Figure 1.15
Impact of boiling on the level of wort proteins (Evans, unpublished).

beyond a certain point, the ELISA developed by Evans and Hejgaard (1999) was less able to measure these LTP1 forms. That is, it was still present in the beer but not measurable with this ELISA. However, van Nierop et al. (2004) also concluded that conformational changes associated with LTP1 denaturation can be deleterious for foam stability, especially when fatty acids are present (Table 1.2). Thus the practical investigations of van Nierop et al. (2002, 2004) strongly suggest that the lipid binding role of LTP1 is of greater importance to foam stability than its direct role as a foam promoting protein. Hence the level of LTP1 measured in beer with the Evans and Hejgaard (1999) ELISA is a very useful measure of potential beer foam stability. In practice, brewers desiring more stable and resilient foam should attempt to minimize the severity of wort boiling within the constraints of their kettle design and the achievement of the other objectives of wort boiling including hop isomerization, protein precipitation and flavor enhancement.

Understanding the impact of wort boiling on the measurement of LTP1 enables a revisiting of the question; do higher malt levels of LTP1 result in beer with higher levels of LTP1? Figure 1.16 compares the level of malt vs. beer LTP1 for pilot brewed (rolling boil, ~102°C) to those from a small scale brewing experiment (bottle in boiling water bath, ~101°C). Where the boil was more rigorous as with most of the pilot scale brews, the level of beer LTP1 is relatively low. Conversely with the less severe small scale wort boil, higher levels of beer LTP1 can be measured. The scatter plot also suggests that, within these two groups, malts with higher levels of LTP1 yield beer with higher levels of

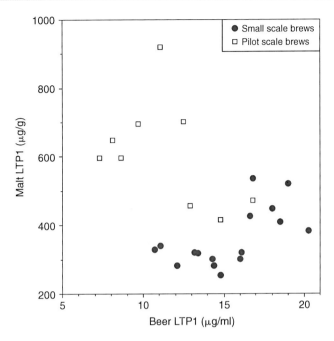

Figure 1.16
Scatter plot comparing LTP1 from malt with the content in the resulting beer by pilot or small scale brews (data extracted and re-evaluated from the experiments described in Evans et al., 1999c).

LTP1. Thus in conclusion, the final level of LTP1 beer and its foam promoting status is an interaction between its level in malt and the vigor of the kettle boil.

One question is what is the influence of turbid or lipid rich worts on eventual beer foam stability? Certainly mash filters and other brewing practices are known to increase lipid content in sweet wort (Kühbeck et al., 2006). To some extent these lipids are removed with the trub during boiling. However, more turbid worts, typically associated with higher levels of lipid result in improved fermentation vigor (Bamforth, 2002) and better foam stability (Kühbeck et al., 2006a,b,c). Such worts would result in improved fermentations due to more vigorous yeast that efficiently remove lipid from the wort as part of their metabolism. Such healthy yeast would also be less likely to secrete harmful lipids or proteinase A into beer.

Fermentation, maturation and beyond

The fundamental role of yeast during fermentation is to attenuate the wort to produce alcohol, although it also produces flavors, using fatty acids, amino acids, micro- and macro-nutrients (i.e. Zn, biotin) to achieve this function. It follows that any brewing process factor that stresses the yeast during or after fermentation may result in the secretion of foam destabilizing components such as proteases, lipids or in extreme cases the lysis of yeast cells which would spill these and other undesirable components into the beer. The obvious starting point is to pitch high quality viable and vital yeast.

Much of the historical information regarding the influence of fermentation on foam stability has been comprehensively reviewed in Bamforth (1985). This review concluded that foam stability could be reduced by:

- Over foaming of beer during fermentation resulting in loss of foam promoting components induced by the large release of CO_2.
- Yeast strains differing in foam inhibiting activity.
- Lipid secretion from stressed yeast.
- The inappropriate or over use of anti-foams.
- Non-flocculent yeast strains, which produce beer with less stable foam than flocculent strains.
- Fermentation conditions that promote yeast activity such as the addition of exogenous enzymes such as amyloglucosidase, excessive wort aeration, excessive levels of trub in the wort, increased temperature and pitching rate.
- Too rapid changes (deceases) in fermentation temperature stressing yeast.
- Beer filtration, particularly membrane filtration (Donn Hawthorne, Fosters Group, personal communication), removing foam promoting proteins.
- Achieving haze stability by the use of proteases such as papain, which not only hydrolyze haze active proteins but also foam promoting proteins (Siebert and Lynn, 1997). Conversely, isinglass finings were suggested to have a positive benefit by removing lipids (Bamforth, 1985; Archibald, 1988).

Much of our recently improved understanding of the influence of this stage on the brewing process has resulted from investigating the effects of high gravity brewing and the secretion of yeast-derived proteases. This identified much useful information that should be considered for optimization of conventional brewing practices. Conveniently, Stewart et al. (2006) recently provided a useful synopsis of the main effects of high gravity brewing on beer foam, including extraction and retention of foam promoting proteins (discussed above), pasteurization and centrifugation. Obviously the act of pitching yeast into high gravity wort, is likely to more severely stress the yeast and result in reduced foam stability. Similarly, the use of centrifugation, although enabling the rapid separation of yeast and other particles from the beer, will lower yeast quality and reduce foam stability because of the increased temperature, gravitational and shear forces. Conversely, Kondo et al. (1995) advocated the use of centrifugation to remove the yeast from wort as soon after sugar assimilation as practicable. This was consistent with the recommendations of several investigators who found that foam stability was improved if yeast was separated from beer as early as possible (Ormrod et al., 1991; Haukeli et al., 1993; Kondo et al., 1998).

The seemingly most potent and insidious contribution yeast makes to reduce foam stability is the excretion of proteinase A. This protease slowly degrades hydrophobic foam promoting proteins, particularly those at around 10 kDa (Shimizu et al., 1995; Wang et al., 2005), which has recently been confirmed to be LTP1 (Leisegang and Stahl, 2005; He et al., 2006). Kondo et al. (1998) linked low yeast vitality levels with increased proteinase A levels and correspondingly poorer foam stability. Poor yeast vitality was increased by the yeast being under stress due to nitrogen starvation, and high levels of alcohol, CO_2 and pressure (Kondo et al., 1995, 1998). In addition, it has been found that different yeast strains either excrete or leach when dead, different levels of proteinase A, with top fermenting strains apparently producing more protease (Muldjberg et al., 1993; Kondo et al., 1995, 1998). The impact of proteinase A, as would be expected from elementary biochemistry, is exacerbated by storing beer at room temperature or above (Nielsen and Hyobye-Hanson, 1988; Kondo et al., 1995; Lusk et al., 2003). Consequently, any measures the brewer can implement to minimize thermal load during warehousing and distribution should have benefits for foam stability, not to mention flavor stability (Bamforth, 2004a,b). However, the thermal treatment of pasteurization can inactivate proteinase A and is advocated as a solution to reduce its impact on foam stability by Stewart et al. (2006).

Palliative options: of gas compostion, widgets and other devices

Nitrogen vs. CO_2

While most beers are naturally and conventionally carbonated, the use of a portion of nitrogen gas has been shown to substantially improve foam stability. In part the benefit of nitrogen is that it has a lower partial pressure compared to CO_2, which results in the production of smaller bubbles (Figure 1.17; Carroll, 1979; Fisher et al., 1999). However, this physical characteristic of nitrogen makes the formation of bubbles less likely than with CO_2. Therefore nitrogen is typically used in conjunction with widgets (see next section) or as

Figure 1.17
Comparison of the foam bubble size of nitrogen vs. CO_2.

"bar gas" (typically 25% nitrogen, 75% CO_2) for the draught dispense of beer via the delivery orifice that ensures the proper level of bubble formation. The smaller nitrogenated bubbles are also more stable because nitrogen has a lower aqueous solubility compared to CO_2, leading to less gas diffusion and an inhibition of disproportionation (Carroll, 1979; Mitani et al., 2002; Bamforth, 2004a). The addition of nitrogen into the beer also changes the foam's mouthfeel to a "creamy" texture, however the lower CO_2 content leads to beer with less "prickle" (acidic and bubble collapse) on the palate (Carroll, 1979) making the beer taste flat and watery. Other investigators have found that in comparing blind-folded with sighted judging of conventionally carbonated beers the visual impact of the foam head on a glass of beer is more important than the tactile impact on beer flavor and mouthfeel (Langstaff and Lewis, 1993). Consequently, the acceptability of nitrogen as a palliative option will depend on the beer style and consumer preferences.

Widgets and other foam promotors

Other strategies for promoting foam stability involve the use of extraneous devices added to the package or included in the glass. The most technically simple is the use of nucleated glassware such as the "headkeeper' style (Parish, 1997), which improves foam stability by promoting bubble nucleation, hence foam replenishment. Of more widespread use for some styles of beer is the widget (in-package foam generating device) that is either attached to the base of the package or floats within it (Figure 1.18). These characteristics reduce the appeal of the widget in the more desirable marketing vehicle, namely the bottle, presumably because the sight of a widget floating or attached to the bottom of a bottle is somewhat disconcerting for the consumer. In cans, the widget can be commonly found in the less carbonated stout and bitter ale style beers of the British Isles (Browne, 1996; Brown, 1997) but has not found acceptance in other regions or beer styles. The widget, combined with the inclusion of nitrogen gas was designed to emulate the draught style pub presentation in small pack. Widgets work as nitrogen foam nucleating devices. In some cases the widget can also form bubbles from its nucleated surfaces (Figure 1.18B). The widget also provides a degree of theatre on package opening with a characteristic rumble of the gas being released through the widget vent (Figure 1.18C(i) to produce a slight fob at the package opening (Figure 1.18D) and pour associated with the rise of a multitude of tiny bubbles (Figure 1.18E) to form a creamy head. A wide range of widget designs have been developed (Browne, 1996; Brown, 1997). The more modern floating widgets (spherical), produce the foam in the ideal place, at the top of the beer. The reliance on nitrogen links the widget to both the benefits and disadvantages of nitrogenated foam as described above. The main downside of widgets is that they are considered by the industry to be expensive both in terms of capital and consumable costs.

PGA and chemical enhancement

Certain additions can result in the improvement of foam. The most widely used is propylene glycol alginate (PGA) (Jackson et al., 1980; O'Reilly, 1996). These

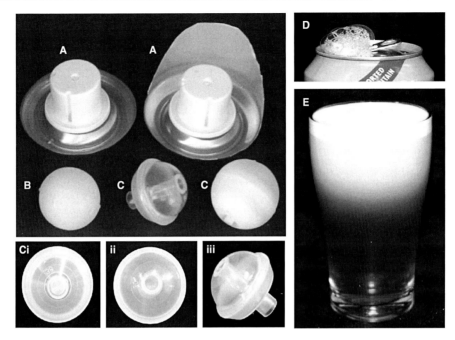

Figure 1.18
A selection of foam generating widgets. (A) Nitrogen injected, with nucleated surface and fused to can bottom. (B) Floating, nitrogen injected with nucleated surface. (C) Floating with nitrogen injected (i) Bottom of widget showing vent, (ii) Top of widget, (iii) Side view of widget. (D) Small amount of fob formed at the opening of widgeted can. (E) Initial pour of widgeted can providing the theatre of numerous small bubbles rising to form head.

authors conclude that PGA primarily protects against the deleterious impacts of lipids. The addition of PGA is reputed to increase foam stability by around 5–10% and many brewers rely on it to provide satisfactory foam quality for some of their products (Evans and Sheehan, 2002). It is also likely that it acts as a foam stabilizing agent in its own right. In this regard, the foam stability results reported by Lusk et al. (2003) showed that PGA could also compensate for the loss of protein due to papain treatment. These authors also suggested that the presence of PGA could contribute to foam density. Similarly, patented exogenous lipid binding proteins vectored via yeast (Kunst et al., 1997) could be added to beer to mop up foam damaging lipids. These preparations are/likely to be effective at removing lipids from beer given time (~24+ hrs) for them to bind the lipids (Roberts et al., 1978). As a consequence of this time-dependence, such remedies will not be effective in binding foam destroying lipids that enter beer at dispense or from the consumer (i.e. lips, lipstick, crisps or chips). There are however, some undesirable consequences for adding PGA in that it may be perceived as a "foreign chemical" by consumers, is expensive costing around US$6/t of malt to add to beer, and has the disadvantage of promoting the formation of undesirable storage haze (Evans and Sheehan, 2002). The supplementation of endogenous foaming protein with whipping proteins has been investigated, but this is an expensive option and is likely to be worthwhile only for those beers seriously deficient in foam potential (Bamforth and Cope, 1987).

... of clean lines, glasses and avoiding detergents and other such nasties

The last critical point in providing the consumer with the desired foam quality is dispensing the beer into its serving container, typically but not always glass. At this stage there are factors that are not always controllable by the brewer such as when they do not own the retail outlet or even more problematic, dispense by the consumer into glasses that may not have been suitably cleaned. With these challenges the brewer's only recourse is to use concerted and well-targeted education campaigns. The necessity for proper cleaning procedures for draught dispense has long been known and its tenets are well summarized by Rees (1976) who also stated the obvious truism that "cleanliness is next to Godliness when it comes to dispensing beer." Such attention to ensure the proper cleaning and subsequent removal of detergent residue is not only necessary for good beer foam quality but flavor. The more universal requirement, for all beer dispense practioners, is that the glass which receives beer is clean. Recently, Stillman (2006) with his "fit to fill" criteria of the glass being free rinsing, visually bright, odor free, disinfected and, cool and dry provided an exemplary discourse for all on how a glass should be cleaned.

Not all glasses are however the same when it comes to beer foam quality. There are of course obvious differences in glass thickness and shape that range between those that are "straight up" (i.e. Pils glass), to the "chalice/cup-shaped" (i.e. Trappist style beer glass) or those that are more "tulip/thistle-shaped" (i.e. wheat beer glass), so shaped so as to concentrate the beer aromas. The thickness of the glass and its surface to volume ratio (i.e. low with tulip/thistle-shaped) will impact upon how quickly the beer warms up and loses CO_2, which will impact on the rate of disproportionation (Delvaux et al., 1995). It also follows that the material that the glass or container is made out of is important for foam quality, in particular cling (Fischer and Sommer, 1999). In general, the more hydrophobic the surface (i.e. plastic such as PE), the larger the area that is covered by foam but this only forms as a thin film. Conversely, lacing covers less area but the amount adhering is greater with more hydrophilic materials (i.e. glass). The amount of lacing will of course be influenced negatively if lipid or detergent residues are on the surface of the container. In addition, surface roughness slightly improves foam adhesion on plastic surfaces. For these reasons different glass compositions of the widely used glass would be expected to influence the lacing characteristics of beer. As already mentioned, the use of nucleated glassware will improve foam stability by promoting creaming (Parish, 1997). Overall, the provision of a clean glass (or container) with the appropriate shape, thickness and surface for the style of beer is critical for the right presentation of the foam quality foam desired. The addition of an attractive brand logo is the final touch to build the brand image consumers will recognize and desire for their favorite beer.

Troubleshooting?

By definition, troubleshooting problems with beer foam can be somewhat involved. The brewer does of course need to ensure their methods for foam quality analysis are adequate and reflect the experience of their customers in the trade. As a guide to beer foam troubleshooting, there are fortunately several excellent articles that provide a systematic framework for investigating problems (Gromus, 1999; Douma et al., 1999; Bamforth, 2000b). In the main, troubleshooting of problems that stem from raw materials (malt and hops), brewhouse practice and subsequent handling of yeast and beer (i.e. storage and dispense) are essentially covered in the preceding sections. Most likely, issues to do with raw materials relating to foam promoting proteins and hop acids are likely to cause relatively incremental but important changes in foam quality. For proteins, previous sections have relied heavily on the use of specific ELISAs for foam promoting proteins such as protein Z and LTP1 (i.e. Evans and Hejgaard, 1999). As they were developed and used as research tools, they are not widely available and are expensive to develop. In lieu, the use of one of the widely available and relatively inexpensive total protein estimation assays, either CBB or PRM, either on their own (see Evans and Sheehan, 2002) or perhaps in conjunction with hydrophobic interaction chromatography (Bamforth et al., 1993; Bamforth, 1995) should be sufficient for almost all quality control or troubleshooting purposes. Two case studies investigating foam quality issues associated with changes in the level of beer foam promoting proteins for example are provided by Douma et al. (1999). Control of the level of hops and their utilization in beer, along with carbonation are routinely undertaken in breweries thus require no further discussion here. Similarly, issues regarding hop composition or type are probably best addressed with suppliers.

The most likely cause of foam complaints is going to be lipids and detergents. Where raw material variation is likely to result in incremental changes in foam quality, lipids and detergents are likely to cause large decreases in foam stability and quality. A relatively simple first step in exposing problems with lipids (and presumably detergents) in beer would be to "cleanse" the beer using a BSA affinity column and evaluate if the treatment resulted in an improvement in foam stability (Dickie et al., 2001). Where a significant improvement was observed, a problem with lipids is likely. Gromus (1999) investigated the influence of a number of negative factors towards the end of the brewing process on foam stability that are summarized below:

- Cleaning and disinfecting agents (particularly tenside residues and small amounts of sodium hydroxide).
- Type of bottle cleaner and its maintenance to ensure low levels of cleaning residue in bottles.
- Glue and label residues.
- New glass bottles with excessive levels of "cold end" coatings suppress foam stability.
- Can/lid lubricant residues of lipidic nature.

Gromus (1999) also noted that filter sheets (supplier dependent), particularly on initial use, can bind foam promoting proteins to reduce foam stability. The time course for this binding was intensively investigated and shown to be non-specific for both haze and foam promoting proteins (Robinson et al., 2004).

Summary

The brand image of a beer is inexorably linked to the quality of the foam on that beer after dispense and is a quality indicator that can be easily applied by all consumers. This is not surprising due to the visual appeal of the foam, its subtle role as a conduit for beer aromas and its contribution to the beer mouth feel. Similarly it is noted that consumers from different nations, regions or even genders have different preferences for the foam on their beer. This is divided by visual cues that include the foam stability, cling, strength, creaming, bubble size and whiteness. Thus Bamforth's (1999) erstwhile boss who said "that generations of biochemists have done less for beer foam than the widget" is alienating a substantial portion of potential consumers from their beer brands because such foam does not meet their expectations. Foam quality is not just about "quick" fixes such as the inclusion of widgets, ever greater levels of tetra hop or gas composition but attention to the beer-making process from grass to glass (malting variety breeding to dispense). Brewers do have solid options in manipulating the quality and quantity of malt foam positive proteins and selection of hop acids, the interaction of which provides the basis for foam stability and quality. Brewers also have a range of palliative options such as additives, gas composition, widgets and methods for dispense that can be used if suitable to the style of beer being produced. The main game is to use these options to optimize the foam quality of their brands and to consistently meet the expectations of the consumers that the brewer is targeting.

Acknowledgments

Dr Evans would like to thank the Australian Grains Research and Development Corporation and the Australian brewing industry for funding. The authors would like to thank Lance Lusk (Miller Brewing, USA), Dana Sedin (Coor-Molsen, USA), Karl Larkenburgess (Anheuser Busch, USA), Harry Craig (DSM, UK), Louise Robinson (Lion Nathan, Australia), Anne Surrel (ISAB, France) for provision of comments, photographs and materials.

References

American, S., of Brewing Chemists ASBC (1992) Methods of analysis. American Society of Brewing Chemists: Pilot Knob, MN.

Andrews, J. and Axcell, B. C. (2003) Wort boiling-evaporating the myths of the past. *Master Brewers Association of the American Technical Quarterly*, 40, 249–254.

Anger, H.-M., Garo, L. and Scholz, M. (2002) A new process for foam stability measurement. *Foam-tester from Lg-Automatic (Denmark). Brauwelt International*, 20, 146–150.

Anness, B. J. and Reed, R. J. R. (1985) Lipids in wort. *Journal of the Institute of Brewing, 91,* 313–317.

Archibald, H. W. (1988) Beer foam getting ahead. *The Brewer, 74,* 295–300.

Archibald, H. W., Weiner, J. P. and Taylor, L. (1988) Observations on factors affecting beer foam characteristics. *The Brewer, 74,* 349–362.

Asano, K. and Hashimoto, N. (1976) Contribution of hop bitter substances to head formation of beer. *Report of the Research Laboratories of Kirin Brewery Co., Ltd, 19,* 9–16.

Asano, K. and Hashimoto, N. (1980) Isolation and characterization of foaming propteins of beer. *Journal of the American Society of Brewing Chemists, 38,* 129–137.

Asano, K., Shingawa, K. and Hashimoto, N. (1982) Characterization of haze-forming proteins of beer and their roles in chill haze formation. *Journal of the American Society of Brewing Chemists, 40,* 147–154.

Asano, K., Ohtsu, K., Shimagawa, K., and Hoshimoto (1984) Affinity of proanthrocyanidins and their oxidation products for haze-forming proteins of beer and the formation of chill haze. *Agricultural Biological Chemistry, 48,* 1139–1146.

Back, W., Forster, C., Krottenthaler, M., Lehmann, J., Sacher, B. and Thum, B. (1999) New research findings on improving taste stability. *Brauwelt International, 17,* 394–405.

Bamforth, C. W. (1985) The foaming properties of beer. *Journal of the Institute of Brewing, 91,* 370–383.

Bamforth, C. W. (1995) Foam: Method, myth or magic. *The Brewer, 81,* 396–399.

Bamforth, C. W. (1999) Bringing matters to a head: The status of research on beer foam. *European Brewing Convention Monograph, XXVII, Amsterdam,* 10–23.

Bamforth, C. W. (2000a) Perceptions of beer foam. *Journal of the Institute of Brewing, 106,* 229–238.

Bamforth, C. W. (2000b) Beer quality series: Foam. *Brewers' Guardian, 129* (3), 40–43.

Bamforth, C. W. (2002) How important is wort clarity? *Brewers' Guardian, 131,* 26–28.

Bamforth, C. W. (2004a) The relative significance of physics and chemistry for beer foam excellence: Theory and practice. *Journal of the Institute of Brewing, 110,* 259–266.

Bamforth, C. W. (2004b) A critical control point analysis for flavour stability of beer. *Master Brewers Association of the Americas Technical Quarterly, 41,* 97–103.

Bamforth, C. W. and Cope, R. (1987) Egg albumen as a source of foam polypeptide in beer. *Journal of the American Society of Brewing Chemists, 45,* 27–32.

Bamforth, C. W. and Jackson, G. (1983) Aspects of foam lacing. *Proceedings of the European Brewery Convention Congress, London, 19,* 331–338.

Bamforth, C. W. and Kanauchi, M. (2003) Interactions between polypeptides derived from barley and other beer components in model foam systems. *Journal of the Science of Food and Agriculture, 83,* 1045–1050.

Bamforth, C. W. and Milani, C. (2004) The foaming of mixtures of albumin and hordein hydrolysates in model systems. *Journal of the Science of Food and Agriculture, 84,* 1001–1004.

Bamforth, C. W., Butcher, K. N. and Cope, R. (1989) The interrelationships between parameters of beer quality. *Ferment, 2,* 54–58.

Bamforth, C. W., Canterranne, E., Chandley, P. and Onishi, A. (1993) The molecular interaction of beer foam. *Proceedings of the European Brewing Convention Congress, Oslo, 24,* 331–340.

Bamforth, C. W., Knapp, G. R. and Smythe, J. E. (2001) The measurement of hydrophobic peptides in beer using the fluorochrome 1-anilino-8-naphthalenesulfonate. *Food Chemistry, 75,* 377–383.

Bateson, J. B. and Leach, A. A. (1969) Nitrogen studies of wort in relationship to beer quality. *Proceedings of the European Brewing Convention Congress, Interlaken, 12,* 161–171.

Bech, L. M., Vaag, P., Heinemann, B. and Breddam, K. (1995) Throughout the brewing process barley lipid transfer protein 1 (LTP1) is transformed into a more foam-promoting form. *Proceedings of the European Brewing Convention Congress, Brussels*, 25, 561–568.

Bech, L. M., Sorenson, S. B., Vaag, P., Muldbjerg, M., Beenfeldt, T., Leah, R. and Breddam, K. (1999). Beverage and a method of preparing it. US Patent #5993865.

Beecher, B., Smidansky, E. D., See, D., Blake, T. K. and Giroux, M. J. (2001) Mapping and sequence analysis of barley hordoindolines. *Theoretical and Applied Genetics*, 102, 833–840.

Birtwistle, S. E., Hunson, J. R. and MacWilliam, I. C. (1962) Use of unmalted wheat flour in brewing. *Journal of the Institute of Brewing*, 68, 467–470.

Blom, J. (1937) Investigations of Foam. *Journal of the Institute of Brewing*, 43, 251–262.

Blum, P. H. (1969) Lipids in malting and brewing. *Brewers Digest*, 44, 58–63.

Bradford, M. M. (1976) A rapid and sensitive method for the quantitization of microgram quantities of protein utilizing the principle of protein-dye binding. *Annals of Biochemistry*, 72, 248–254.

Brey, S. E., Bryce, J. H. and Stewart, G. G. (2002) The loss of hydrophobic polypeptides during fermentation and conditioning of high gravity and low gravity brewed beer. *Journal of the Institute of Brewing*, 108, 424–433.

Brierley, E. R., Wilde, P. J., Oniski, A., Hughes, P. S., Simpson, W. J. and Clarke, D. C. (1996) The influence of ethanol on the foaming properties of beer protein fractions: A composition of rudin and microconductivity methods of foam assessment. *Journal of the Science Food and Agriculture*, 70, 531–537.

Brown, D. (1997) Nitrogen and its foaming relationship with widgets. *The Brewers*, 83, 25–32.

Browne, J. J. C. (1996) What are widgets? *The Brewers*, 82, 498–503.

Carroll, T. C. N. (1979) The effect of dissolved nitrogen gas on beer foam and palate. *Master Brewers Association of the Americas Technical Quarterly*, 16, 116–119.

Clark, D. C. (1991) New approaches for studying beer foam. *Ferment*, 4, 370–374.

Clark, D. C., Wilde, P. J. and Marion, D. (1994) The protection of beer foam against lipid-induced destabilization. *Journal of the Institute of Brewing*, 100, 23–25.

Compton, S. J. and Jones, C. G. (1985) Mechanism of dye response and interference in Bradford protein assay. *Annals of Biochemistry*, 151, 369–374.

Constant, M. (1992) A practical method for characterising poured beer foam quality. *Journal of the American Society of Brewing Chemists*, 50, 37–47.

Cooper, D. J., Stewart, G. G. and Bryce, J. H. (1998a) Hydrophobic polypeptide extraction during high gravity mashing-experimental approaches for its improvement. *Journal of the Institute of Brewing*, 104, 283–287.

Cooper, D. J., Stewart, G. G. and Bryce, J. H. (1998b) Some reasons why high gravity brewing has a negative effect on head retention. *Journal of the Institute of Brewing*, 104, 83–87.

Cooper, D. J., Husband, F. A., Mills, E. N. C. and Wilde, P. J. (2002) Role of beer lipid-binding proteins preventing lipid destabilization of foam. *Journal of Agricultural and Food Chemistry*, 50, 7645–7650.

Curioni, A., Pressi, G., Furegon, L. and Peruffo, A. D. B. (1995) Major proteins of beer and their precursors in barley: Electrophoretic and immunological studies. *Journal of Agricultural and Food Chemistry*, 43, 2620–2626.

Cvengrochova, M., Sepelova, G. and Smogrovicova, D. (2003) Influence of pre-isomerised hop on taste and foam stability. *Monatsschrift für Brauwissenschaft*, 56, 206–209.

Dale, C. J. and Young, T. W. (1987) Rapid methods for determining the high molecular weight polypeptide components of beer. *Journal of the Institute of Brewing*, 93, 465–467.

Delcour, J. A., Vanhamel, S., Moerman, E. and Vancraenenbroeck, R. (1987) Use of proanthocyanin-free malt Galant and/or physio-chemical stabilization treatments for the production of chill-proof beers. *Master Brewers Association of the Americas Technical Quarterly*, 24, 21–27.

Delvaux, F., Deams, V., Vanmachelen, H., Neven, H. and Derdelinckx, G. (1995) Retention of beer flavours by the choice of appropriate glass. *Proceedings of the European Brewery Convention Congress, Brussels*, 25, 533–542.

Dickie, K. H., Cann, C., Norman, E. C., Bamforth, C. W. and Muller, R. E. (2001) Foam negative materials. *Journal of the American Society of Brewing Chemists*, 59, 17–23.

Diffor, D. W., Lickens, S. T., Rehberger, A. J. and Burkhardt, R. J. (1978) The effect of isohumulone/isocohumulone ratio on beer head retention. *Journal of the American Society of Brewing Chemists*, 36, 63–65.

Douliez, J. P., Michon, T., Elmorjani, K. and Marion, D. (1999) Structure, biological and technological functions of lipid transfer proteins and indolines, the major lipid transfer proteins from cereal kernels. *Journal of Cereal Science*, 32, 1–20.

Douma, A. C., Mocking-Bode, H. C. M., Kooijman, M., Stolzenbach, E., Orsel, R., Bekkers, A. C. A. P. A. and Angelino, S. A. G. F. (1997) Identification of foam-stabilizing proteins under conditions of normal beer dispense and their biochemical and physiochemical properties. *Proceedings of the European Brewing Convention Congress, Maastrict*, 26, 671–679.

Douma, A. C., Bos, M. A., Mocking-Bode, H. C. M. and Angelino, S. A. G. F. (1999) Beer foam trouble-shooting. *European Brewery Convention Monograph*, 27, 84–93.

Eglinton, J. K., Langridge, P. and Evans, D. E. (1998) Thermostability variation in alleles of barley *beta*-amylase. *Journal of Cereal Science*, 28, 301–309.

Evans, D. E. and Hejgaard, J. (1999) The impact of malt derived proteins on beer foam quality. Part I. The effect of germination and kilning on the level of protein Z4, protein Z7 and LTP1. *Journal of the Institute of Brewing*, 105, 159–169.

Evans, D. E. and Sheehan, M. C. (2002) Do not be fobbed off, the substance of beer foam, a review. *Journal of the American Society of Brewing Chemists*, 60, 47–57.

Evans, D. E., MacLeod, L. C. and Lance, R. C. M. (1995) The importance of protein Z to the quality of barley and malt for brewing. *Proceedings of the European Brewery Convention Congress, Brussels*, 25, 225–232.

Evans, D. E., Nischwitz, R., Stewart, D. C., Cole, N. and MacLeod, L. C. (1999a) The influence of malt foam-positive proteins and non-starch polysaccharides on beer foam quality. *European Brewing Convention Monograph*, XXVII, Amsterdam, 114–128.

Evans, D. E., Ratcliffe, M., Jones, B. L. and Barr, A. R. (1999b) Variation and genetic control of foam-positive proteins in Australian barley varieties. *Proceedings of the Australian Barley Technical Symposium*, Melbourne, Victoria, Vol. 9, pp. 3.6.1–3.6.6, http://www.cdesign.com.au/proceedings%5Fabts1999/.

Evans, D. E., Sheehan, M. C. and Stewart, D. C. (1999c) The impact of malt derived proteins on beer foam quality. Part II: The influence of malt foam-positive proteins and non-starch polysaccharides on beer foam quality. *Journal of the Institute of Brewing*, 105, 171–177.

Evans, D. E., Sheehan, M. C., Tolhurst, R. L., Skerritt, J. S., Hill, A. and Barr, A. R. (2003) Application of immunological methods to differentiate between foam-positive and haze active proteins originating from malt. *Journal of the American Society of Brewing Chemists*, 61, 55–62.

Evans, D. E., Collins, H. M., Eglinton, J. K. and Wilhelmson, A. (2005) Assessing the impact of the level of diastatic power enzymes and their thermostability on the hydrolysis of starch during wort production to predict malt fermentability. *Journal of the American Society of Brewing Chemists, 63,* 185–198.

Evans, D. E., Surrel, A., Sheehy, M., Stewart, D. and Robinson, L. H. (2008) Comparison of foam quality and the influence of hop a-acids and proteins by five foam analysis methods. *Journal of the American Society of Brewing Chemists, 65,* 1–10.

Ferreira, I. M. P. L. O., Jorge, K., Nogueira, L. C., Silva, F. and Trugo, L. C. (2005) Effects of the combination of hydrophobic polypeptides, iso-α-acids and malt-oligosaccharides on beer foam stability. *Journal of Agricultural and Food Chemistry, 53,* 4976–4981.

Fischer, S. and Sommer, K. (1999) Cling of beer foam to different surfaces. *Proceedings of the European Brewing Convention Congress, Cannes, 27,* 183–189.

Fisher, S., Hauser, G. and Sommer, K. (1999) Influence of dissolved gases on foam. *European Brewing Convention Monograph, XXVII, Amsterdam,* 37–46.

Fukuda, K., Saito, W., Arai, S. and Aida, Y. (1999) Production of a novel proanthrocyanidin-free barley line with high quality. *Journal of the Institute of Brewing, 105,* 179–183.

Furukubo, S., Shoboyaski, M., Fukui, N., Isoe, A. and Nakatami, K. (1993) A new factor which effects the foam adhesion of beer. *Master Brewers Association of the Americas Technical Quarterly, 30,* 155–158.

Gibson, C. E., Evans, D. E. and Proudlove, M. O. (1996) Protein Z4 and beer foam. *Ferment, 9,* 81–84.

Giese, H. and Hopp, H. E. (1984) Influence of nitrogen on the amount of hordein, protein Z and β-amylase messenger RNA in developing endosperms of barley. *Carlsberg Research Communications, 49,* 365–383.

Gromus, J. (1999) Negative influences on foam. *European Brewery Convention, Monograph,, 27,* 69–83.

Hallgren, L., Rosendale, I. and Rasmussen, J. N. (1991) Experiences with a new foam stability analyser. *Journal of the American Society of Brewing Chemists, 49,* 76–78.

Hangsted, C. and Erdal, K. (1991) Head hunting. *Proceedings of the European Brewing Convention Congress, Lisbon, 23,* 449–456.

Hao, J., L, Q., Dong, J., Yu, J., Fan, W. and Chen, J. (2006) Identification of the major proteins in beer foam by mass spectometry following sodium dodecyl sulphate-polyacrylamide gel electrophoresis. *Journal of the American Society of Brewing Chemists, 64,* 166–174.

Haukeli, A. D., Wulff, T. O. and Lie, S. (1993) Practical experiments to improve foam stability. *Proceedings of the European Brewing Convention Congress, Oslo, 24,* 365–372.

He, G.-Q., Wang, Z.-Y., Liu, Z.-S., Chen, Q. H., Ruan, H. and Schwartz, P. B. (2006) Relationship of proteinase activity, foam proteins and head retention in unpasteurized beer. *Journal of the American Society of Brewing Chemists, 64,* 33–38.

Hejgaard, J. (1977) Origin of dominant beer protein immunochemical identity with β-amylase-associated protein from barley. *Journal of the Institute of Brewing, 83,* 94–96.

Hejgaard, J. and Kaersgaard, P. (1983) Purification and properties of the major antigenic beer protein of barley origin. *Journal of the Institute of Brewing, 89,* 402–410.

Hejgaard, J., Bjorn, S. and Nielsen, G. (1984) Localization to chromosomes of structural genes for the major protease inhibitors of barley grains. *Theoretical and Applied Genetics, 68,* 127–130.

Hii, V. and Herwig, W. C. (1982) Determination of high molecular weight proteins in beer using coomassie blue. *Journal of the American Society of Brewing Chemists, 40,* 46–50.

Hirota, N., Kuroda, H., Takoi, T., Kaneko, T., Kaneda, H., Yoshida, I., Takashio, M. T., Ito, K. and Takeda, K. (2005) Development of a novel barley with improved beer foam and flavour stability – the impact of lipoxygenase-1-less barley in the brewing industry. *Proceedings of the European Brewing Convention Congress, Prague, 30,* 46–51.

Hirota, N., Kuroda, H., Takoi, T., Kaneko, T., Kaneda, H., Yoshida, I., Takashio, M. T., Ito, K. and Takeda, K. (2006) Development of a novel barley with improved beer foam and flavour stability – the impact of lipoxygenase-1-less barley in the brewing industry. *Master Brewers Association of the Americas Technical Quarterly, 43,* 131–135.

Hollemans, M. and Tonies, A. R. J. M. (1989) The role of specific proteins in beer foam. *Proceedings of the European Brewing Convention Congress, Oslo, 22,* 561–568.

Honno, E., Furukubo, S., Kondo, H., Ishibaski, Y., Fukui, N. and Nakatani, K. (1997) Improvement of foam lacing of beer. *Master Brewers Association of the Americas Technical Quarterly, 34,* 299–301.

Hug, H. and Pfenninger, H. (1980) Improvement of beer filterability using heat stable β-glucanase. *Brauerei Rundschau, 91,* 61–65.

Hughes, P. S. (1997) Characterising beer foam quality. *Brewers' Guardian, 126,* 33–36.

Hughes, P. (2000) The significance of iso-α-acids for beer quality. *Journal of the Institute of Brewing, 106,* 271–276.

Hughes, P. S. and Simpson, W. J. (1995) Interactions between hop bitter acids and metal cations assessed by ultra-violet spectrophotometry. *Cerevisiae Biotechnology, 20,* 35–39.

Hughes, P. S. and Wilde, P. J. (1997) New techniques for the evaluation of interactions in beer foams. *Proceedings of the European Brewing Convention Congress, Maastricht, 26,* 525–534.

Hughes, P., Mills, C., Kauffman, J., Bierley, E., Dickie, K., Proudlove, M., Onishi, A. and Wilde, P. (1999) The foaming and interfacial behavior of beer polypeptides: The effect of hydrophobicity. *European Brewery Convention Monograph, XXVII,* Amsterdam, 129–140.

Hung, J. K. S., Wallin, C. E. and Bamforth, C. W. (2005) An evaluation of an automated procedure for measuring beer foam stability. *Master Brewers Association of the Americas Technical Quarterly, 42,* 178–183.

Ishibashi, Y., Terano, Y., Fukui, N., Honbou, N., Kakui, T., Kawasaki, S. and Nakatami, K. (1996) Development of a novel method for determining beer foam and haze proteins by using immunochemical method – ELISA. *Journal of the American Society of Brewing Chemists, 54,* 177–182.

Ishibashi, Y., Kakni, T., Terano, Y., Hon-no, E., Kogin, A. and Nakatani, K. (1997) Application of ELISA to quantitative evaluation of foam-active protein in the malting and brewing process. *Journal of the American Society of Brewing Chemists, 55,* 20–23.

Jackson, G. and Bamforth, C. W. (1982) The measurement of foam-lacing. *Journal of the Institute of Brewing, 88,* 378–381.

Jackson, G. and Wainright, T. (1978) Melanoidins and beer foam. *Journal of the American Society of Brewing Chemists, 36,* 192–195.

Jackson, G., Roberts, R. T. and Wainright, T. (1980) Mechanism of beer foam stabilization by propylene glycol alginate. *Journal of the Institute of Brewing, 86,* 34–37.

Jegou, S., Douliez, J.-P., Molle, D., Boivin, P. and Marion, D. (2001) Evidence of the glycation and denaturation of LTP1 during the malting and brewing process. *Journal of Agricultural and Food Chemistry, 49,* 4942–4949.

Jende-Strid, B. (1997) Proanthocyanidin-free malting barley – A solution of the beer haze problem. *Proceedings of the European Brewing Convention Congress, Madrid, 21,* 101–108.

Jones, B. L. (2005) Endoproteinases of barley and malt. *Journal of Cereal Science, 42,* 139–156.

Jones, B. L. and Marinac, L. A. (1995) Barley LTP1 (PAP1) and LTP2 are inhibitors of green malt cysteine endoproteinases. *Journal of the American Society of Brewing Chemists, 53*, 194–195.

Kaersgaard, P. and Hejgaard, J. (1979) Antigenic beer macromolecules, an experimental survey of purification methods. *Journal of the Institute of Brewing, 85*, 103–111.

Kakui, T., Ishibashi, Y., Kunihige, Y., Isoe, A. and Nakatani, K. (1999) Application of enzyme-linked immunosorbent assay to quantitative evaluation of foam-active protein in wheat beer. *Journal of the American Society of Brewing Chemists, 57*, 151–154.

Kalla, R., Shimamoto, K., Potter, R., Nielsen, P. S., Linnestad, C. and Olsen, O. A. (1994) The promotor of barley aleurone-specific gene encoding a putative 7KDa lipid transfer protein. *Plant Journal, 6*, 849–860.

Kano, Y. and Kamimura, M. (1993) Simple methods for determination of the molecular weight distribution of beer proteins and their application to foam and haze studies. *Journal of the American Society of Brewing Chemists, 51*, 21–28.

Kauffman, J. A., Mills, E. N. C., Brett, G. M., Fido, R. J., Tatham, A. S., Shewry, P. R., Onishi, A., Promtlor, M. and Morgan, R. A. (1994) Immunological characterization of barley peptides in larger foam. *Journal of Science Food and Agriculture, 66*, 345–355.

Kihara, M., Kaneko, T. and Ito, K. (1998) Genetic variation of *beta*-amylase thermostability among varieties of barley, *Hordeum vulgare* L., and relation to malting quality. *Plant Breeding, 117*, 425–428.

Klopper, W. J. (1973) Foam stability and foam cling. *Proceedings of the European Brewing Convention Congress, Salzburg, 14*, 363–371.

Knapp, G. R. and Bamforth, C. W. (2002) The foaming properties of proteins isolated from barley. *Journal of the Science Food and Agriculture, 82*, 1276–1281.

Kobayashi, N., Kaneda, H., Kano, Y. and Koslimo, S. (1994) Behaviour of lipid hydroperoxides during mashing. *Journal of the American Society of Brewing Chemists, 52*, 141–145.

Kobayashi, N., Segawa, S., Umemoto, S., Kuroda, H., Mitani, Y., Watari, J. and Takashio, M. (2002) A new method for evaluating foam-damaging effect by free fatty acids. *Journal of the American Society Brewing Chemists, 60*, 37–41.

Kondo, H., Shibano, Y., Fukui, N., Nakatani, K., Oda, K. and Amachi, T. (1995) Development of a novel and sensitive method for measurement of proteinase A in beer. *Proceedings of the Europeran Brewing Convention Congress, Brussels, 25*, 669–676.

Kondo, H., Yomo, H., Fukukubo, S., Kawasaki, Y. and Nakatani, K. (1998) Advanced method for measuring proteinase A in beer. *Proceedings of the 25th Convention, The Institute of Brewing, Asia Pacific Section, Perth, 25*, 119–124.

Kreis, M. and Shewry, P. R. (1992) The control of protein synthesis in developing barley seeds. In: *Barley: Genetics, Biochemistry, Molecular Biology, and Biotechnology* (P. R. Shewry ed.), pp. 319–333. C.A.B. International: London.

Kühbeck, F., Back, W. and Krottenthaler, M. (2006a) Influence of lauter turbidity on composition, fermentation performance and beer quality – a review. *Journal of the Institute of Brewing, 112*, 215–221.

Kühbeck, F., Back, W. and Krottenthaler, M. (2006b) Influence of lauter turbidity on wort composition, fermentation, performance and beer quality in large-scale trials. *Journal of the Institute of Brewing, 112*, 222–231.

Kühbeck, F., Schutz, M., Thiele, F., Krottenthaler, M. and Back, W. (2006c) Influence of lauter turbidity and hot trub on wort composition, fermentation, and beer quality. *Journal of the American Society of Brewing Chemists, 64*, 16–28.

Kunst, A., Lalor, E., Schmedding, D. J. and Schie, B. J. A. (1997) A glyco-protein preparation for use as a stabilizer of beverages. European Patent #9705272.

Kuroda, H., Kobayashi, N., Kaneda, H., Watari, J. and Takashio, M. (2002) Characterization of factors that transform linoleic acid into di and trihydroxyoctadecenoic acids in mash. *Journal of Bioscience and Bioengineering*, 93, 73–77.

Langstaff, S. A. and Lewis, M. J. (1993) Foam and the perception of beer flavour and mouth feel. *Master Brewers Association of the Americas Technical Quarterly*, 30, 16–17.

Le Mestre, M., Colas, B., Simatos, D., Closs, B., Courthaudon, J. L. and Lorient, D. (1990) Contribution of proteins flexibility to the foaming properties of casein. *Journal of Food Science*, 55, 1445–1447.

Leach, A. A. (1968) Nitrogenous components of worts and beer brewed from all-malt and malt plus wheat flour grists. *Journal of the Institute of Brewing*, 74, 183–192.

Leeson, T. J., Velissarion, M. and Lyddiatt, A. (1990) Biochemical and physical analysis of beer-roles for macromolecular species in foam stabilization at dispense, pp. 194–206. In: *Food Polymers, Gels and Colloids* (E. Dickinson, ed.), Royal Society of Chemistry.

Leiper, K. A., Stewart, G. G. and McKeown, I. P. (2003) Beer polypeptides and silica gel. Part II: Polypeptides involved in foam formation. *Journal of the Institute of Brewing*, 109, 73–79.

Leisgang, R. and Stahl, U. (2005) Alteration in proteinase sensitivity of a foam protein during the malting and brewing process. *Proceedings of the European Brewery Convention Congress, Prague*, 30, 642–645.

Letters, R. (1992) Lipids in brewing, friend or foe? *Ferment*, 5, 268–274.

Lewis, M. J. and Lewis, A. S. (2003) Correlation of beer foam with other beer properties. *Master Brewers Association of the Americas Technical Quarterly*, 40, 114–124.

Lewis, M. J. and Serbia, J. W. (1984) Aggregation of protein and precipitation of polyphenol in mashing. *Journal of the American Society of Brewing Chemists*, 42, 40–43.

Lewis, M. J., Krumland, S. C. and Muhleman, D. J. (1980) Dye-binding method for measurement of protein in wort and beer. *Journal of the American Society of Brewing Chemists*, 38, 37–41.

Lopez, M. and Edens, L. (2005) Effective prevention of chill haze in beer using an acid proline-specific endoproteinase from *Aspergillis niger*. *Journal of Agricultural and Food Chemistry*, 53, 7944–7949.

Lusk, L. T., Goldstein, H. and Ryder, D. (1995) Independent role of beer proteins, melanoidins and polysaccharides in foam formation. *Journal of the American Society of Brewing Chemists*, 53, 93–103.

Lusk, L. T., Ting, P., Goldstein, H., Ryder, D. and Navarro, A. (1999) Foam tower fractionation of beer proteins and bittering acids. *European Brewing Convention Monograph*, XXVII, Amsterdam, 166–187.

Lusk, L. T., Duncombe, G. R., Kay, S. B., Navarro, A. and Ryder, D. (2001a) Barley β-glucan and beer foam stability. *Journal of the American Society Brewing Chemists*, 59, 183–186.

Lusk, L. T., Goldstein, H., Watts, K., Navarro, A. and Ryder, D. (2001b) Monitoring barley lipid transfer protein levels in barley, malting and brewing. *Proceedings of the European Brewery Convention Congress, Budapest*, 28, 663–672.

Lusk, L. T., Cronan, C. L., Ting, P. L., Seabrooks, J. and Ryder, D. (2003) An evolving understanding of foam bubbles based upon beer style development. In: *Proceedings of the European Brewery Convention Congress*, Dublin, vol. 29, paper #79.

Lynch, D. M. and Bamforth, C. W. (2002) Measurement and characterisation of bubble nucleation in beer. *Journal of Food Science*, 67, 2696–2701.

Maeda, K., Yokoi, S., Kamada, K. and Kamimura, M. (1991) Foam stability and physiochemical properties of beer. *Journal of the American Society of Brewing Chemists*, 49, 14–18.

Melm, G., Tung, P. and Pringle, A. (1995) Mathematical modeling of beer foam. *Master Brewers Association of the Americas Technical Quarterly*, 32, 6–10.

Mills, E. N. C., Kauffman, J. A., Morgan, M. R. A., Field, J. M., Hejgaard, J., Proudlore, M. O. and Onishi, A. (1998) Immunological study of hydrophobic polypeptides in beer. *Journal of Agricultural and Food Chemistry*, 46, 4475–4483.

Mitani, Y., Joh, M., Segawa, S., Shinotsuka, K. and Ohgaki, K. (2002) Dynamic behavior of carbon dioxide gas related to formation and diminution of beer foam. *Journal of the American Society Brewing Chemists*, 60, 1–9.

Morris, K. S. and Hough, J. S. (1987) Lipid–protein interactions in beer and beer foam brewed with wheat flour. *Journal of the American Society Brewing Chemists*, 45, 43–47.

Muldjberg, M., Meldal, M., Breddam, K. and Sigsgaard, P. (1993) Protease activity in beer and correlation to foam. *Proceedings of the European Brewery Convention Congress*, 357–364.

Mulroney, A. R., Wenn, R. V., Ootwyn, J. and Williamson, R. R. (1997) The measurement of beer foam using a new laser-based video device. *Proceedings of the European Brewing Convention Congress, Maastricht*, 26, 615–622.

Mundy, J. and Rogers, J. C. (1986) Selective expression of a probable amylase/proteinease inhibitor in barley aleurone cell: Comparison to the barley amylase/subtilisin inhibitor. *Planta*, 169, 51–63.

Narziss, L. (1993) Modern wort boiling. *Proceedings of the Institute of Brewing Convention, Somerset West, South Africa*, 195–212.

Narziss, L., Reichender, E. and Barth, D. (1982a) Investigations on the effect of glycoproteidases on foam properties of beer. *Monatsschrift fur Brauwissenschaft*, 35, 275–283.

Narziss, L., Reichender, E. and Barth, D. (1982b) Concerning the influence of high molecular protein fractions and glycoproteins on beer foam with particular emphasis on technological procedures. *Monatsschrift fur Brauwissenschaft*, 35, 213–223.

Narziss, L., Miedaner, H., Graft, H., Eichhorn, P. and Lusting, S. (1993) Technological approach to improve flavour stability. *Master Brewers Association of the Americas Technical Quarterly*, 30, 48–53.

Nielsen, G., Johansen, H., Jemsen, J. and Hejgaard, J. (1983) Localization on barley chromosome 4 of genes coding for β-amylase (Bmy 1) and protein Z (Paz 1). *Barley Genetics Newsletter*, 13, 55–57.

Nielsen, H.a. H.-H. T (1988) Damage to beer foam by enzymes from yeast. *Brauwelt International*, 19, 441–443.

Nishida, Y., Tada, N., Inui, T., Kageyama, N., Furukubo, S., Takaoka, S. and Kawasaki, Y. (2005) Innovative control technology of malt components by use of a malt factionation technique. *Proceedings of the European Brewery Convention Congress, Prague*, 30, 93–100.

Onishi, A. and Proudlove, M. O. (1994) Isolation of beer foam polypeptides by hydrophobic interaction chromatography and their partial characterization. *Journal of Science Food and Agriculture*, 65, 233–240.

Onishi, A., Canterranne, E., Clarke, D. J. and Proudlove, M. O. (1995) Barley lipid binding proteins: Their role in beer stabilization. *Proceedings of the European Brewery Convention Congress, Brussels*, 25, 553–560.

Onishi, A., Proudlove, M. O., Dickie, K., Mills, E. N. C., Kauffman, J. A. and Morgan, M. R. A. (1999) Monoclonal antibody probe for assessing beer foam stabilizing proteins. *Journal of Agricultural and Food Chemistry*, 47, 3044–3049.

Ono, M., Hashimoto, S., Kakudo, Y., Nagami, K. and Kumada, J. (1983) Foaming and beer flavor. *Journal of the American Society of Brewing Chemists*, 41, 19–23.

Ormrod, I. H., Lalor, E. F. and Sharpe, F. R. (1991) The release of yeast proteolytic enzymes into beer. *Journal of the Institute of Brewing*, 97, 441–443.

Osbourne, T. B. (1924) *The Vegetable Proteins*, 2nd ed. Longmans, Green and Co: London.

Ovesna, J., Machova-Polakova, K., Kucera, L., Vaculova, K. and Miltova, J. (2006) Evaluation of Czech spring malting barleys with respect to β-amylase allele incidence. *Plant Breeding, 125,* 236–242.

O'Donnell, D. C. (1987) Pilot brewing of proanthrocyanidin-free malt. *Proceedings of the Australian Barley Symposium, Sydney,* 211–221.

O'Reilly, J. and Taylor, R. (1996) Solubility PGA in beer head retention. *Brewers' Guardian, 125,* 22–24.

Palmer, J. J. (2006) How the mash works. In: *How to Brew: Everything you need to know to brew beer right the first time,* pp. 141–152. Brewers Association: Boulder, CO.

Paris, M. and Eglinton, J. K. (2002) Genotyping single nucleotide polymorphisms for selection of barley β-amylase alleles. *Plant Molecular Biological Report, 20,* 149–159.

Parish, M. (1997) New research proves effectivness of "Head Keeper" nucleated glassware. *Brewers Digest, 71,* 26–27.

Perrocheau, L., Bakan, B., Boivin, P. and Marion, D. (2006) Stability of barley and malt lipid transfer protein 1 (LTP1) towards the heating and reducing agents: Relationships with the brewing process. *Journal of Agricultural and Food Chemistry, 54,* 3108–3113.

Polakova, K., Laurie, D., Vaculova, K. and Ovesna, J. (2003) Characterization of b-amylase alleles in 79 varieties with pyrosequencing. *Plant Moecular Biological Report, 21,* 439–447.

Powling, A., Islam, A. K. M. R. and Shepard, K. W. (1981) Isozymes in wheat-barley hybrid derivative lines. *Biochemical Genetics, 19,* 225–232.

Prins, A. (1988) Principles of foam stability. In: *Advances in Food Emulsions & Foams* (E. Dickinson and G. Stainsby eds), pp. 91–122. Elsevier Applied Science: London.

Prins, A. and van Marle, J. T. (1999) Foam formation in beer: Some physics behind it. *Beer Foam Quality,* pp. 26–36, vol. Monograph 27. Amsterdam, The Netherlands; Fachverlag Hans Carl, Nurnberg.

Rasmussen, J. N. (1981) Automated analysis of foam stability. *Carlsberg Research Communication, 46,* 25–36.

Rees, T. C. (1976) Cleaning procedures in the public house. *Brewers' Guardian, 105,* 32–36.

Roberts, R. T. (1975) Glycoproteins and beer foam. *European Brewery Convention Congress, 15,* 453–464.

Roberts, R. T. (1976) Interaction between beer protein and isohumulone. *Journal of the Institute of Brewing, 82,* 282.

Roberts, R. T., Keeney, P. J. and Wainwright, T. (1978) The effects of lipids and related materials on beer foam. *Journal of the Institute of Brewing, 84,* 9–12.

Robinson, L. H., Evans, D. E., Kaukovirta-Norja, A., Vilpola, A., Aldred, P. and Home, S. (2004) The interaction between brewing conditions and malt protein quality and their impact on beer colloidal stability. *Master Brewers Association of the Americas Technical Quarterly, 41,* 353–362.

Ronteltap, A. D. (1989) Beer foam physics. PhD Thesis, Agricultural University, Wageningen.

Ronteltap, A. D., Hollemans, M., Bisperink, C. G. J. and Prins, A. (1991) Beer foam physics. *Master Brewers Association of the Americas Technical Quarterly, 28,* 25–32.

Ross, S. and Clark, G. L. (1939) On the measurement of foam stability with special reference to beer. *Wallerstein Communications, 6,* 46–54.

Roza, J. R., Wallin, C. E. and Bamforth, C. W. (2006) A comparison between instrumental measurement of head retention/lacing and perceived foam quality. *Master Brewers Association of the Americas Technical Quarterly, 43,* 173–176.

Rudin, A. D. (1957) Measurement of the foam stability of beers. *Journal of the Institute of Brewing, 63,* 506–509.

Rudin, A. D. (1958) Effect of nickel on the foam stability of beers in relation to their isohumulone contents. *Journal of the Institute of Brewing*, 64, 238–239.

Rudin, A. D. and Hudson, J. R. (1958) Significance of isohumulone and certain metals in gushing beers. *Journal of the Institute of Brewing*, 64, 317–318.

Sarker, D. K., Wilde, P. J. and Clarke, D. C. (1995) Control of surfactant-induced destabilization of foams through polyphenol-mediated protein–protein interactions. *Journal of Agricultural and Food Chemistry*.

Sarker, D. K., Wilde, P. J. and Clark, D. C. (1998) Enhancement of protein foam stability by formation of wheat arabinoxylan-protein crosslinks. *Cereal Chemistry*, 75, 493–499.

Schwarz, P. B. and Pyler, R. E. (1984) Lipoxygenase and hydroperoxide isomerase activity of malting barley. *Journal of the American Society of Brewing Chemists*, 42, 47–53.

Segawa, S., Yamashita, S., Mitani, Y. and Takashio, M. (2002) Analysis of detrimental effect on head retention by low molecular weight surface-active substance using surface excess. *Journal of the American Society Brewing Chemists*, 60, 31–36.

Sheehan, M. C. and Skerritt, J. H. (1997) Identification and characterisation of beer polypeptides derived from barley hordeins. *Journal of the Institute of Brewing*, 103, 297–306.

Shimizu, C., Yokoi, S., Shigyo, T. and Koshimo, S. (1995) The mechanism controlling the decrease in beer foam stability using proteinase A. *Proceedings of the European Brewing Convention Congress*, Brussels, 25, 569–576.

Siebert, K. J. and Knudson, E. J. (1989) The relationship of beer molecular weight protein and foam. *Master Brewers Association of the Americas Technical Quarterly*, 26, 139–146.

Siebert, K. J. and Lynn, P. Y. (1997) Mechanisms of beer colloidal stabilization. *Journal of the American Society of Brewing Chemists*, 55, 73–78.

Siebert, K. J. and Lynn, P. Y. (2005) Comparison of methods for measuring protein in beer. *Journal of the American Society of Brewing Chemists*, 63, 163–170.

Simpson, W. J. and Hughes, P. S. (1994) Stabilization of foams by hop derived bitter acids. Chemical interactions in beer foam. *Cerevisiae Biotechnology*, 19, 39–44.

Skands, B., Lavrsen, I. and Bonfils, L. (1999) The use of a robot in assessing foam quality. *European Brewing Convention Monograph*, XXVII, Amsterdam, 62–68.

Skiver, K., Leah, R., Muller-Uri, Olsen, F. L. and Mundy, J. (1992) Structure and expression of the barley lipid transfer protein gene Ltp1. *Plant Molecular Biology*, 18, 585–589.

Slack, P. T. and Bamforth, C. W. (1983) The fractionation of polypeptides from barley and beer by hydrophobic interaction chromatography: The influence of their hydrophobicity on foam stability. *Journal of the Institute of Brewing*, 89, 397–401.

Smith, R. J. Davidson, D., and Wilson J. H. (1998) Natural foam stabilizing and bittering compounds derived from hops. *Journal of the American Society of Brewing Chemists*, 56, 52–57.

Smythe, J. E., O'Mahony, M. A. and Bamforth, C. W. (2002) The impact of appearance of beer on its perception. *Journal of the Institute of Brewing*, 108, 37–42.

Sorensen, S. B., Bech, L. M., Muldberg, M., Beenfeldt, T. and Breddam, K. (1993) Barley lipid transfer protein 1 is involved in beer foam formation. *Master Brewers Association of the Americas Technical Quarterly*, 30, 135–145.

St John-Coghlan, P., Woodrow, J., Bamforth, C. W. and Hinchliffe, E. (1992) Polypeptides with enhanced foam potential. *Journal of the Institute of Brewing*, 98, 207–213.

Stewart, D. C., Hawthorne, D. and Evans, D. E. (1998) Cold sterile filtration: A small scale filtration test and investigation of membrane plugging. *Journal of the Institute of Brewing*, 104, 321–326.

Stewart, G. G., Mader, A., Chlup, P. and Miedl, M. (2006) The influence of process parameters on beer foam stability. *Master Brewers Association of the Americas Technical Quarterly*, 43, 47–51.

Stillman, C. (2006) Is the glass fit to fill-glass washing – the final critical control point of beer quality. *Brewer and Distiller International*, 2, 19–21.

Stowell, K. C. (1985) The effect of various cereal adjustments on the head retention properties of beer. *Proceedings of the European Brewing Convention Congress, Helsinke*, 20, 507–513.

Tada, N., Inui, T., Kageyama, N., Takaoka, S. and Kawasaki, Y. (2004) The influence of malt acrospires on beer taste and foam quality. *Master Brewers Association of the Americas Technical Quarterly*, 41, 305–309.

Thurston, P. A., Quain, D. E. and Tubb, R. S. (1982) Lipid metabolism and regulation of volatile ester synthesis in *Saccharomyces cerevisiae*. *Journal of the Institute of Brewing*, 80, 90–94.

Todd, P. H., Held, R. W. and Guzinski, J. G. (1996) The development and use of modified hop extracts in the art of brewing. *Master Brewers Association of the Americas Technical Quarterly*, 33, 91–95.

Townsend, A.-A. and Nakai, S. (1983) Relationships between hydrophobicity and foaming characteristics of food proteins. *Journal of Food Science*, 48, 588–594.

van Nierop, S. N. E., Evans, D. E., Axcell, B. C. and Cantrell, I. C. (2002) Studies on beer foam proteins in a commercial brewing process. *Proceedings of the 27th Convention, The Institute of Brewing*, Asia Pacific Section, Adelaide, paper 14.

van Nierop, S. N. E., Evans, D. E., Axcell, B. C., Cantrell, I. C. and Rautenbach, M. (2004) The impact of different wort boiling temperatures on the beer foam stabilizing properties of lipid transfer protein 1. *Journal of Agricultural and Food Chemistry*, 52, 3120–3129.

Vaag, P., Bech, M. L., Cameron-Mills, V. and Svendsen, I. (1999) Characterization of a beer protein originating from barley. *Proceedings of the European Brewing Convention Congress, Cannes*, 27, 157–166.

Vaag, P., Bech, M. L., Cameron-Mills, V. and Sorensen, M. B. (2000) 17 kDa foam protein. International patent: PCT/IB99/01597.

von Wettstein, D., Jende-Strid, B., Ahrenst-Larsen, B. and Erdal, K. (1980) Proanthrocyanidin-free barley prevents the formation of beer haze. *Master Brewers Association of the Americas Technical Quarterly*, 17, 16–23.

Vundla, W. and Torline, P. (2007) Steps towards the formation of a model foam standard. *Journal of the American Society of Brewing Chemists*, 65, 21–25.

Wackerbauer, K., Meyna, S. and Theiss, L. (2005) LOX activity and different fractions of hydroxy fatty acids as well as their molecular stability in barley and malt. *Proceedings of the European Brewing Convention Congress, Prague*, 30, 721–736.

Walstra, P. (1989) Principles of foam formation and stability. In: *Foams: Physics, Chemistry and Structure*, pp. 1–15 (A. J. Wilson, ed.). Berlin, Heidelberg.

Wang, Z.-Y., He, G.-Q., Liu, Z.-S., Ruan, H., Chen, Q.-H. and Xiong, H.-P. (2005) Purification of yeast proteinase A from fresh beer and its specificity for foam proteins. *International Journal of Food Science and Technology*, 40, 1–6.

Weiss, A., Schonberg, Ch., Mitter, W., Biendl, M., Back, W. and Krottenthaler, M. (2002) Sensory and analytical characterisation of reduced and isomerised hop extracts and their influence and use in beer. *Journal of the Institute of Brewing*, 108, 236–242.

Wilde, P. J., Clarke, D. C. and Marion, D. (1993) Influence of competitive adsorption of a lysopalmitoyl-phosphalidyldoline on the functional properties of puroindoline A lipid binding protein isolated from wheat flour. *Journal of Agricultural and Food Chemistry*, 41, 1570–1576.

Wilde, P. J., Husband, F. A., Cooper, D., Ridout, M. J., Muller, R. E. and Mills, E. N. C. (2003) Destabilisation of beer foam by lipids: structural and interfacial effects. *Journal American Society of Brewing Chemists*, 61, 196–202.

Williams, K. M., Fox, P. and Marshall, T. (1995) A comparison of protein assays for the determination of the protein concentration of beer. *Journal of the Institute of Brewing*, 101, 365–369.

Wilson, R. J. H., Roberts, T. R., Smith, R. J., Bradley, L. L. and Moir, M. (1999) The inherent foam stabilizing and lacing properties of some minor, hop-derived constituents of beer. *European Brewing Convention Monograph*, XXVII, Amsterdam, 188–207.

Yabuuchi, S. and Yamashita, H. (1979) Gas chromatographic determination of trihydroxyactadecenoic acids in beer. *Journal of the Institute of Brewing*, 85, 216–218.

Yang, G. and Schwarz, P. B. (1995) Activity of lipoxygenase isoenzymes during malting and mashing. *Journal of the American Society of Brewing Chemists*, 53, 45–49.

Yasui, K., Yakoi, S., Shigyo, T., Tamaki, T. and Shinotsuka, K. A. (1998) Customer-orientated approach to the development of a visual and statistical foam analysis. *Journal of the American Society of Brewing Chemists*, 56, 152–158.

Yokoi, S., Maeda, K., Xiao, R., Kamada, K. and Kaminura, M. (1989) Characterization of beer proteins responsible for the foam of beer. *Proceedings of the European Brewing Convention Congress, Zurich*, 22, 593–600.

Yokoi, S., Yamashita, N., Kunitake, N. and Koshimo, S. (1994) Hydrophobic beer proteins and their function in beer foam. *Journal of the American Society of Brewing Chemists*, 52, 123–126.

2
Beer flavor
Paul Hughes

Introduction

The flavor attributes of beer are critical to its overall acceptance by consumers. Whilst beer has been defined as the second most bland drink after milk (presumably excluding water!), beer flavor is the ultimate expression of a rather complex series of production processes. Additionally, with the exception of a handful of examples[1] there are few practical options for the substantial removal of components from beer during production, which limits the opportunities for remedial action. This implies that, for product consistency, a fine degree of control is required. This is particularly relevant for beer flavor, where the typical dynamic range of flavor perception ranges from sub-ppt (low picomolar) to high-ppm (high millimolar) concentrations, or 11–12 orders of magnitude.

In this chapter, we start by reiterating the flavor unit (FU) concept originally expounded by Meilgaard. We then survey beer flavor as the interaction of malting and beer production processes with brewing raw materials. There are two other broad sources of flavors in beer: those that develop during beer storage, and those that are not normally present in beer that are derived from accidental contact with contaminated materials, the so-called taints. This approach, though, considers beer in terms of its specific components, so we close this chapter by appraising beer flavor in more holistic (i.e. consumer relevant) terms, and outline approaches for the assessment of beer flavor attributes.

The flavor unit

The comparison of analytical data for flavor compounds in beers is difficult to understand in sensory terms without an appreciation of the flavor thresholds of each of the analytes. Thus, whilst 15 ppb of hydrogen sulfide is sensorially significant, for dimethyl sulfide the flavor impact is much less or even absent at

---

[1]Most losses of components post wort separation are either through precipitation during wort boiling or evaporation during wort boiling and fermentation. Further removal is limited to the application of palliatives to remove polyphenols and polypeptides or enzymic treatment to modify the properties of macromolecules.

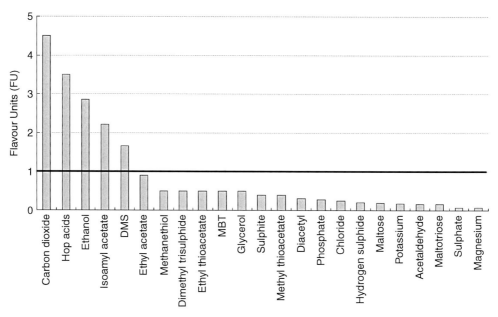

Figure 2.1
An incomplete Flavor Unit (FU) plot for a typical European lager beer. Few compounds are present at above flavor threshold. The introduction of taints or off-flavors can substantially distort this plot. For instance, modest exposure of beer to light will result in the development of FUs of MBT (3-methyl-2-butene-1-thiol) that exceed the total FUs for the rest of the beer flavor impact characters.

the same concentration. Meilgaard (1975) proposed the correction of analytical figures with flavor threshold estimates to derive the flavor unit (FU). The result is a set of analytical data that is scaled according to our ability to perceive it.

$$\text{Flavour Units} = \frac{\text{Concentration of flavour compound}}{\text{Flavour threshold}} \quad (2.1)$$

(*Note*: FU is dimensionless and therefore the units of analysis and threshold must be the same).

There are limitations to such an approach, not least of which is the variation of individuals' thresholds but, nonetheless, it does help to evaluate the significance of changes in flavor compound concentration. For instance, viewing analytical data in this way (Figure 2.1) shows clearly that there are typically few compounds that exceed their flavor threshold in beer and implies a total FU score for a given product. Such a score can be substantially distorted by the introduction of certain off-flavors or taints.

Brewing raw materials and beer flavor

Quantitatively, the most important raw materials used for the production of beer are the carbohydrate sources, that is barley (usually malted), adjuncts such as maize, wheat and sorghum. Less commonly oats, and, in some instances, flavored sugars (e.g. chip sugar) also confer flavor on beer when they are pressed into service. However, in terms of raw materials, it is the use of hops that uniquely defines the direct raw material contribution to beer flavor. Hops imbue beer with its characteristic bitterness and in many cases can provide an appreciable hoppy aroma to the final product. This is dependent on the way in which hops are used and, indeed, there are now a plethora of hop products available to the brewer to allow downstream adjustment of final hoppy flavors in beer independent of bitterness.

Malted barley

Malted barley is the major raw material for most beers produced world-wide. The vast majority of this is white malt, which is so-called because it is kilned primarily to remove the water absorbed during germination, rather than to generate color. The act of kilning also helps to reduce the levels of flavor-negative green notes which are characteristic of unkilned malt and due to the presence of aldehydes such as hexanal. The degree of color formation during kilning positively correlates to the formation of increasing amounts of flavor-active Maillard reaction products, which are inevitable given that the main aim of the malting process is to release fermentable carbohydrates and free amino nitrogen for subsequent fermentation. Thus the use of specialty malts, which are applied in smaller proportions, adds both color and flavor to the final beer (Table 2.1).

The heat-promoted chemical reactions that occur during malt kilning are complex, including the thermal degradation of phenolic acids, the caramelization of sugars, Maillard reactions (including the Strecker degradation of amino acids),

Table 2.1
Typical flavors derived from specialty malts (Hughes and Baxter, 2001)

Malt	Colour (°EBC)	Typical flavour attributes
Pale ale	4.5–4.8	Biscuity
Caramalt	25–35	Sweet, nutty, cereal, toffee
Crystal	100–300	Malty, toffee, caramel
Amber	40–60	Nutty, caramel, fruity
Chocolate	900–1200	Mocha, treacle, chocolate
Black	1250–1500	Smoky, coffee
Roasted barley	1000–1550	Burnt, smoky

and thermal degradation of oxygenated fatty acids derived both chemically and enzymatically from lipids (Figure 2.2). Thus a range of volatile compounds are formed, such as fatty acids, aldehydes, alcohols, furans, ketones, phenols, pyrazines and sulfur compounds. Compounds such as furaneol, maltol and

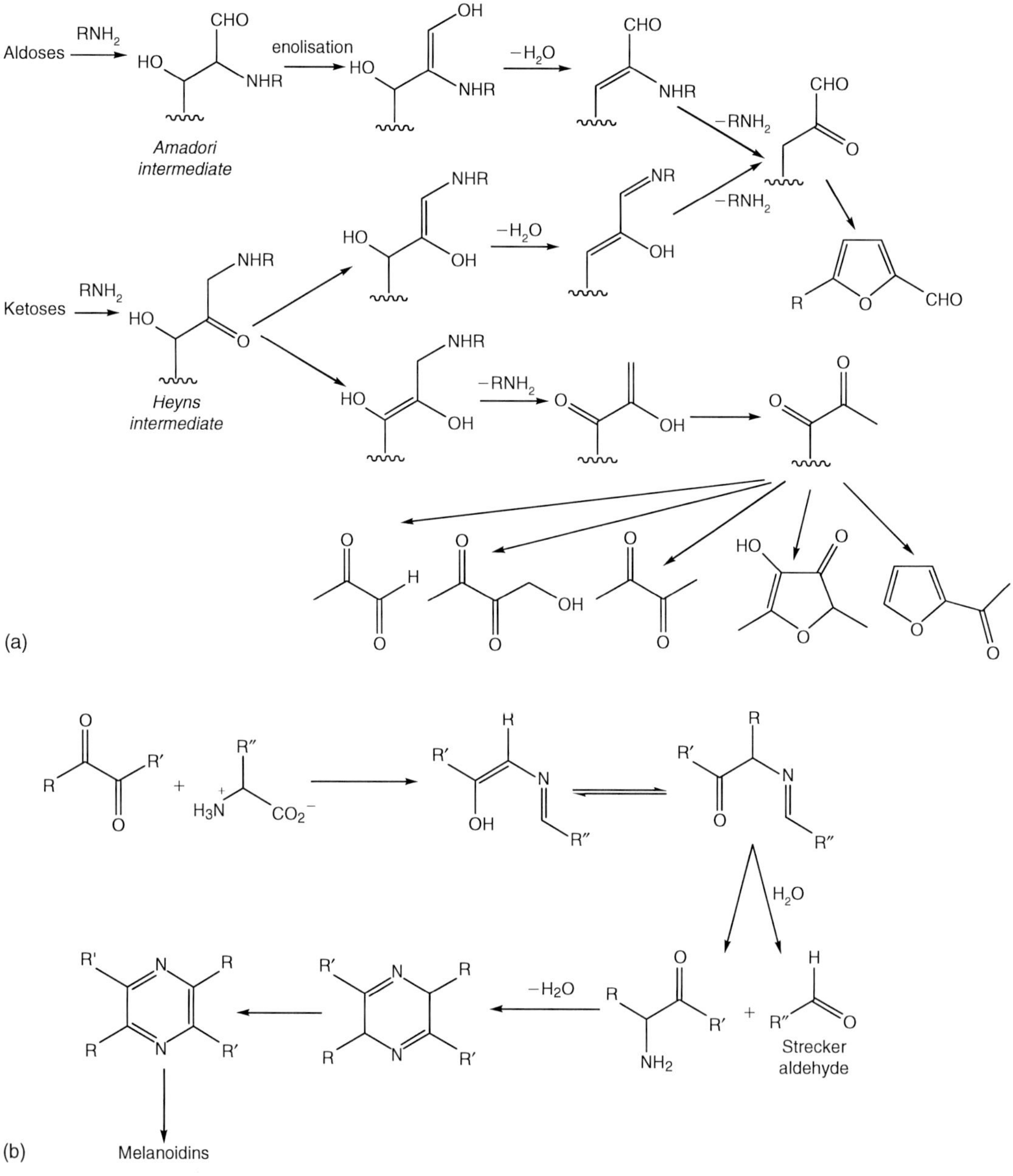

Figure 2.2
An overview of the Maillard reaction cascade (Hughes and Baxter, 2001). These reaction pathways are affected by various parameters, not least temperature, pH and water-activity.

isomaltol (Table 2.2) are highly flavor-active and can contribute to beer flavor. During the course of the Maillard reaction cascade, the oxygen of furan rings may be replaced by sulfur or nitrogen moieties, leading to the formation of the corresponding thiophenes and pyrroles. Other malt-derived heterocycles identified in malt include thiazoles, thiazolines, pyridines, pyrrolizines and pyrazines. Some pyrazines, such as the dimethylpyrazines, can occur at levels above two flavor units in some cases, thereby substantially affecting beer flavor (Herent et al., 1997). These heterocycles have a diverse range of flavor attributes, and have been variously described as malty, oxidized, sweet, meaty and burnt.

The presence of these compounds in malt indicates that they are likely to occur in sweet wort. However, many are lost during the boiling stage either by evaporation or chemical breakdown and during fermentation, by the action of yeast. Of those that survive into beer, dimethyl sulfide (DMS) is one of the most significant, particularly for lager beers.

A significant component of malt is lipid – typically 3% (w/w) of the dry grain (Palmer, 2006). Most of this material, being relatively insoluble in aqueous media, is lost with the brewers' grains or precipitated with the solids (trub) which are formed both during the boil and during the subsequent cooling operations. The major fatty acid component is linoleic acid, which can yield many C_5–C_{12} saturated and unsaturated aldehydes, ketones and alcohols, some of which are highly flavor-active. This is discussed further below.

Table 2.2
Flavor-active Maillard compounds found in malts (Hughes and Baxter, 2001)

Compound	Structure	Flavor attributes
Furaneol		Toffee, caramel
Maltol		Caramel
Isomaltol		Caramel, burnt sugar

Hop compounds

Much of the bitterness of beer is due to the presence of the iso-α-acids. These are a mixture of six major components, which are three stereoisomeric pairs of compounds derived from each of the three hop α-acids (Figure 2.3a). These α-acids are isomerized to the iso-α-acids during the wort boiling stage, although in practice the final utilization is around 30–40%. In contrast, isomerization by hop processing outwith the brewery (Figure 2.3a) gives yields in excess of 90%. There are probably minor differences in the bitterness of the individual compounds. Nevertheless, whilst hops come in a wide range of varieties, current thinking is that variety has little effect on bitterness intensity or quality. There are some who consider the cohumulone content of the α-acids to have poorer quality bitterness, but to date this has not been conclusively demonstrated. It is likely though that the utilization of the isocohumulones is significantly higher than the other iso-α-acids, which can be attributed to their lesser hydrophobicity, so they are more readily retained in solution.

Hops also contain the β-acids. Although these do not undergo isomerization in the sense of α-acids, they do form bitter degradation products – the hulupones – when present during wort boiling (Figure 2.3c). In practice, this contribution to bitterness is considered to be minor.

The mechanism of bitterness perception is an active and fascinating area of current research. Whilst it is outside the scope of this chapter to review this in detail, it is worth pointing out that the bitterness of the individual hop acids broadly mirrors the hydrophobicity model put forward by Gardner (1978, 1979). Essentially, there appears to be a strong correlation between bitterness intensities and hydrophobicities across a diverse range of hop-derived bitter substances. Certainly the more hydrophobic tetrahydroiso-α-acids are considered to be appreciably more bitter when present in beer, whilst the less hydrophobic ρ-iso-α-acids are appreciably less bitter.

The addition of hops to beer not only imparts bitterness but also a unique "hoppy" aroma. There has been some debate as to whether this arises from the breakdown of the non-volatile hop components, such the α- and β-acids, or from the oil constituents. It is now generally accepted that the latter is the predominant source of hoppy aroma in beers. Beers vary substantially in the intensity and quality of hoppy flavor. Intensity is dependent on brewing practice and dosing level, while quality is mostly dependent on the varieties used. The difficulties in understanding the provenance of hoppy flavors in beer is due to the complexity of hop oil, containing in excess of 300 components (Moir, 1994, Table 2.3). Before considering what constitutes hoppy aroma in beer, it is worth reviewing the composition of hop oil itself.

Of this oil, around 40–80% is hydrocarbon, which, with the exception of some simple aliphatic hydrocarbons, is mainly terpene in origin. Monoterpenes are represented by acyclic, monocyclic and bicyclic compounds, but myrcene is the most abundant and can account for up to 30% of the total oil. There is a group of at least 40 sesquiterpenes, which again can be either acyclic, mono-, bi- or tricyclic. The major components from this group are usually humulene and caryophyllene (Figure 2.4). Only two diterpenes have been identified in hop

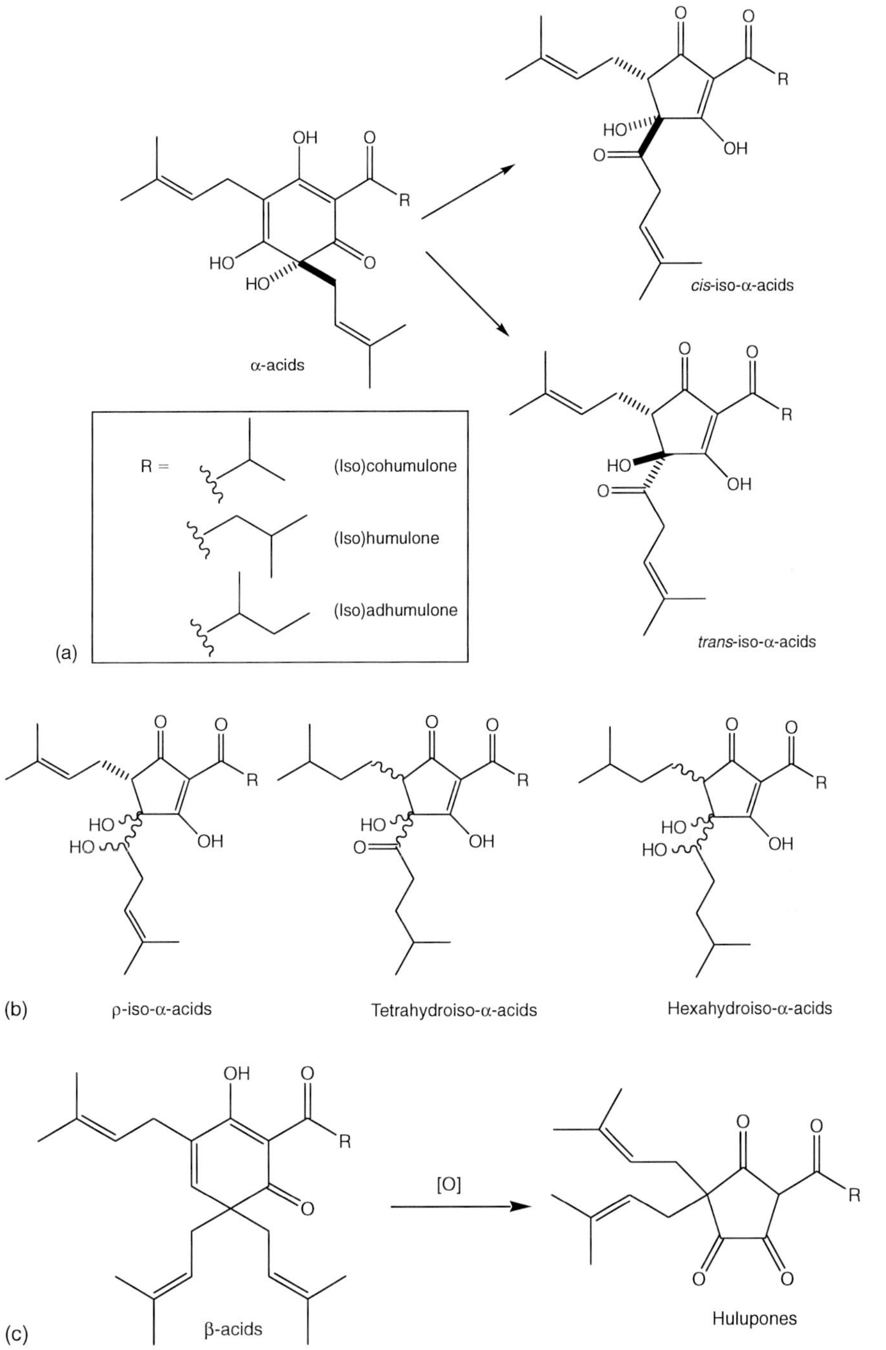

Figure 2.3
Hop acid contributors to beer bitterness. (a) The formation of the bitter iso-α-acids from the non-bitter α-acids; (b) the three groups of chemically modified iso-α-acids; (c) the formation of hulupones from oxidative degradation of the β-acids.

Table 2.3
Classification of hop oil components by chemical identity (Moir, 1994)

Classification	Approximate number identified
Hydrocarbons	60
Aldehydes	20
Ketones	50
Esters	70
Acids	10
Alcohols	70
Oxygen heterocycles	30
Sulfur compounds	30

oil – *m*- and *p*-camphorene – both of which are Diels-Alder adducts formed from two molecules of myrcene.

Some saturated and unsaturated aldehydes have been reported as hop oil constituents, as well as the monoterpenoid citral a and b (geranial and neral respectively) and citronellal (Figure 2.5). Aldehydes, particularly when unsaturated, are often highly flavor-active (see below), but in fact can be reduced to their corresponding alcohols during fermentation. Indeed, the presence of yeast aldo- and ketoreductase activities during fermentation has been demonstrated (Laurent et al., 1995).

There are a large number of methyl ketones which have been reported to occur in hop oil. A homologous series from heptan-2-one to heptadeca-2-one is evident in most hop varieties, together with a number of branched and unsaturated ketones. There is also some evidence for the presence of alkane-2,4-diones in the hop oil of one variety, Wye Target (Moir, 1994). The highly flavor-active nor-carotenoids, β-ionone and β-damascenone, have been identified in hop oil and found in beer at levels which indicate that they may be important contributors to hop flavor in beer.

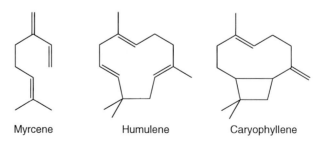

Myrcene   Humulene   Caryophyllene

Figure 2.4
The major terpene hydrocarbons in hop oil.

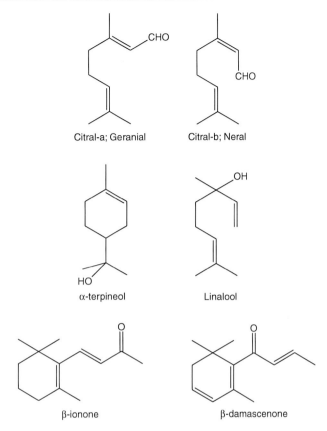

Figure 2.5
Some relevant oxygenated hop oil components. Linalool in particular is considered to be an indicator of the intensity of hoppy aroma.

The esters identified in hop oil include a homologous series of methyl esters, from hexanoate to dodecanoate, as well as a number of branched-chain and unsaturated methyl esters. Most of the hop oil esters are hydrolyzed during fermentation, and transesterification in the presence of increasing levels of ethanol has been shown to occur. Methyl esters of some conjugated acids, such as methyl geranate, resist hydrolysis and therefore are detectable in beer.

Some simple aliphatic carboxylic acids are sometimes present in trace quantities in hop oil, but the presence of branched-chain acids (such as 2-methylbutanoic acid and 4-methyl-3-pentenoic acid), has been shown to be due to the oxidation of the α- and β-acids. There is a wide range of alcohols present in hop oil, from series of straight- and branched-chain alkan-1-ols, to terpene alcohols. Linalool is the major terpene alcohol found in hops and can account for up to 1% of the total oil. Isomeric compounds include geraniol (largely esterified in hop oil) and α-terpineol. Moir (1994) suggested that a useful distinction could be made between terpene alcohols which can be considered to be the end-products of biosynthetic pathways (e.g. linalool) and allylic terpene alcohols (such as β-humulene-1-ol) which are terpene oxidation products and can be prepared by rearrangement of their epoxide counterparts. Levels of the former tend to fall during hop storage while levels of the latter increase.

The chemical lability of the sesquiterpene hydrocarbons is evident from their behavior during hop storage. Thus the levels of these hydrocarbons decrease on storage, whilst levels of epoxide auto-oxidation products increase. Humulene-4,5-epoxide has been identified in a commercial beer, and a number of humulene diepoxides have been detected in both hops and beer (Figure 2.6).

Some cyclic ethers have been detected in hop oil, including the *cis*- and *trans*-linalool oxides. These are presumably formed by the cyclization of linalool. Such components are efficiently transferred to wort during the wort boiling process and, together with other compounds such as karahana ether and hop ether are considered to be important hop flavor compounds. The 3(2H)-furanones, found in both hop oil and beer, show structural homology which indicates that they are derived by the degradation of the α- or β-acids.

Hops also contain a range of sulfur compounds. Green hops contain two series of S-alkyl thioesters, the only volatile sulfur compounds present. Their levels are essentially unchanged during kilning and resist both hydrolysis and transesterification. There are a number of literature reports of various sulfur compounds found in hops which are artefacts formed by the reaction of terpene

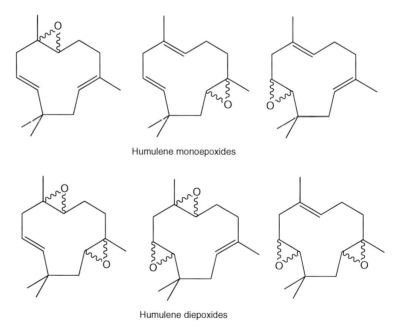

Figure 2.6
The humulene mono- and diepoxides. Whilst it is thought that these compounds only play a minor role in the expression of hoppy aroma in beer, these compounds can undergo a further series of complex reactions, which may in their own right result in the formation of flavor-active compounds (Deinzer and Yang, 1994).

hydrocarbons with residual sulfur which has been used as a fungicide to protect against powdery mildew damage. This usually does not result in any flavor problems, but a more serious issue of the presence of free sulfur on hops is the ability of yeast to generate sulfury and burnt off-flavors during fermentation when the hops are used for late or dry hopping.

From what has been said about the complexity, volatility and chemical lability of various groups of hop oil components, it is perhaps unsurprising that hoppy aroma in beer is far from understood. Indeed, it is likely that compounds which are considered to be hop-derived that occur in beer do not actually occur in hops themselves. A number of hop-derived compounds have been measured in beers, together with their thresholds (Table 2.4). There is a need though to treat such information with caution, as the analytical values are heavily dependent on the beer type and processing conditions employed, and the way in which the analysis was carried out. Furthermore, threshold values tend to be product specific and rely on the purity of the compounds which have been assessed.

The exact make-up of hoppy flavor in beer depends on where the hops or hop products are employed. Thus, to impart bitterness, it is usual to add hops, or more commonly hop pellets, at the beginning of the boil. The hoppy aroma which results from this is referred to as kettle hop aroma. Some workers have reported the loss of 95% of the total hop oil within 5 minutes of addition of the hops to a boil. Thus it is clear that little hop oil components survive the boil, although other workers have considered the hop acid degradation product 2-methyl-3-buten-2-ol to be a kettle hop character. In addition to this is a vast array of oxygenated sesquiterpenes, virtually all of which are below their flavor thresholds. For the production of lager beers, it is common practice to

Table 2.4
Flavor thresholds of typical hop aroma compounds found in beer

Compound	Range of reported levels ($\mu g/l$)	Flavour threshold in beer ($\mu g/l$)
Linalool	1–470	27, 80
Linalool oxides	nd–49	–
Citronellol	1–90	–
Geraniol	1–90	36
Geranyl acetate	35	–
α-Terpineol	1–75	2000
Humulene epoxide I	nd–125	10
Humulene epoxide II	1.9–270	450
Humulenol	1–1150	500, 2500

add a portion of the hops (typically 20%) towards the end of the boil. While this does not allow sufficient time for the effective conversion of the α-acids to their bitter isomerized counterparts, it does permit the extraction, steam fractionation and chemical modification of part of the hop oil present. Late hop flavor in lager beers is often described as floral and spicy, consisting primarily of monoterpene alcohols such as linalool and geraniol, together with the norcarotenoids and cyclic ethers. For the production of ales, it is common practice to add whole hop cones or whole hop pellets to the cask. The resulting dry hop character is relatively simple, comprising of a combination of monoterpenes such as myrcene, aliphatic esters and linalool.

Whilst traditionally beer was produced with whole hop cones, this practice is relatively rare today, with the use of standardized pellets, extracts which can provide bitterness, hoppy flavor, or both and highly refined extracts for the modification of hoppy flavors at the end of beer production. Finally, hoppy aroma is a hop variety characteristic. Thus while hops for providing bitterness are sold on the basis of their α-acid content, specific varieties prized for the quality of their aroma often command premium prices. The current status of research however means that the ability of a hop variety to impart a good aroma cannot readily be measured analytically and still requires brewing trials and sensory evaluation to ensure that this is indeed the case.

Impact of beer production processes

Ultimately it is the combination of production processes and raw materials that define the flavor of beer. Beer production is a complicated process, relying on three biotransformation steps-mashing, fermentation and maturation – on to which is superimposed the chemical changes that arise particularly during wort boiling and wort clarification. In this part, we consider only the flavor attributes that arise by "design," rather than taints and off-flavors, which are considered separately below.

Mashing and sweet wort separation

These critical operations are essential for the preparation of the medium for subsequent fermentation and for providing the precursors to many of the process-derived flavor attributes in beer. The development of free amino nitrogen and reducing sugars during mashing provides a minor set of flavor precursors that can undergo the Maillard reaction, principally during wort boiling, although decoction processes are likely to promote Maillard reaction chemistry. The other key reaction cascade that is initiated during mashing is the development of lipid oxidation products. Whilst the impact of mashing conditions on final beer flavor stability is contentious, there is little doubt that either auto- or lipoxygenase-induced oxidation results in the degradation of malt lipids. The most notable set of reactions is that which starts with the oxidation of linoleic acid to generate the corresponding hydroperoxides (Figure 2.7). The

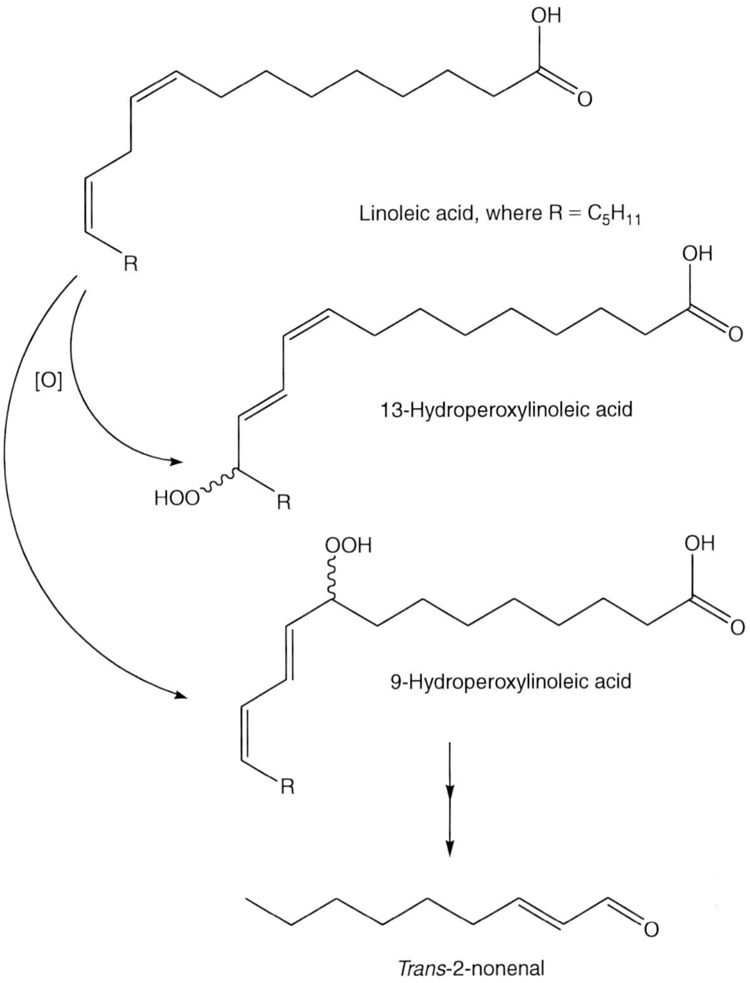

Figure 2.7
Auto- or lipoxygenase-induced oxidation results in the degradation of malt lipids. The most notable set of reactions is that which starts with the oxidation of linoleic acid to generate the corresponding hydroperoxides.

9-hydroperoxide is considered to ultimately degrade to form the highly flavor-active *trans*-2-nonenal (t-2-N). This compound, with a distinct cardboard or fresh linen aroma, is typically perceptible at sub-ppb levels. Put another way, conversion of all the linoleic acid in an all-malt mash into *trans*-2-nonenal would result in more than 10 million FU of trans-2-nonenal! Of course, this is not what happens in reality, but it serves to illustrate the fact that, for highly flavor-active species, the chemistry of their formation need not be efficient or readily apparent.

Wort boiling and wort clarification

In terms of flavor, perhaps the most notable impacts on beer flavor are:

- The dissolution of α-acids and their isomerization to the iso-α-acids
- The breakdown of S-methylmethionine (SMM) to DMS and the subsequent evaporation of the latter
- The loss of hop oil components through both evaporation and chemical modification.

The impact of trub formation on beer flavor is not completely clear but is most likely to be indirect. Thus the losses of polyphenols and polypeptides may have consequences for the subsequent flavor stability of beer in-pack, and the chelation of cations can impact on fermentation performance, but overall the three processes defined above are the critical processes that occur during boiling to affect beer flavor.

The dispersal of the α-acids in boiling wort is dependent on the hop products used. Wilson and Roberts (2006) have shown that α-acids are more rapidly extracted from pellets than from extracts, presumably due to the better dispersion of pellet material relative to resinous extracts. The resulting isomerization results in a mixture of iso-α-acids that is predominantly of the *cis*-configuration (ca 68%). The *cis*-isomers tend to be more stable in packaged beer (De Cooman et al., 2000) and demonstrate distinct sensory and physicochemical properties from their *trans*-counterparts (Hughes, 2000). The resulting bitterness expressed by the iso-α-acids has a significant time-dependency (Pangborn et al., 1983).

The degradation of SMM, an amino acid residue found in barley proteins, to DMS (Figure 2.8) can have a significant impact on beer flavor and is the most significant flavor compound to be derived from malted barley. Typically the aroma of DMS is described as sweetcorn, tinned tomato or baked beans. The level of DMS can be either above or below threshold and it is often a characteristic of a brand. As SMM degradation can occur both in wort boiling and whirlpool rest, a key control point for DMS in beer is control particularly of the whirlpool rest, as the opportunity for DMS evaporation is less.

It is well established that hop oil compounds are substantially lost by evaporation even over short periods of time (typically more than 90% in less than

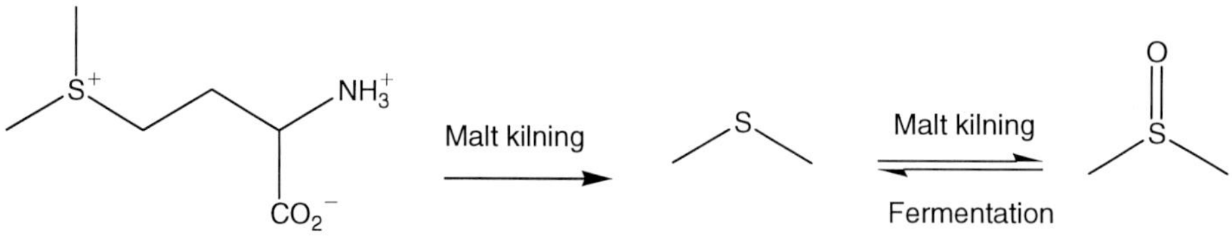

Figure 2.8
S-methylmethionine is prone to thermal degradation during malt kilning, wort boiling and whirlpool operations to yield DMS. Further oxidation during malt kilning can also result in the formation of DMSO, which can be subsequently reduced to DMS during fermentation.

10 minutes). Indeed, for hop products that contain both oil and bitter fractions, substantial loss of the hop oil is highly desirable, otherwise the resulting hoppy flavor can be overwhelming in the final product. Thus, even when using isomerized pellets, it is important to add them sufficiently early to allow not only the dispersion of the iso-α-acids, but also the evaporative losses of hop oil components.

Fermentation

Without doubt it is the fermentation of wort that is decisive in final beer flavor expression. The two major products of fermentation – ethanol and carbon dioxide – are both above threshold for many beers, and confer mouthfeel, body, warmth and liveliness to the final product. It is rare that either is out of specification, as ethanol is subject to regulatory control through taxation and carbon dioxide can be readily adjusted. However, ethanol and carbon dioxide are not the only products of a brewery fermentation. A whole array of volatile compounds are formed, the profile of which are a function both of the yeast strain used and the composition of the boiled wort that is pitched. Here, the principal volatiles (esters, higher alcohols, aldehydes, diacetyl and sulfur compounds) are considered in turn.

The esters are the most important group of fermentation-derived flavor-active components. They impart fruity, floral and solvent-like flavors and aromas to beers. The most important esters include ethyl acetate (solvent-like, fruity), isoamyl acetate (sweet, banana), isobutyl acetate (banana, fruity), ethyl caproate (apple) and 2-phenylethyl acetate (rose, honey). As might be expected, the most abundant esters are those of organic acids with ethanol, given that ethanol is the most abundant alcohol in beer. (For a detailed overview of how esters are thought to form during fermentation, see Boulton and Quain (2006) and Casey (2007).) Suffice to say that ester synthesis regulation is complicated, and seems to be influenced by other pathways that affect the flux of acetyl coenzyme A, not least of which is lipid synthesis. Clearly, the impact of esters on beer flavor is such that the control of ester synthesis is essential for reproducible beer flavor.

Of the 40 or so higher alcohols identified in beers, the most important in terms of flavor are *n*-propanol, isobutanol, 2-methylbutanol and 3-methylbutanol. These can be considered to be more potent analogues of ethanol, imparting warming as well as "alcoholic" flavor and aroma. 2-Phenylethanol is also important in some beers, imparting a rose or floral aroma. In a rare consideration of flavor interactions, Hegarty et al. (1995) showed that higher levels of 2-phenylethanol suppressed the perception of DMS in final beer. The higher alcohols, as well as affecting beer flavor in their own right, are also precursors of esters, by way of their reaction with acylated coenzyme A, particularly acetyl coenzyme A for the ultimate formation of acetate esters.

The formation of aldehydes during fermentation is intimately bound up with the formation of higher alcohols. Thus the pathways for the formation of ethanol, *n*-propanol, isobutanol and 2- and 3-methylbutanols proceed *via* their corresponding aldehydes. Aldehydes are much more flavor-active than their corresponding alcohols (Table 2.5), and are generally considered to contribute negative flavor attributes to beer. The major aldehyde in beer, acetaldehyde,

Table 2.5
Aldehydes commonly occurring in beer (after Hughes and Baxter, 2001)

Compound	Typical levels in beer (mg/l)	Threshold (mg/l)	Flavor units	Flavor
Acetaldehyde	2–20	25	0.08–0.8	Green, paint
trans-2-Butenal	0.003–0.02	8.0	0.00	Apple, almond
2-Methylpropanal	0.02–0.5	1.0	0.02–0.5	Banana, melon
C_5 aldehydes	0.01–0.3	ca 1.0	0.01–0.3	Grass, apple, cheese
trans-2-Nonenal	0.00001–0.002	0.0001	0.09–18	Cardboard
Furfural	0.01–1.0	200	0.0	Papery, husky
5-Methylfurfural	<0.01	17	0.0	Spicy
5-Hydroxymethylfurfural	0.1–20	1000	0.0–0.02	Aldehyde, stale

confers an emulsion paint or green apple aroma and taste to beer. Other aldehydes can be responsible for flavor defects in aging beer, either by higher alcohol oxidation, or Strecker degradation of certain amino acids (see below).

The presence of vicinal diketones (VDKs; diacetyl and its less flavor-active homologue pentane-2,3-dione) in beer is often, but not always, considered to be a negative contributor to beer flavor. Diacetyl typically has a butterscotch flavor, and thus is also associated with the perception of sweetness. VDK's also has a substantial impact on mouthfeel, conferring a perception of thickness on the palate. The formation of diacetyl is considered to be derived from precursor α-acetohydroxy acids, themselves intermediates in the biosynthesis of valine and isoleucine. During fermentation, extracellular VDKs are metabolized by yeast cell reductases which reduce either one or both of the carbonyl moieties. These reduced metabolites are much less flavor-active than the VDKs themselves, and so their persistence in beer can be tolerated.

Perhaps some of the most easily recognizable flavors in beers are sulfur-containing. These can be classified as thiols, sulfides, disulfides, trisulfides, thioesters and sulfur heterocycles (Table 2.6). Often the most important sulfur compound in beer is DMS. Residual DMSO that arises from SMM oxidation during malt kilning can be reduced by yeast DMSO reductase, particularly after the cessation of yeast growth. The thiols are not desired in beer unless their levels are under rigorous control. The parent of all thiols, hydrogen sulfide, is flavor-active at around 15 ppb and, at controlled levels, can make a positive contribution to beer flavor. It is formed by the pathways that assimilate sulfur from sulfate and sulfite, which in turn is involved in the biosynthesis of the sulfur-containing amino acids (methionine and cysteine) and the aliphatic alcohol amino acids (serine and threonine; Boulton and Quain, 2006).

Two other sulfides are worth considering here. Methional (4-thiapentanal) and methionol (4-thiapentan-1-ol) are derived from by the Strecker degradation of methionine and have flavors reminiscent of raw and cooked potato respectively.

Table 2.6
Classification of flavor-active sulfur compounds that occur in beer

Class	Examples	Flavour descriptors
Inorganic	Hydrogen sulfide Sulfur dioxide	Rotten eggs Struck match
Thiols	Methanethiol 3-Methyl-2-butene-1-thiol	Putrefaction Lightstruck, "skunky"
Sulfides	Dimethyl sulfide	Sweetcorn
Disulfides	Dimethyl disulfide	Rotten vegetable
Trisulfides	Dimethyl trisulfide	Rotten vegetable, onion
Thioesters	Ethyl thioacetate	Cabbage

Methional is the more flavor-active, and can have a significant impact on the flavor expression of ageing beer.

Aliphatic thiols, particularly methanethiol, are highly flavor-active and, with flavor attributes such as rotting vegetable, are to be avoided. The formation of methanethiol is thought to be mediated by S-adenosylmethionine, but may also form by the photodegradation of methionine in the presence of riboflavin. Because of its flavor potency (1 FU < 1 ppb) modest quantities of methanethiol will have a substantial effect on beer flavor quality. The most infamous thiol that is found in beer though is 3-methyl-2-butene-1-thiol (MBT). This compound, flavor-active at low ppt levels, is formed by the light-induced degradation of hop-derived iso-α-acids (see below).

The di- and trisulfides are represented mainly by dimethyl disulfide and dimethyl trisulfide respectively. Both are highly flavor-active with typical sulfury flavors. The mechanisms for their formation though are not completely clear. Thioesters (principally methyl and ethyl thioacetates) are less flavor-active and have less offensive flavor attributes, such as cooked vegetable. Their formation is thought to be analogous to the formation of esters, that is the reaction between active acetate and thiols. Finally, the heterocyclic sulfur compounds are a diverse group. They may be derived from Maillard reaction chemistry, or from thiamine (vitamin B_1) degradation. These compounds have relatively low volatility and are generally less well-characterized than other sulfur compounds. However, flavor attributes such as roasted and meaty are common for such compounds.

Maturation and finishing

The major impact that maturation has on beer flavor is the reduction of diacetyl to acetoin and butane-2,3-diol as mentioned earlier. This process can be accelerated by enzyme treatment with Maturex® (Novozymes, 2008), which converts α-acetolactate directly into acetoin and reduces or eliminates the need for maturation. Post-maturation, impacts on flavor are relatively slight, with the exception of specific flavor additions (e.g. bitterness and hoppy aroma adjustment). Physical stabilizers, such as PVPP and silica gels should not contribute either positively or negatively to beer flavor. Body feed kieselguhr can be prone to picking up contamination, and must therefore be transported under closely controlled conditions (e.g. dedicated shipping containers). The early applications of membrane filtration were suspected of reducing mouthfeel as the pores of the membrane become partially occluded through operation, removing macromolecules that contribute to the perception of fullness.

In-pack flavor changes

The composition of beer is complex, so it is not surprising that the finished product bears compounds that are mutually reactive. It is important to recognize that, as there are no reported instances of the development of components that are negative from a consumer well-being perspective, the major concern for a brewer is the consistency of his/her product in the market and the maintenance of optimal quality.

Perhaps the most infamous in-pack flavor change is that induced by the exposure of beer to light. Native iso-α-acids are prone to photochemical degradation by both visible and ultra-violet light which results in the formation of 3-methyl-2-butene-1-thiol (MBT; Figure 2.9). This component, with flavor

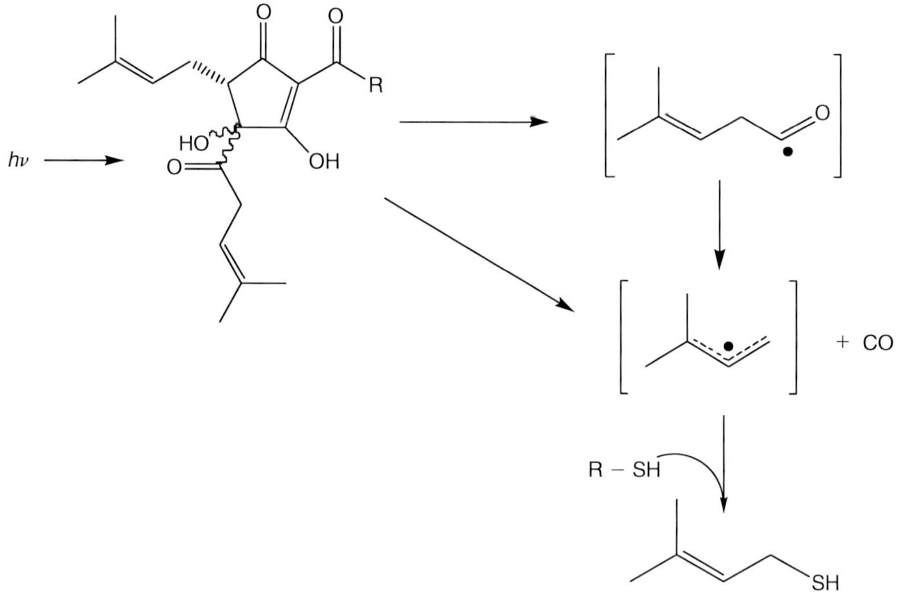

Figure 2.9
The formation of 3-methyl-2-butene-1-thiol (MBT) by light-induced degradation of the hop-derived iso-α-acids.

thresholds reported from 1 to 35 ng/l is reported to have a skunky aroma (the compound is similar to, but not the same as, the major impact compound emitted by skunks). The critical factor though is its low flavor threshold, so that even brief exposures to light can render a beer unpleasant or undrinkable. Packaging beer in green or clear glass exacerbates the problem, whilst the use of opaque packaging or brown glass affords sufficient protection to the contained beer. Brewers are making increasing use of 100% substitution of the native iso-α-acids with chemically modified counterparts, with good effect. The success is highly dependent on the rigour with which the brewery is cleaned between "conventional" and light-stable brews. Even separate trap filters are considered necessary to minimize the risk of contamination of the light-stable beer with native iso-α-acids.

There are other components that form more slowly in beer but nonetheless can have a profound impact on final beer flavor. The formation of *trans*-2-nonenal (E-2-N) as a result of linoleic acid oxidation (Figure 2.7) is well known and again, it is the flavor threshold of this compound (ca 0.1 μg/l) which dictates that very little needs to be formed to have an undesirable impact. Its flavor can be described as cardboard or fresh linen, but either way is not generally desired in beer.

If t-2-N is considered to be the most important impact character of beer flavor stability, then the next "tier" of flavor impact compounds is likely to be the so-called Strecker aldehydes – 2-methylpropanal (from valine), 2- and 3-methylbutanal (from isoleucine and leucine respectively), methional (from methionine) and phenylacetaldehyde (from phenylalanine). These aldehydes increase during beer ageing and have a range of flavor attributes, such as honey and cooked potato. In principle then, control of residual free amino nitrogen in final product could be a useful strategy for the retardation of beer flavor changes in-pack.

A number of other compounds have been observed to change level during beer ageing, many of which though are considered to be markers, rather than actual flavor-active entities in their own right. So, for instance, furfural forms during beer ageing, but never reaches its flavor threshold. Its value lies in its reflection of the time-temperature history of the product in question. The formation of marker esters, such as mono- and diethyl succinate and ethyl nicotinate similarly reflects beer ageing but without having any significant influence on final beer flavor directly.

One sulfur compound that has not been considered explicitly above is sulfur dioxide. This is not particularly flavor-active, but can occur at around 1 FU or higher. It is derived during fermentation as part of the sulfate assimilation pathway, but can also be added either consciously, or indirectly as it is used as a preservative in additives such as finings and primings. Sulfur dioxide has recently attracted attention due to its status as an allergen, which has meant that it has come under regulatory control. Probably the most significant impact that sulfur dioxide has on beer flavor is that it helps beer to retain its freshness in-pack. It is thought that this is due to its ability to form bisulfite addition products with aldehydes (reducing their flavor activity) and by acting as an antioxidant, either as is or as an addition product with acetaldehyde.

Taints and off-flavors

The overall flavor expression of beer is expected to derive from a combination of the production process and judicious selection of ingredients. However, on occasion, components that are foreign to typical beers can be found, or, alternatively, compounds are present at levels far above their expected level. The focus here is on the presence of adventitious materials that elicit a taste response, rather than other components that can have implications for consumer well-being (e.g. mycotoxins, nitrosamines). The distinction made here is that taint compounds are normally absent from beer, whereas off-flavors are caused by the presence of components that are usually present (often at substantially less than one FU), but in fact are present at levels above one FU and therefore have a significant if not substantial impact on beer flavor expression.

Taints are, on the whole, extremely flavor-active, and have been extensively reviewed by Bennett (1996). The flavor potency of taints can be very high so that they can be missed by all but the most diligent of analyses. Thus it is common to attempt to characterize a suspected taint by first assessing the sensory properties of the contaminated beer. This can provide a clue as to the analytical target and therefore steer the analytical approach to confirm the presence of a given taint.

Perhaps the most well-known are those derived from phenol, present either as halogenated (chloro- or bromo-) phenols or anisoles (i.e. O-methylated phenols). With typical flavor thresholds at or below 1 ng/l, very small quantities of these substances are required to adversely affect beer flavor. Their presence in beer is considered to be due to the historical use of pentachlorophenol to preserve wood, protecting it from moulds, mildew and termites. It has also been detected in food contact packaging materials, such as boards and paper. The chlorination of phenol also yields the antifungal tetra- and trichlorophenols, and it is the migration of these compounds into foodstuffs and beer that results in typical chlorophenolic taints. Descriptions of such taints are often listed as trichlorophenol (TCP) or hospital/disinfectant-like. Whilst these compounds are toxic, their high level of flavor activity essentially protects the consumer from overconsumption. Whilst the use of pentachlorophenol has waned, sporadic occurrences of such taints can still arise. For instance treated wood that is converted to sawdust can find its way back into the brewery as packaging for equipment, such as pumps. Migration of pentachlorophenol from the sawdust to the pump represents a taint risk if the pump is not very thoroughly cleaned prior to installation. Chlorinated anisoles are quite distinct in their sensory characteristics, exhibiting musty or, in wine-parlance "corked" aroma and taste and are formed by bacteria-mediated methylation of the corresponding chlorophenol.

Other contact materials can contribute to taints in final product. Metallic taints can be picked up from filter aids, although this is rarer as most if not all brewers set rigorous iron specifications for kieselguhr to minimize the impact of iron-induced oxidation on beer flavor during storage. Oil and grease taints can originate from lubricants and the so-called cardboard-like "Labox" taint is thought to arise from volatiles absorbed by can lacquers.

Holistic flavor perception

It is common practice to relate specific flavors to one or more chemical components. Indeed, many renditions of the flavor profile incorporate chemical terminology such as DMS, diacetyl and butyric. Whilst this is helpful in understanding the origin of a whole range of flavors, and for defining analytical targets for flavor control, there are a number of flavor expressions that seem to defy this straight-forward mapping (Figure 2.10).

As mentioned earlier, Hegarty et al. (1995) demonstrated that the presence of 2-phenylethanol has an antagonistic effect on DMS perception, indicating that even a simple regression model to relate concentration to perceived intensity is fraught with difficulty.

It is perhaps no coincidence that those flavor expressions which prove difficult to define can also prove difficult to train sensory panellists on, due to the lack of unequivocal training standards. Perhaps the most apparent examples are the expressions of hoppy aroma and beer ageing. In addition to the specific challenges implicated by Figure 2.10, other phenomena which have at least some basis in flavor perception, such as satiety and drinkability, prove to be sufficiently abstract to raise the question as to whether they are actually real constructs or not (Thomson and Bailey, 2006).

Currently there are no models that can fully accommodate the modeling of all but the simplest of sensory responses, so it would seem that for the foreseeable future, sensory testing will be a mandatory requirement for product monitoring and control. Such models might incorporate some form of accounting for the non-linearity of human responses to flavors and activities to accommodate matrix effects.

Analytical/flavour correlation		
Low	*Moderate*	*High*
Characteristics		
Multicomponent; Suspicion of undiscovered impact compounds and strong matrix effects	Multicomponent; Reasonable handle on analytics. Additive models probably insufficient	Single/few measurable components. Weak matrix effects
Examples		
Beer ageing Hoppy character	Estery Bitterness	Diacetyl Lightstruck

Figure 2.10
Success in the prediction of beer flavor performance from analytical data is dependent on the character being evaluated. The challenge is particularly onerous for the flavor attributes towards the left of the figure as presented.

Summary

Here, the flavor of beer has been considered from the points of view of its raw materials and the impact of processing on them. The classical approach for assessing beer flavor attributes is to consider each component separately or in small groups of very similar species. This approach works well for distinct entities such as diacetyl, DMS and bitterness, but is more problematic for less readily defined flavor attributes. A fruitful area of future flavor research is to understand the interactions between flavors more fully, either through empirical experimentation, or by exploiting the recent developments in the area of taste and olfactory receptor genetics.

References

Bennett, S. J. E. (1996) Off-flavours in alcoholic beverages. In: *Food Taints and Off-Flavours* (M. J. Saxby ed.), 2nd edition, pp. 290–320. Chapman & Hall: London.

Boulton, C. and Quain, D. (2006) *Brewing Yeasts and Fermentation*. Blackwell: Oxford. pp. 113–141

Casey, G. P. (2007) A journey in brewing science and the ASBC. In: *Brewing Chemistry and Technology in the Americas* (P. W. Gales ed.), pp. 99–231. ASBC: St Paul, MN.

De Cooman, L., Aerts, G., Overmeire, H. and De Keukeleire, D. (2000) Alterations of the profiles of iso-α-acids during beer ageing, marked instability of *trans*-iso-α-acids and implications for beer bitterness consistency in relation to tetrahydroiso-α-acids. *Journal of the Institute of Brewing*, 106, 169–178.

Deinzer, M. and Yang, X. (1994) EBC Monograph 22. Symposium on Hops, Zoeterwoude, Getränkefachverlag Hans Carl, Nürnberg, 181–187.

Gardner, R. J. (1978) Lipophilicity and bitter taste. *Journal of Pharmacy and Pharmacology*, 30, 531.

Gardner, R. J. (1979) Lipophilicity and the perception of bitterness. *Chemical Senses and Flavour*, 4, 275–286.

Hegarty, P. K. Parsons, R., Bamforth, C. W. and Molzahn, S. W. (1995) Phenyl ethanol – a factor determining lager character. *Proceedings of the 25th EBC Congress*, Brussels, IRL Press: Oxford, pp. 515–522.

Herent, M.-F., Vanthournhout, C., Gijs, L. and Collin, S. (1997) Influence de la composition en heterocycles azotes de malts speciaux sur le profile aromatique de la biere. *Proceedings of the 26th EBC Congress*, Maastricht, pp. 167–174.

Hughes, P. S. (2000) The significance of iso-α-acids for beer quality. *Journal of the Institute of Brewing*, 106, 271–276.

Hughes, P. S. and Baxter, E. D. (2001) *Beer: Quality, safety and nutritional aspects*. Royal Society of Chemistry: Cambridge. 138 pp.

Laurent, M., Geldorf, B., Van Nedervelde, L., Dupire, S. and Debourg, A. (1995) Characterization of the aldoketoreductase yeast enzymatic system involved in the removal of wort carbonyls during fermentation. *Proceedings of the 25th EBC Congress*, Brussels, IRL Press: Oxford, pp. 337–344.

Meilgaard, M. C. (1975) Flavour chemistry of beer – Part I: Flavour interactions between principal volatiles. *MBAA Technical Quarterly*, 12, 107–117.

Moir, M. (1994) Hop aromatic compounds, EBC Monograph XXII, Symposium on Hops, Zoeterwoude, The Netherlands, Getränkefachverlag Hans Carl, Nürnberg, pp. 165–180.

Novozymes, www.novozymes.com, accessed January 5, 2008.

Palmer, G. H. (2006) Barley and malt. In: *Handbook of Brewing*, 2nd edition (F. G. Priest and G. G. Stewart, eds), Boca Raton, FL.

Pangborn, R. M., Lewis, M. J. and Yamashita, Y. F. (1983) Comparison of time-intensity with category scaling of bitterness of iso-alpha-acids in model systems and in beer. *Journal of the Institute of Brewing*, 89, 349–355.

Thomson, D. H. and Bailey, P. (2006) Psychology of drinkability, EBC Monograph 34 (CD-ROM), Drinkability Symposium, Edinburgh, Getränkefachverlag Hans Carl, Nürnberg, 64–77.

Wilson, R. J. H. and Roberts, T. R. (2006) Hops. In: *Handbook of Brewing*, 2nd edition (F. G. Priest and G. G. Stewart, eds), Boca Raton, FL.

3
The flavor instability of beer

Charles W. Bamforth and Aldo Lentini

Perhaps the biggest remaining quality challenge for brewers is the achievement of flavor stability. The factors determining flavor robustness in beer are extremely complex.

There is no universally accepted terminology for the flavor changes that occur in beer and certainly no surety that any two beers will age in exactly the same way, to give the same flavor notes in identical proportions. "Gently"-flavored lagers and strong ales are not predisposed to the selfsame flavor changes. If we are to generalize in pursuit of simplification, then one of the first changes is a perceptible decline in bitterness, and the beer may be perceived as harsh. There will also be a decline in fruity/estery and floral notes. Some beers will develop a ribes (blackcurrant buds, tomcat urine) aroma and most beers are claimed to develop a wet paper or cardboard character. Bready, sweet, toffee-like, honey, earthy, straw, hay, woody, winey and sherry-like are all notes that have been reported (Drost et al., 1971; Meilgaard, 1972; Dalgliesh, 1977; Whitear, 1981).

There is a difference between the flavor perception of beer aged "naturally" (remembering that this may vary in regions with extremes in climate) and in beer which has been "forced" aged. Thus Kaneda et al. (1995) say that a beer aged at 25°C develops primarily caramel-like characteristics, whereas at 30°C or 37°C the cardboard notes are emphasized.

To add to the complexity of the situation, it is not entirely clear that beers displaying a pronounced age character necessarily meet with disfavor in the marketplace. Thus Guinard and co-workers (2001) showed that when judged under branded conditions, imported beers were found by an expert panel to display stale characteristics, but they were nonetheless preferred to domestic beers. When the beers were not brand-identified, there was an equal preference for the imported and domestic beers. This highlights that branding may be at least as significant as inherent flavor characteristics in consideration of beer choice. Stephenson and Bamforth (2002) also demonstrated that branding was of major significance in beer, however when a given brand is identified there is a preference for the fresh version of that beer.

Achieving flavor stability is a major challenge, especially as what happens to the beer in between packaging and consumption is often out of the control of the brewer. It has even been suggested that the aged character should be maximized in beer before it leaves the brewery, on the basis that no further flavor change will occur (Torline et al., 1999). This chapter makes the fundamental assumption, however, that most brewers do desire to minimize flavor change and that their beers should be inherently fresh. The basic credo is that all packagings of a given brand of beer should taste identical, such that a consumer knows precisely how a beer will taste when purchased.

In theory, any perceptible change in flavor that renders a beer different from that expected for the beer in question amounts to flavor instability. For the most part discussion is of carbonyl compounds. It was Hashimoto (1966) who first reported on the substantial increase in the level of carbonyl compounds in ageing beer. Thereafter, Palamand and Hardwick (1969) first described the development of E-2-nonenal (cardboard-like aroma), which above all other compounds is the one most frequently referred to in the context of staling. However, many other compounds may change in their amount, taking them either above or below their flavor threshold and thus registering as a change in perceived flavor. As many as 600–700 substances contributing to the flavor of beer can be detected by the human taste and olfactory system, some at extremely low concentrations. Table 3.1 lists some of the compounds that have been associated with flavor deterioration in beer. There is a tremendous diversity here, albeit a preponderance of carbonyl compounds, so small wonder that agents which bind carbonyls can strip the aged character from beer (Hashimoto, 1972b; Bamforth, 2000).

Table 3.1
Compounds formed during beer storage

Class	Compounds
Aldehydes	Acetaldehyde E-2-Nonenal E-2-Octenal E,E-2,4-Decadienal E,E-2,6-Nonadienal 2-Methylbutanal 3-Methylbutanal Benzaldehyde 2-Phenylacetaldehyde 3-(Methylthio) propionaldehyde
Ketones	E-β-Damascenone diacetyl 3-Methyl-2-butanone 4-Methyl-2-butanone 4-Methyl-2-pentanone 2,3-Pentanedione

(Continued)

Table 3.1
(Continued)

Class	Compounds
Cyclic acetals	2,4,5-Trimethyl-1,3-dioxolane 2-Isopropyl-4,5-dimethyl-1,3-dioxolane 2-Isobutyryl-4,5-dimethyl-1,3-dioxolane 2-Sec butyl-4,5-dimethyl-1,3-dioxolane
Heterocylic compounds	Furfural 5-Hydroxymethylfurfural 5-Methylfurfural 2-Acetylfuran 2-Acetyl-5-methylfuran 2-Propionylfuran Furan Furfuryl alcohol Furfuryl ethyl ether 2-Ethoxymethyl-5-furfural 2-Ethyoxy-2,5-dihydrofuran Maltol Dihydro-5,5-diemethyl-2(3H)-furanone 5,5-Dimethyl-2(5H)-furanone 2-Acetylpyrazine 2-Methoxypyrazine 2,6-Dimethylpyrazine Trimethylpyrazine Tetramethylpyrazine
Ethyl esters	Ethyl-3-methylbutyrate Ethyl-2-methylbutyrate Ethyl-2-methylpropionate Ethylnicotinate Diethyl succinate Ethyl lactate Ethyl phenylacetate Ethyl formate Ethyl cinnamate
Lactones	γ-Nonalactone γ-Hexalactone
S-compounds	Dimethyl trisulphide 3-Methyl-3-mercaptobutylformate

From Vanderhaegen et al. (2006).

There is substantial literature on the flavor instability of beer. Alas, much of it is of dubious value and can display serious deficiencies, particularly for its lack of robustness in sensory techniques (Meilgaard, 2001) and an over-reliance on chemical and instrumental data. Ultimately, the only test of what is and what is not relevant for the enhancement of beer shelf life is whether flavor stability can be unequivocally demonstrated to occur when perceived organoleptically.

The challenges of studying such a complex phenomenon and the attendant seeming inconsistencies in the results obtained, have led to considerable disagreement in the literature on the relative importance of various chemical changes and process stages to flavor instability. The apparent failure of brewers to achieve the degree of flavor robustness they want in the finished product, despite the major advances made in control of parameters such as oxygen levels in the final package, has led them to search further back in the process for the reasons and, they hope, the solutions. Thus, there has been great focus on the brew house in recent years and even some suggestions that flavor instability is built into the system as early as the malt house. The apparent improvements that have been made by focusing on malting and wort production appear to be relatively minor.

Another major shortcoming with research on flavor stability is the aforementioned over-emphasis on E-2-nonenal. As long ago as 1981, van Eerde and Strating showed that whilst this compound increases within days in beers aged at 40°C, there was no equivalent increase in 4 months of storage at 20°C. Narziss et al. (1980, 1999), Foster et al. (2001), Schieberle and Komarek (2003) and Vesely et al. (2003) reported similar findings.

Factors impacting the shelf life of beer

The extent to which a foodstuff such as beer will age in the marketplace can be described by the formula given by Singh and Cadwallader (2003).

$$rQ = \varphi(Ci, Ej) \tag{3.1}$$

where
rQ = rate of quality deterioration
Ci = compositional factors (e.g. content of reactive species, catalysts, inhibitors, pH, etc.)
Ej = environmental factors (e.g. temperature, light, mechanical stress)
φ = proportionality constant

Many factors contribute to the ageing of beer and can be basically divided into *intrinsic* factors (i.e. compositional ones) and *extrinsic* factors (i.e. events and conditions outwith the beer but to which the beer is exposed). Although the formula makes no attempt to weigh the various parameters, it is clear that a change in any one of them will impact flavor stability.

Of all the intrinsic factors, the one most extensively studied is oxygen.

Oxygen

Oxygen accounts for 21% of the gases in dry air and is thus plentifully available to react with wort and beer if air is allowed access. The concentration that will dissolve in wort and beer is dependent upon:

- The partial pressure of oxygen above the liquid: higher pressures (and proportion of oxygen in the gas phase) give a higher oxygen concentration in solution

Table 3.2
Solubility of oxygen in worts of increasing strength

Gravity (Plato)	Solubility of oxygen (ppm) at 1 atmosphere pressure and 10°C	Solubility of oxygen (ppm) at 1 atmosphere pressure and 20°C
Deionized water	10.9	8.9
9	8.7	7.1
12	8.3	6.8
15	8.0	6.6
18	7.8	6.4

- The temperature: higher temperatures afford less oxygen in solution
- The quantity of other materials dissolved in the liquid, which in turn relates to the strength of the wort and beer: more competing solutes lead to less oxygen in solution.

The oxygen concentration in de-ionized water is 0.34 mM (10.9 ppm) at 10°C and 0.28 mM (8.9 ppm) at 20°C. Because of the high concentration of dissolved materials in wort, solubility of oxygen is less at successively higher strengths (Table 3.2). For worts at lower atmospheric pressures (higher altitudes) the concentration of dissolved oxygen will be proportionately less. At a given atmospheric pressure, the solubility of oxygen in beer will also be less than in pure water, but rather greater than in wort.

Oxygen is much more soluble (seven to eightfold) in organic solvents than in water. This has seldom been taken into consideration when considering oxidation in brewing systems: the oxygen concentration in a localized lipid environment (e.g. mash and trub particles) may be somewhat greater than in the bulk aqueous phase.

Reactive Oxygen Species

It was Bamforth and Parsons (1985) who first drew attention to the role of active oxygen species rather than ground-state oxygen in potentiating flavor damage in beer.

The principle basis for the toxicity of oxygen is via its conversion to reactive "free radical" forms, or reactive oxygen species (ROS) as they are often now collectively termed, because not all damaging species produced from oxygen are radicals.

A free radical is a species with an independent existence containing one or more unpaired electrons. Free radicals can be formed either by loss or acquisition of electrons from non-radicals.

An input of energy to the oxygen molecule can "flip" the spin of one of the outer orbital electrons in oxygen, generating the highly reactive *singlet oxygen*.

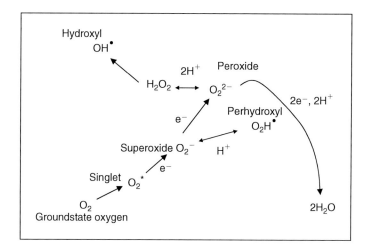

Figure 3.1
The activation of oxygen through the addition of electrons.

Addition of one extra electron to ground-state oxygen forms the *superoxide* radical, O_2^-. As superoxide needs to gain only one electron to complete its outermost orbital, it is more reactive than ground state oxygen. One further electron leads to *peroxide*, O_2^{2-}, which is not a radical, but is more reactive than ground state oxygen. Essentially the two atoms in the ground state oxygen are held together by two bonds, in superoxide by one-and-a-half bonds and in peroxide by only one bond. This makes the bond in hydrogen peroxide relatively fissile and the input of energy (e.g. as light) effects a splitting.

$$H_2O_2 \rightarrow 2\ OH^\bullet \quad (3.2)$$

The *hydroxyl* radical produced is immensely reactive.
A summary of the ROS is given in Figure 3.1.

Transition metal ions

With the exception of zinc, the metals in row one of the d-block of the Periodic Table possess unpaired electrons (i.e. they too are radicals). Iron (II), then can donate an electron to oxygen, with the attendant formation of iron (III) and superoxide.

$$Fe\ (II) + O_2 \rightarrow Fe\ (III) + O_2^- \quad (3.3)$$

Copper can both receive electrons from and donate electrons to superoxide.

$$Cu\ (II) + O_2^- \rightarrow Cu\ (I) + O_2 \quad (3.4)$$

$$Cu\ (I) + O_2^- \rightarrow Cu\ (II) + O_2^{2-} \quad (3.5)$$

$$Net:\ 2\ O_2^- \rightarrow O_2^{2-} + O_2 \quad (3.6)$$

In other words the copper is acting as a catalyst – it ultimately remains unchanged and is available to continue this dismutation reaction, provided it has access to superoxide.

Of the other metals in this category that are likely to be found in brewing systems, manganese can enter into this type of radical reaction, but zinc cannot.

Fenton reaction

It's over 100 years since Fenton first demonstrated the reaction

$$Fe\,(II) + H_2O_2 \rightarrow complex \rightarrow Fe\,(III) + OH^\bullet + OH^- \tag{3.7}$$

Furthermore, these reactions occur

$$OH^\bullet + H_2O_2 \rightarrow H_2O + H^+ + O_2^- \tag{3.8}$$

$$Fe\,(III) + O_2^- \rightarrow Fe\,(II) + O_2 \tag{3.9}$$

Analogous reactions occur with copper.

And so it can be seen from reactions of this type (Equations 3.2–3.8) that metal ions such as iron and copper are effective in stimulating the formation and multiple inter-conversions of radicals from oxygen.

Sources of hydrogen peroxide

Whilst hydrogen peroxide can be produced by the metal ion-catalyzed activation of oxygen, it might also be noted that it can be produced by the oxidation during mashing of sulfhydryl groups in malt proteins (Muller, 1997). The proteins cross-link through the disulfide bridges formed (thereby forming complexes that retard wort separation), and simultaneously hydrogen peroxide is produced.

Sulfur compounds

Sulfur is of further relevance in the context of radical damage. Just as oxygen can be activated into more potent forms, so too can the next element in its group, sulfur. For instance

$$Cu\,(II) + RSH \rightarrow RS^\bullet + Cu\,(I) + H^+ \tag{3.10}$$

Indeed, the co-presence of –SH and oxygen can lead to

$$RS^\bullet + O_2 \rightarrow RSO_2^\bullet \tag{3.11}$$

Superoxide and hydroxyl also react with RSH to form $RS^\bullet$.

These various sulfur radicals are extremely reactive, too, and can promulgate the formation of oxygen radical species if they encounter more oxygen.

In a complex "soup" such as wort or beer, then, there are a myriad of opportunities for radicals to form. There will be many more than those identified here.

Radicals produce radicals

The reaction shown in Equation 3.11 is just one example of a radical producing another through reaction with a non-radical species. Another example is the formation of a 1-hydroxyethyl radical from ethanol, through the reaction of the hydroxyl radical with ethanol.

$$C_2H_5OH + OH^\bullet \rightarrow CH_3CH^\bullet OH + H_2O \qquad (3.12)$$

The hydroxyethyl radical is probably one of the most abundant in beer, owing to the high concentration of ethanol present (Andersen et al., 2000). These authors imply that this is the radical primarily detected in electron spin resonance spectroscopy (esr) measurements on beer and suggest that this radical reacts with ground state oxygen to generate perhydroxyl, which is the species that is primarily causing the damage in the beer. The hydroxyethyl radical will also degrade to produce acetaldehyde (Andersen and Skibsted, 1998).

Thus these radical products can react with other species (and with one another). In other words, once a radical has been produced it can dissipate its energy in the formation of other radicals, which in turn pass on their energy through reactions with other species; viz, a chain reaction is set in motion, sometimes known as *propagation* (Figure 3.2). It is only when two radicals interact to form a non-radical species that the chain is terminated.

The extent to which these various reactions occur is a function of

- the rate of formation of the very first radicals, which in turn depends upon the concentration of the species destined to become a radical (e.g. oxygen) and the concentration of activating species (noting that these activating species may not always be in forms where they are "active" – e.g. a metal ion such as copper may not be capable of converting peroxide to hydroxyl if it is sequestered

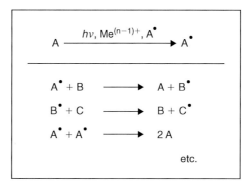

Figure 3.2
Radical propagation.

by a chelating agent such as an amino acid or if it is hidden away within an insoluble particle. The state and availability of metal ions is referred to as their *speciation*.)
- the concentration of other species capable of reacting with the first radicals, in turn, to become radicals themselves
- prevailing local conditions (e.g. pH, temperature)
- the relative rate constants for the various reactions under these conditions
- the presence of any catalytic species (including enzymes).

Clearly there is a vast complexity of reactions and interactions that can occur, particularly in a complex aqueous milieu such as wort and beer. All we can hope to do is to assess the probability of individual reactions occurring.

Let us take a very simple scenario: the opportunity for oxygen to be "activated" in beer.

Let us assume that we have a concentration of oxygen in beer of 0.3 ppm, which might be considered a typical achievement for many brewers for beer in final pack (although some of course aspire to lower oxygen concentrations than this). This is equivalent to a concentration of oxygen of 9.4 µM. Furthermore, let us suppose that free (unchelated, non-bound) iron is present in the beer at just 0.01 ppm (0.18 µM).

The rate of a chemical reaction is a function of the concentration of the reactants multiplied by a rate constant: faster reactions have higher rate constants.

The rate constant for the reaction of oxygen with iron (II) is $1.3 \times 10^6 \, M^{-1} s^{-1}$. And so, with the prevailing concentrations of iron and oxygen, the rate of activation of oxygen to superoxide in reaction 3.3 (see earlier) would be

$$\text{rate} = (1.3 \times 10^6) \times (9.4 \times 10^{-6}) \times (0.18 \times 10^{-6})$$
$$= 2.2 \times 10^{-6} \, M^{-1} s^{-1}$$

The superoxide radicals formed may have various fates. The likelihood of these various fates depends on which molecules they encounter, which in turn depends on the concentrations of these molecules in solution. Thus in a beer, the greatest likelihood is that the first molecule encountered will be one of water, the next most likely being one of ethanol, and so on. Because more than a single superoxide radical will be produced in a given locale (i.e. there will be a high localized concentration), there is also a high probability that these radicals will react together. In fact this happens far more rapidly if one or both of adjacent superoxides are protonated to form the perhydroxyl radical.

$$O_2^- + H^+ \rightarrow O_2H^\bullet \quad (3.13)$$

The pK_a of this equilibrium is 4.8. Thus at beer pH the majority of the superoxide will be in the perhydroxyl form, enabling these reactions to occur.

$$HO_2^\bullet + O_2^- + H^+ \rightarrow H_2O_2 + O_2 \quad k = 8 \times 10^7 \, M^{-1} s^{-1} \quad (3.14)$$

$$HO_2^\bullet + HO_2^\bullet \rightarrow H_2O_2 + O_2 \quad k = 8 \times 10^5 \, M^{-1} s^{-1} \quad (3.15)$$

The hydrogen peroxide formed is even more reactive than is superoxide (perhydroxyl), and in turn it could react with the iron in the beer, according to the Fenton reaction (see Equation 3.7) for which the rate constant is $76\,M^{-1}\,s^{-1}$.

Consequently, we arrive at the hydroxyl radical, one of the most reactive species known, and one capable of reacting with a myriad of species, including ethanol, to produce the hydroxyethyl radical (Equation 3.12). The rate constant for the reaction of hydroxyl with ethanol is $7.2 \times 10^8\,M^{-1}\,s^{-1}$ (at pH 7).

Hydroxyl and perhydroxyl (and also singlet oxygen) are capable of reacting with an unsaturated fatty acid, such as linoleic acid, and thereby setting in motion the chain reaction suggested by many to lead to stale flavor development (Figure 3.3). The authors do not have to hand the rate constant for the reaction of hydroxyl with linoleic acid, but it is likely to be of the same order of magnitude as for the reaction with lecithin, viz. $10^8\,M^{-1}\,s^{-1}$. The rate constant for the reaction of perhydroxyl with linoleic acid is five orders of magnitude lower ($1.18 \times 10^3\,M^{-1}\,s^{-1}$). Other radicals capable of triggering the peroxidation of unsaturated fatty acids include $RS^\bullet$ (see above).

The reader by now will have realized the complexity of the situation – and should not lose sight of the fact that this is only the tip of the iceberg! The flux of species through the myriad of reactions will depend on the rate constants and concentration of the various species. For instance, the rate constant for the dismutation reaction of superoxide/perhydroxyl is almost five orders of magnitude greater than that for the reaction of perhydroxyl with linoleic acid. It is only if the linoleic acid concentration is substantially higher than the local concentration of perhydroxyl that the latter reaction would be favored. Many suppositions must enter into a discussion of a highly simplified scenario such as this, but it is evident that the rate-limiting step (the one with the lowest rate constant) is for the conversion of peroxide to hydroxyl (Fenton reaction).

Oxidation of unsaturated fatty acids is the only route by which staling materials (principally carbonyls) arise in beer via radical reactions (see later). It is also important to stress that unsaturated fatty acids and iso-α-acids, etc., do not need to be degraded to any great extent in these radical reactions in order to cause a flavor change in beer: the carbonyl staling products have very low

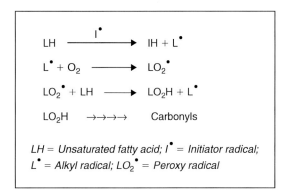

Figure 3.3
Autocatalytic oxidation of unsaturated fatty acids.

flavor thresholds. Such carbonyls can have flavor thresholds of 0.1 ppb or less. Although the concentrations of the reactants in the various reactions we have discussed above are relatively low and, despite the proportionately low rate constants of reactions such as the Fenton reaction, only a very limited amount of oxidation needs to take place to generate perceptible stale character (see Bamforth (1999) for a calculation).

The impact of temperature

Apart from oxygen, the most dramatic impact on flavor stability of any parameter, on the route from barley to a beer in the customer's hand, is the temperature of storage of the beer. According to Arrhenius' Law

$$k_{t+10} = 2 \sim 3k_t \qquad (3.16)$$

where k_t is the rate of a chemical reaction at a given temperature and k_{t+10} is the rate of that reaction when the temperature is raised by 10°C. If the temperature of the beer is raised by 10°C, the rate of reactions, including those responsible for staling, is increased two to threefold.

This huge impact of temperature is illustrated in Figure 3.4, in which a factor of 3 in Equation 3.16 is used. There is a strong likelihood that this is valid, because accelerated ageing regimes are generally founded on a supposition that 1 day at 60°C equates to 4 weeks at 30°C and the shape of the plot bears this out. Perusal of Figure 3.4 reveals that if beer can be held at, say, 10°C as

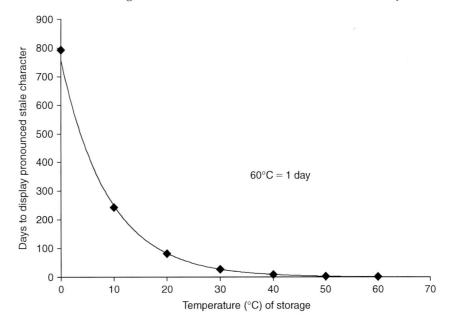

Figure 3.4
The impact of temperature on flavor deterioration in beer (from Bamforth 2004b).

opposed to room temperature then this will "buy" about 5 months of shelf life. Equally, increasing the temperature of storage from 20°C to 35°C (e.g. beer being held in winter as opposed to summer in a garage in Davis, California) has a vast impact on the development of stale character. It is only too apparent how critical the logistics of distribution are and how any attempt to bring the temperature down post-packaging is highly advisable.

The chemistry of flavor change in beer

Many flavor active compounds present in uninfected beer are capable of changing their levels during storage in the final package.

Compounds may

(a) decrease in level leading to flavor deterioration by loss of a desirable character
(b) increase in level leading to a flavor deterioration by an increase in an undesirable character.

In turn, category b compounds may arise

(i) because they are produced *de novo* in a chemical reaction
(ii) by the release of pre-formed material that is bound up in the beer with a "holding agent" that prevents their flavor from being expressed
(iii) because conditions have changed in a beer which makes the likelihood of either type of change [(i) or (ii)] more likely, for example a change in redox conditions.

The relevant chemistry underpinning the changes described in (a), (b), (i), (ii) and (iii) is as follows, remembering that, while all of these reactions are feasible and at some time or another have been proposed as contributors to flavor change, they may have varying degrees of actual relevance.

In passing we might note that there has been an over-emphasis on the compound E-2-nonenal in the literature. To imply that a solitary compound is primarily responsible for ageing is naive.

Enzymic oxidation of unsaturated fatty acids

Lipoxygenase (LOX) catalyses the oxidation of polyunsaturated fatty acids, notably linoleic acid, to hydroperoxides (Doderer et al., 1992). In turn these are substrates for hydroperoxide isomerase (Schwarz and Pyler, 1984; Zimmerman and Vick, 1970) and hydroperoxide lyase (Kuroda et al., 2003). An ensuing sequence of non-enzymic reactions leads to the production of unsaturated carbonyl compounds, including E-2-nonenal (formerly known as *trans*-2-nonenal). It is argued that hydroperoxides produced upstream in malting and brewing survive into the finished beer and progressively decay to release stale character. LOX is produced in the barley embryo during germination (Boivin et al., 1996), during which hydroperoxides are also produced (Bamforth et al., 1993). However, the latter are not measurable after kilning and their fate is unknown. LOX is a very heat-sensitive enzyme and is substantially destroyed during kilning (Bamforth et al., 1991), especially during more stringent ale regimes. It will survive mashing at lower temperatures, but is rapidly destroyed at 65°C (Boivin et al., 1996). It has been argued that if this enzyme has any relevance whatsoever in mashing, then it can only be at the point of initial striking of malt with brewing water, at which point alone there seems to be sufficient substrate and enzyme for the enzyme to act (Biawa and Bamforth, 2002). Taking mashing pH from 5.5 to 5.0 halves lipoxygenase activity (Doderer et al., 1991). Barleys lacking lipoxygenase have been developed and are claimed to result in beers more resistant to staling (Hirota et al., 2005).

Non-enzymic oxidation of unsaturated fatty acids

Linoleic acid is susceptible to oxidation even in the absence of enzymes. The reaction is autocatalytic and needs only a small amount of initial "trigger" to start the cascade of radical reactions. The first agents needed to start the chain reaction may typically be oxygen radicals such as hydroxyl and perhydroxyl (the protonated form of superoxide that is the prevalent at beer pH (Kaneda et al., 1997)). Linoleic acid, oxygen and activating metal ions such as iron may be at relatively minuscule levels in beer, yet still sufficient to allow the staling sequence to occur (Bamforth, 1999; see earlier discussion). Other workers (e.g. Lermusieau et al., 1999; Noel et al., 1999a,b) have proposed that nonenal production does not occur in the finished beer, but rather that 30% occurs during mashing and 70% during wort boiling.

Oxidation of iso-α-acids

Unhopped beers seldom develop an oxidized flavor, which suggests a likely role for the iso-α-acids as precursors of staling compounds (Hashimoto et al., 1979). In model systems it has been shown that volatile carbonyls (alkenals and alkadienals with chain lengths of between 6 and 12 carbon atoms) can be produced from a solution of bitter substances, higher alcohols and melanoidins. The *trans* isomers are more prone to degradation than are the *cis* isomers (De Cooman et al., 2000; Araki et al., 2002). Reduced side-chain iso-α-acids do not give staling carbonyls (Hashimoto, 1988).

Oxidation of higher alcohols

Alcohols in beer can be converted to their equivalent aldehydes through the mediation of melanoidins, with the oxidized carbonyl groups on the latter acting as electron acceptors (Hashimoto, 1972a). Devreux et al. (1981) suggest that the reaction is inhibited by polyphenols and requires light, so is of secondary significance in beer. Meanwhile, Irwin et al. (1991) argue that the efficiency of conversion is so small as to make the pathway irrelevant.

Strecker degradation of amino acids

Amino acids can react with α-dicarbonyl compounds, such as the intermediates in browning reactions. The amino acid is converted into an aldehyde with one fewer carbon atom (Blockmans et al., 1975; Hashimoto and Kuroiwa, 1975). Polyphenols may have a catalytic role (Blockmans et al., 1979).

Aldol condensations

In the aldol condensation, separate aldehydes or ketones react to form larger carbonyl species. This is a plausible route through which E-2-nonenal may be produced, by a reaction between acetaldehyde and heptanal (Hashimoto and Kuroiwa, 1975). Proline may act as a catalyst.

Acetal formation

Cyclic acetals can be formed by the condensation of 2,3-butanediol with carbonyls such as acetaldehyde (Peppard and Halsey, 1982).

Binding of carbonyls by sulfur dioxide

Sulfite is capable of forming addition complexes with carbonyl containing compounds, the resultant "adducts" display no perceptible flavor at the concentrations likely to be found in beer (Barker et al., 1983). It has been suggested that carbonyls produced upstream bind to the sulfite produced by yeast, thereby carrying through into the finished beer, to be progressively released as SO_2 is consumed in other (as yet unknown) reactions (Ilett and Simpson, 1995). It has been suggested that the greater significance of sulfite for protecting against staling is through its role as an antioxidant (Kaneda et al., 1994). In this regard, Dufour et al. (1999) indicate that SO_2-carbonyl binding actually occurs through the C=C of the unsaturated aldehyde, rather than at the carbonyl group and, as such, is non-reversible.

Binding of carbonyls by amino groups in proteinaceous species

A similar scenario is understood to occur with carbonyl compounds entering into reversible Schiff base formation with amino groups, including proteinaceous species in the grist (Lermusieau et al., 1999).

Reduction of carbonyl compounds by yeast

Yeast is capable of reducing carbonyl compounds (Peppard and Halsey, 1981). These of course include the well appreciated reactions leading from acetaldehyde to ethanol and diacetyl to acetoin and butanediol. But many other aldehydes and ketones produced upstream will be reduced, leading to the belief by some that upstream production of such carbonyls is unimportant. Various enzymes may be involved in this reduction (Collin et al., 1991; Debourg et al., 1993, 1994; Laurent et al., 1995).

Release of flavor active compounds by enzymes from yeast

Yeast is known to release a range of enzymes with potential impact on product quality. Included amongst these are the glycosidases, the substrates for which include complexes of carbohydrate with several significant hop aroma components (Biendl et al., 2003). If these enzymes remain in beer (e.g. if beer is not pasteurized) then conceptually there may be a progressive change in hop character over time. Chevance et al. (2002) showed β-glucosidase enhanced the release of (E)-β-damascenone in beer.

Oxygen radical scavenging by polyphenols and melanoidins

Oxygen radicals will react with a diversity of species and trigger a cascade of ensuing radical events in the process stream and in beer. Radical scavengers, which halt this cascade by trapping radicals without forming fresh radicals, may include polyphenols (Owades and Jakovac, 1966) and melanoidins (Hayase et al., 1986). Polyphenols may not only scavenge superoxide (Yuting et al., 1990) and hydroxyl (Husain et al., 1987) and the peroxy radicals formed in the autocatalytic oxidation of unsaturated fatty acids (Torel et al., 1986), but may also protect against staling by chelating metal ions and inhibiting LOX (Boivin et al., 1975). Again we find conflicting evidence in the literature. Andersen et al. (2000) could find no impact of polyphenols on free radical scavenging in wort and beer, whereas Kaneda et al. (1995) and McMurrough et al. (1996) suggested that up to 60% of the reducing power in beer comes from this source. Regarding Maillard reaction products, Andersen et al. (2000) found them to be pro-oxidants, whereas Bright et al. (1999) and Coghe et al. (2003) described the benefit to flavor stability of these materials, especially from more roasted malts.

Hydrogen peroxide removal by peroxidases

A key player in the process of oxidation in the brew house and finished beer is hydrogen peroxide. It is detectable in beer and implicated in radical formation. It is also produced during mashing, and during the cross-linking of thiol groups in gel proteins (Muller, 1997). In turn, the peroxide is removed by a reaction with polyphenols in reactions catalyzed by the peroxidases (Clarkson et al., 1982). Thus the risk of radical generation in mashing depends substantially on the relative ability of peroxidases to out-compete non-enzymic systems that trigger the formation of hydroxyl from peroxide in the Haber-Weiss reaction.

Vicinal diketone release in beer from incompletely eliminated precursors

If acetolactate and acetohydroxybutyrate are not completely converted during fermentation and maturation to diacetyl and pentanedione, thereby allowing yeast to consume them, then they will survive into beer and progressively degrade in the final package (Inoue and Yamamoto, 1971).

Sulfur compounds

Reducing agents, such as the amino acid cysteine, will progressively release dimethyl sulfide from dimethyl sulfoxide in final package (Bamforth, 1985). Residual yeast (in naturally conditioned products) will also do this and will reduce sulfur dioxide to hydrogen sulfide (Walker and Simpson, 1994). Compounds responsible for the ribes character, 3-methyl-3-mercaptobutyl formate (Schieberle, 1991) and 4-mercapto-4-methyl-penta-2-one (Tressl et al., 1980) are also produced on storage. Peppard (1978) and Gijs et al. (2002) have reported the development of DMTS in beer from various precursors.

Changes in ester levels

Stenroos (1973) and Neven et al. (1997) reported a decrease in the level of iso-amyl acetate during the storage of beer. However, a range of other esters (see Table 3.1) increase in quantity during storage (Bohnan, 1985b; Gijs et al., 2002; Lustig et al., 1993; Miedaner et al., 1991; Williams and Wagner, 1978).

An evaluation of processes from barley to beer in the context of flavor instability

Table 3.3 lists the potential impact of all stages from field to package on flavor instability.

Some critical comments

One of the biggest challenges in any discussion of flavor instability is the acute shortage of good sensory data to support many of the claims that have been made, particularly for oxygen control upstream. This is illustrated in Figure 3.5. Most commonly, the intensity of stale flavor is reported on a scale of the

Table 3.3
Process impacts on flavor instability (derived from Bamforth, 2004a)

Raw material or process stage	Parameter	Hypothesis
Barley selection	Potential for staling precursors	Barleys differ in their propensity to develop lipoxygenase (LOX)
	2-Row vs. 6 row	6-Row has higher enzyme potential, more polyphenol potential
	Winter vs. spring	Winter has more polyphenol potential
	Low proantho-cyanidin varieties	Less antioxidant potential through lower polyphenol levels?
	Lipoxygenase-null varieties	Absence of lipoxygenase so no contribution of enzyme-catalyzed lipid oxidation
Steeping and germination	Development of redox enzymes and the extent to which they can act	Factors promoting modification also encourage LOX in terms of levels and extent to which it produces hydroperoxides
		Peroxidases (POD) also increase in level
		Developing sugar and amino acids as melanoidin precursors
Kilning	Destruction of LOX	Higher kilning temperatures destroy LOX
	Destruction of hydroperoxides	Hydroperoxides are also lost on kilning – but where do they go?
	Production of melanoidins	Melanoidins and intermediates leading to them may have antioxidant properties. However they may also promote the oxidation of higher alcohols
Malt storage	Loss of LOX	LOX levels diminish on storage, thereby leaving less oxidation potential in the grist
Milling	Hammer vs. roller	Finer milling leads to greater extraction of all components – including LOX and lipid
	Wet vs. dry	Greater opportunity for leaching of undesirables in wet system – also commencement of LOX reaction
	Embryo preservation	Milling procedures that avoid embryo damage will leave lipid and LOX in an unextractable form and they will go into the spent grains
Mashing-in	Oxygen	It is the point at which milled grist is mixed with water that the LOX risk is greatest because it is here that substrate concentrations are high enough for LOX and it is not yet destroyed

(Continued)

Table 3.3
(Continued)

Raw material or process stage	Parameter	Hypothesis
Mashing	Temperature regime	Low mashing temperatures allow more LOX survival. Heat damage during decoction mashing
	Number of vessels	Increased risk of oxygen pick up with more vessels – for example in decoction mashing
	Oxygen	Oxygen as substrate for LOX, but also reacting non-enzymatically with –SH groups in proteins to make peroxides, which act as substrates for POD. Also oxygen radical formation
	pH	Lower mashing pH reduces rate of LOX. However superoxide radical more damaging in protonated forms at lower pH
	Vessel fabric	Leaching of Cu from copper vessels. Cu promotes oxygen radical formation
Solid adjuncts	Antioxidant potential	Roasted adjuncts contain melanoidins that may act as radical scavengers
	Lipid level	Some cereals, for example rice, have high lipid potential unless polished
Wort separation	System	Modern mash filter affords less turbid worts and reduced lipid level. Brighter worts allow increased levels of SO_2 production by yeast
	Collection of weaker worts	Increased extraction of lipid, but also more tannoids (potential anti-oxidants)
Kettle	Duration	Risk of thermal damage in prolonged boiling (and in whirlpool stand)
	Kettle design	Rolling boils to ensure volatilization of undesirable flavors
	Energy saving approaches	Thermal damage, for example in high temperature/high pressure wort boiling
Liquid adjuncts	Selection and level of use	Sugars and syrups devoid of staling precursors so their use "thins out" stale potential
Hopping	Choice of material	Most polyphenols in whole hops, least in extracts
		Reduced iso-α-acids do not degrade to unsaturated carbonyls
		Cis isomers less prone to degradation – selection of hop products with high *cis* ratio
Hot wort clarification	Removal of trub	Trub impacts yeast performance (vigor – ability of yeast to eliminate carbonyls and to produce sulfur dioxide). Also risk of thermal damage if whirlpool stand is prolonged

Table 3.3
(Continued)

Raw material or process stage	Parameter	Hypothesis
Cooling and oxygenation	Introduction of oxygen	Oxygen "melding" into wort; O_2 impacts cold break formation. Impact on vigor of yeast performance (see hot wort clarification)
Yeast selection	Yeast characteristics (strain, pitching rate, health)	Strains differ in ability to produce SO_2
		Healthy yeast and good pitching rates for efficient removal of carbonyls, including VDKs and acetaldehyde
Cold wort handling pre-fermentation	Removal of cold break	Brighter worts lead to more SO_2 production by yeast
Primary fermentation	Removal of undesirable flavor components	See yeast selection. If fermentation is out of spec then there will be increased survival of VDK and acetaldehyde and insufficient production of SO_2
Secondary fermentation/ warm conditioning	Flavor refinement	Final mopping of VDK's (ensure precursors also fully eliminated)
Cold conditioning	Buffer stock	Beer at lowest practical temp is at its most stable
	Oxygen stripping	Opportunity here to eliminate last traces of oxygen prior to packaging
Filtration and stabilization	Clarification with minimum O_2 and metal pick up	Filtration presents risk of oxygen and iron pick-up. Some argue for avoidance of PVPP because it removes polyphenol antioxidants
Packaging	Lowest oxygen pick-up	The biggest threat to high oxygen levels and therefore staling in beer
Final beer composition	pH	Higher the pH on the scale 4–4.5, the less the level of the most damaging oxygen species
	O_2	The lower the better
	SO_2	Antioxidant and binds staling agents
	Precursors (e.g. acetolactate, DMSO)	If they are present in beer they will potentiate change in final product
	"Carbonyl potential"	The less "bound-up" carbonyl the better
Warehousing	Temperature, time	All chemical reactions proceed much more quickly as temperature increases
Transportation	Temperature	As for warehousing, agitation also promotes breakdown
	Agitation	
	Distance logistics	
Package labeling	Born on or best before dates	Make customer aware of the risks

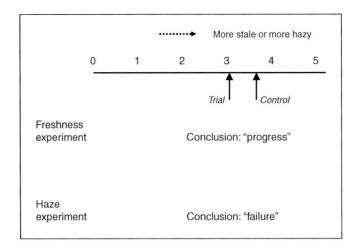

Figure 3.5
Different expectations for flavor stability and colloidal stability trials (from Bamforth, 2004b).

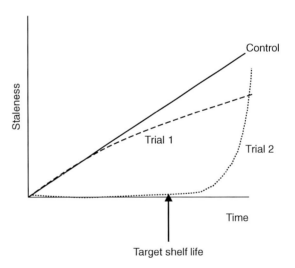

Figure 3.6
Better to gauge flavor stability in terms of time to onset of stale flavor (from Bamforth, 2004b).

type whereby zero indicates no stale character and five signifies intense staling. The researchers may do their trials and then report, say, that the control beer (no precautions) has a staleness value of 3.7 and the trial (with precautions – e.g. inert gas blanketing of brew house vessels) has a value of 3.1. Ergo, the treatment is claimed to be beneficial in moving towards improved flavor stability. Leaving aside the frequent absence of statistical treatment of the data (reliability of tasting protocol, numbers of replications of the trials) and taking the information in good faith and at face value, should we be impressed? Say that the trial was not on flavor stability, but rather on haze stability. Would an improvement in a haze value from 3.7 to 3.1 impress anyone? After all, the beer would still be hazy. In common with haze, we fervently suggest that the criterion should not be intensity of stale character, but rather *time to development of stale character* (Figure 3.6). At the end of the ageing time period, the beer from trial 2 shows worse staling than the beer from trial 1. But the time "bought" in terms of weeks to the onset of staling is vastly improved.

References

Andersen, M. L. and Skibsted, L. H. (1998) Electron spin resonance spin trapping identification of radicals formed during aerobic forced aging of beer. *Journal of Agricultural and Food Chemistry*, 46, 1272–1275.

Andersen, M. L., Outtrup, H. and Skibsted, L. H. (2000) Potential antioxidants in beer assessed by ESR spin trapping. *Journal of Agricultural and Food Chemistry*, 48, 3106–3111.

Araki, S., Takashio, M. and Shinotsuka, K. (2002) A new parameter for determination of the extent of staling in beer. *Journal of the American Society of Brewing Chemists*, 60, 26–30.

Bamforth, C. W. (1985) Cambridge prize lecture: Biochemical approaches to beer quality. *Journal of the Institute of Brewing*, 91, 154–160.

Bamforth, C. W. (1999) Enzymic and non-enzymic oxidation in the brewhouse: A theoretical consideration. *Journal of the Institute of Brewing*, 105, 237–242.

Bamforth, C. W. (2000) Making sense of flavor change in beer. *Master Brewers Association of the Americas Technical Quarterly*, 37, 165–171.

Bamforth, C. W. (2004a) Fresh controversy: Conflicting opinions on beer staling. *Proceedings of the 28th Convention of the Institute and Guild of Brewing, Asia-Pacific Section*, 63–73.

Bamforth, C. W. (2004b) A critical control point analysis for flavor stability of beer. *Master Brewers Association of the Americas Technical Quarterly*, 41, 97–103.

Bamforth, C. W. and Parsons, R. (1985) New procedures to improve the flavor stability of beer. *Journal of the American Society of Brewing Chemists*, 43, 197–202.

Bamforth, C. W., Clarkson, S. P. and Large, P. J. (1991) The relative importance of polyphenol oxidase, lipoxygenase and peroxidases during wort oxidation. *Proceedings of the 23rd European Brewery Convention Congress, Lisbon*, 617–624.

Bamforth, C. W., Muller, R. E. and Walker, M. D. (1993) Oxygen and oxygen radicals in malting and brewing: A review. *Journal of the American Society of Brewing Chemists*, 51, 79–88.

Barker, R. L., Gracey, D. E. F., Irwin, A. J., Pipasts, P. and Leiska, E. (1983) Liberation of staling aldehydes during storage of beer. *Journal of the Institute of Brewing*, 89, 411–415.

Biawa, J.-P. and Bamforth, C. W. (2002) A two-substrate kinetic analysis of lipoxygenase in malt. *Journal of Cereal Science*, 35, 95–98.

Biendl, M., Kollmannsberger, H. and Nitz, S. (2003) Occurrence of glycosidically bound flavor compounds in different hop products. *Proceedings of the 29th European Brewery Convention Congress, Dublin*, pp. 252–259.

Blockmans, C., Masschelein, C. A. and Devreux, A. (1979) Origine de certains composes carbonyls formes au course du viellissement de la biere. *Proceedings of the 17th European Brewery Convention Congress, Berlin*, 279–291.

Boivin, P., Malanda, M., Maillard, M. N., Berset, C., Hugues, M., Forget-Richard, F. and Nicolas, J. (1995) Role des antioxygenes naturels du malt dans la stabilite organoleptique de la biere. *Proceedings of the 25th European Brewery Convention Congress, Brussels*, 159–168.

Boivin, P., Clamagirand, V., Maillard, M. N., Berset, C. and Malanda, M. (1996) Malt quality and oxidation risk in brewing. *Proceedings of the 24th Convention of the Institute of Brewing, Asia Pacific Section*, 110–115.

Bright, D., Stewart, G. G. and Patino, H. (1999) Coors Brewing Company. A novel assay for antioxidant potential of specialty malts. *Journal of the American Society of Brewing Chemists*, 57, 133–137.

Chevance, F., Guyot-DeClerck, C., DuPont, J. and Collin, S. (2002) Investigation of the beta-damascenone level in fresh and aged commercial beers. *Journal of Agricultural and Food Chemistry*, 50, 3818–3821.

Clarkson, S. P., Large, P. J. and Bamforth, C. W. (1992) A two-substrate kinetic study of peroxidase cationic isoenzymes in barley malt. *Phytochemistry*, 31, 743–749.

Coghe, S., Vanderhaegen, B., Pelgrims, B., Basteyns, A-V. and Delvaux, F. R. (2003) Characterization of dark specialty malts: New insights in color evaluation and pro- and antioxidative activity. *Journal of the American Society of Brewing Chemists*, 61, 125–132.

Collin, S., Montesinos, M., Meersman, E., Swinkels, W. and Dufour, J.-P. (1991) Yeast dehydrogenase activities in relation to carbonyl compounds removal from wort and beer. *Proceedings of the 23rd European Brewery Convention Congress, Lisbon*, 409–416.

Dalgliesh, C. E. (1977) Flavour stability. *Proceedings of the 16th European Brewery Convention Congress, Amsterdam*, 623–659.

De Cooman, L., Aerts, G., Overmeire, H. and De Keukeleire, D. (2000) Alterations of the profiles of iso-alpha-acids during beer ageing, marked instability of *trans*-iso-alpha-acids and implications for beer bitterness consistency in relation to tetrahydroiso-alpha-acids. *Journal of the Institute of Brewing*, 106, 169–178.

Debourg, A., Laurent, M., Dupire, S. and Masschelein, C. A. (1993) The specific role and interaction of yeast enzymatic systems in the removal of flavour-potent wort carbonyls during fermentation. *Proceedings of the 24th European Brewery Convention Congress, Oslo*, 437–444.

Debourg, A., Laurent, M., Goossens, E., Borremans, E., Van De Winkel, L. and Masschelein, C. A. (1994) Wort aldehyde reduction potential in free and immobilized yeast systems. *Journal of the American Society of Brewing Chemists*, 52, 100–106.

Devreux, A., Blockmans, C. and Van de Meerrsche, K. (1981) Carbonyl compounds formation during aging of beer. *EBC – Flavor symposium Mongraph VII*, 191–201.

Doderer, A., Kokkelink, I., Van der Veen, S., Valk, B. E. and Douma, A. C. (1991) Purification and characterization of lipoxygenase from germinating barley. *Proceedings of the 23rd European Brewery Convention Congress, Lisbon*, 109–116.

Doderer, A., Kokkelink, I., Van der Veen, S., Valk, B. E., Schram, A. W. and Douma, A. C. (1992) Purification and characterization of two lipoxygenase isoenzymes from germinating barley. *Biochimica et Biophysica Acta*, 1120, 97–104.

Drost, B. W., Eerde, P., Van Hoekstra, S. and Strating, J. (1971) Fatty acids and the staling of beer. *Proceedings of the 13th European Brewery Convention Congress, Estoril*, 9, 451–458.

Dufour, J-P., Leus, M., Baxter, A. J. and Hayman, A. R. (1999) Characterization of the Reaction of Bisulfite with Unsaturated Aldehydes in a Beer Model System Using Nuclear Magnetic Resonance Spectroscopy. *Journal of the American Society of Brewing Chemists*, 57, 138–144.

Foster, R. T., Samp, E. J. and Patino, H. (2001) Multivariate modeling of sensory and chemical data to understand staling in light beer. *Journal of the American Society of Brewing Chemists*, 59, 201–210.

Greenhoff, K. and Wheeler, R. E. (1981) Evaluation of stale flavor and beer carbonyl development during normal and accelerated aging using liquid and high performance liquid chromatography. *Proceedings of the 18th European Brewery Convention Congress, Copenhagen*, 405–412.

Guinard, J. X., Uotani, B. and Schlich, P. (2001) Internal and external mapping of preferences for commercial lager beers: Comparison of hedonic ratings by consumers blind versus with knowledge of brand and price. *Food Quality and Preference*, 12, 243–255.

Hashimoto, N. (1966) Studies on volatile carbonyl compounds in beer. *Report of the Research Laboratory of the Kirin Brewery Company*, 9, 1–10.

Hashimoto, N. (1972a) Oxidation of higher alcohols by melanoidins in beer. *Journal of the Institute of Brewing*, 78, 43–51.

Hashimoto, N. (1972b) Oxidation degradation of isohumulones in relation to beer flavor. Oxidation of higher alcohols by melanoidins in beer. *Report of the Research Laboratory of the Kirin Brewery Company*, 15, 7–15.

Hashimoto, N. (1988) Melanoidin-mediated oxidation: A greater involvement in flavor staling. *Report of the Research Laboratory of the Kirin Brewery Company*, 31, 19–32.

Hashimoto, N. and Kuroiwa, Y. (1975) Proposed pathways for the formation of volatile aldehydes during storage of bottled beer. *Proceedings of the American Society of Brewing Chemists*, 104–111.

Hashimoto, N., Shimazu, T. and Eshima, T. (1979) Oxidative degradation of isohumulones in relation to beer flavor. *Report of the Research Laboratory of the Kirin Brewery Company*, 22, 1–10.

Hirota, N., Kuroda, H., Takoi, K., Kaneko, T., Kaneda, H., Yoshida, I., Takashio, M., Ito, K. and Takeda, K. (2005) Development of novel barley with improved beer foam and flavor stability – the impact of lipoxygenase-1-less barley in brewing industry. *Proceedings of the 30th European Brewery Convention Congress, Prague*, 46–52.

Husain, S. R., Cillard, J. and Cillard, P. (1987) Hydroxyl radical scavenging activity of flavonoids. *Phytochemistry*, 26, 2489–2491.

Ilett, D. R. and Simpson, W. J. (1995) Loss of sulfur dioxide during storage of bottled and canned beers. *Food Research International*, 28, 393–396.

Inoue, T. and Yamamoto, Y. (1971) Decomposition rate of precursors of diacetyl and 2,3-pentanedione during beer fermentation. *Report of the Research Laboratory of the Kirin Brewery Company*, 14, 55–59.

Irwin, A. J., Barker, R. L. and Pipasts, P. (1991) The role of copper, oxygen and polyphenols in beer flavor instability. *Journal of the American Society of Brewing Chemists*, 49, 140–149.

Kaneda, H., Osawa, T., Kawakishi, S., Munekata, M. and Koshino, S. (1994) Contribution of carbonyl-sulfite adducts to beer stability. *Journal of Agricultural and Food Chemistry*, 42, 2428–2432.

Kaneda, H., Kobayashi, N., Furusho, S., Sahara, H. and Koshino, S. (1995) Reducing activity and flavor stability of beer. *Master Brewers Association of the Americas Technical Quarterly*, 32, 90–94.

Kaneda, H., Takashio, M., Tomaki, T. and Osawa, T. (1997) Influence of pH on flavour staling during beer storage. *Journal of the Institute of Brewing*, 103, 21–23.

Kuroda, H., Furusho, S., Maeba, H. and Takashio, M. (2003) Characterization of factors involved in the production of E-(2)-nonenal during mashing. *Bioscience Biotechnology and Biochemistry*, 67, 691–697.

Laurent, M., Geldorf, B., Van Nedervelde, L., Dupire, S. and Debourg, A. (1995) Characterization of the aldoketoreductase yeast enzymatic system involved in the removal of wort carbonyls during fermentation. *Proceedings of the 25th European Brewery Convention Congress, Brussels*, 337–344.

Lermusieau, G., Noël, S., Liégeois, C. and Collin, S. (1999) Nonoxidative mechanism for development of *trans*-2-nonenal in beer. *Journal of the American Society of Brewing Chemists*, 57, 29–33.

McMurrough, I., Madigan, D., Kelly, R. J. and Smyth, M. R. (1996) The role of flavanoid polyphenols in beer stability. *Journal of the American Society of Brewing Chemists*, 54, 141–148.

Meilgaard, M. (1972) Stale flavor carbonyls in brewing. *Brewers Digest*, 47, 48–57.

Meilgaard, M. C. (1975) Flavor chemistry of beer. Part 1: Flavor interaction between principal volatiles. *Master Brewers Association of the Americas Technical Quarterly*, 12, 107–117.

Meilgaard, M. (2001) Effects on flavor of innovations in brewery equipment and processing: A review. *Journal of the Institute of Brewing*, 107, 271–286.

Muller, R. E. (1997) The formation of hydrogen peroxide during oxidation of thiol-containing proteins. *Journal of the Institute of Brewing*, 103, 307–310.

Narziss, L., Miedaner, H. and Graf, H. (1985) Carbonyls and aging of beer. 2. Influence of some technical parameters. *Monatsschrift fur Brauwissenschaft*, 38, 472–477.

Narziss, L., Miedaner, H. and Lustig, S. (1999) The behavior of volatile aromatic substances as beer ages. *Monatsschrift fur Brauwissenschaft*, 52, 164–175.

Noel, S., Liegeois, C., Lermusieau, G., Bodart, E., Badot, C. and Collin, S. (1999a) Release of deuterated nonenal during beer aging from labeled precursors synthesized in the boiling kettle. *Journal of Agricultural and Food Chemistry*, 47, 4323–4326.

Noel, S., Metais, N., Bonte, S., Bodart, E., Peladan, F., Dupire, S. and Collin, S. (1999b) The use of oxygen 18 in appraising the impact of oxidation process during beer storage. *Journal of the Institute of Brewing*, 105, 269–274.

Owades, J. L. and Jakovac, J. (1966) Study of beer oxidation with O^{18}. *Proceedings of the American Society of Brewing Chemists*, 180–183.

Palamand, S. R. and Hardwick, W. A. (1969) Studies on the relative flavor importance of some beer components. *Master Brewers Association of the Americas Technical Quarterly*, 6, 117–128.

Peppard, T. L. and Halsey, S. A. (1981) Malt flavour – transformation of carbonyl compounds by yeast during fermentation. *Journal of the Institute of Brewing*, 87, 386–390.

Schieberle, P. and Komarek, D. (2002) Changes in key aroma compounds during natural beer aging. In: *Freshness and Shelf life of Foods* (K. R. Cadwallader and H. Weenen eds), pp. 70–79. ACS: Washington, DC.

Schwarz, P. B. and Pyler, R. E. (1984) Lipoxygenase and hydroperoxide isomerase activity of malting barley. *Journal of the Institute of Brewing*, 42, 47–53.

Singh, T. K. and Cadwallader, K. R. (2003) The shelf life of foods: An overview. *Freshness and shelf life of foods. ACS Symposium Series*, 836, 2–21.

Stephenson, W. H. and Bamforth, C. W. (2002) The impact of lightstruck and stale character in Beers on their perceived Quality: A consumer study. *Journal of the Institute of Brewing*, 108, 406–409.

Torel, J., Cillard, J. and Cillard, P. (1986) Antioxidant activity of flavonoids and reactivity with peroxy radical. *Phytochemistry*, 25, 383–385.

Torline, P., Dercksen, A. and Axcell, B. (1999) An electronic look at beer flavor instability. *Master Brewers Association of the Americas Technical Quarterly*, 36, 289–292.

Vanderhaegen, B., Neven, H., Verachtert, H. and Derdelinckx, G. (2006) The chemistry of beer aging—a critical review. *Food Chemistry*, 95, 357–381.

Van Eerde, P. and Strating, J. (1981) Trans-2-nonenal. *European Brewery Convention Flavor Symposium, Monograph VII*, 117–121.

Vesely, P., Lusk, L., Basarova, H., Seabrooks, J. and Ryder, D. (2003) Analysis of aldehydes in beer using solid-phase microextraction with on-fiber derivatization and gas chromatography/mass spectrometry. *Journal of Agricultural and Food Chemistry*, 51, 6941–6944.

Wainwright, T. (1974) Control of diacetyl. *The Brewer*, 60, 638–643.

Walker, M. D. and Simpson, W. J. (1994) Control of sulphury flavours in cask conditioned beer. *Brewers' Guardian*, 123(11), 37–38, 40.

Whitear, A. L. (1981) Factors affecting beer stability. *EBC Flavor Symposium, Monograph VII*, 203–210.

Yuting, C., Rongliang, Z., Zhongjian, J. and Yong, J. (1990) Flavonoids as superoxide scavengers and antioxidants. *Free Radical Biology and Medicine*, 19, 19–21.

Zimmerman, D. C. and Vick, B. A. (1970) Hydroperoxide isomerase: A new enzyme of lipid metabolism. *Plant Physiology*, 46, 445–453.

4

Colloidal stability of beer

Kenneth A. Leiper and Michaela Miedl

Summary

This chapter will cover the area of beer stability in relation to the formation of colloidal haze, and the methods available to reduce haze formation. There are six types of beer stability: biological, non-biological haze, foam, flavor staling, light and gushing. The first two are generally referred to as haze and will be discussed here. Biological haze is caused by microbial contamination and is avoidable with care and the use of best practice procedures. Biological stability is covered in detail in the accompanying chapter by Annie Hill. In modern brewing, non-biological haze is more important and forms the main subject of this chapter.

There are three important types of non-biological haze. The most important is caused by the interaction of protein and polyphenol. The other two types, chill haze and hazes caused by other substances will also be covered. Foam stability is covered in the chapter by Evans and Bamforth in this book, but as beer proteins are involved with foam as well as haze, foam stability will be discussed here as well as it is difficult to separate the two areas. Flavor (light and ageing) stability and gushing are outside the scope of this chapter.

Protein-polyphenol haze is normally referred to as "permanent" or "colloidal" haze. This can be a serious problem in brewing, and brewers wish to reduce its effects. This chapter will cover: the importance of the whole brewing process in reducing haze, the proteins and polyphenols involved, their interactions, the products available to reduce their levels, how they operate, and how their efficacies can be measured.

Biological stability

There is a risk in brewing that the beer can become contaminated by microorganisms. Fortunately, beer is an inhospitable environment for microbial growth. It has a low pH (in the range 3.8–4.5), ethanol is present in a range of concentrations (typically 3–6% abv), there is a limited range of nutrients, hop acids are present, the environment is anaerobic and the liquid is carbonated.

Most potential contaminants originate from raw materials. Barley can contain *Fusarium* fungus which can release mycotoxins or cause gushing. It can also carry bacteria, which may contribute nitrosamines and cause filtration problems. Thus, contaminants can cause flavor deterioration, turbidity and potential health problems.

It is important to exclude these contaminants from the brewing process. Modern plant and good hygiene throughout the process will help. Most breweries pasteurize beer to ensure stability, but with good hygiene and efficient filtration the use of this expensive and potentially damaging process can be reduced.

Importance of whole process to ensure stability

Beer stability cannot be ensured by treating beer with one "super-product" which will solve everything. Stability is affected by the whole brewing process, so care must be taken at every stage. This is summarized in Figure 4.1 wherein the major materials and processes are shown. A (+) indicates that a benefit is gained, a (−) indicates a disadvantage and a (?) indicates that any advantage is debatable.

Raw materials are the source of haze material precursors. Malt must be of good quality, be in good condition and have a sufficient level of amylolytic enzymes. No individual barley variety appears to be particularly noted for producing haze precursors, however, batches of barley with high nitrogen levels may be more problematic. It is not certain if nitrogen modification of malt is related to protein stability. The use of adjuncts such as syrup, rice and maize grits dilutes haze-causing substances, although haze-causing proteins are not significantly reduced (Chapon, 1994). Adjuncts derived from barley and wheat may increase the level of haze precursors. Wheat also increases the level of pentosans, but is good for foam stability. Malt with no proanthocyanidins (haze-causing polyphenols) is available. Proteins will be discussed in detail later.

Lower mash pH causes polyphenol precipitation, which is desirable as it removes haze precursors. The benefit of "protein rests" in mashing is debatable as is any advantage in using mash filters in place of lauter tuns. Long run-offs extract more polyphenols especially if the pH is allowed to rise. Excessive use of lauter rakes is undesirable as is sparging at too high a temperature. Shear damage as a result of transfer by piping and pumping, etc. should be avoided as this can contribute to haze (Curin et al., 1989).

A good boil with agitation and turbulence is essential as it removes substances that could survive into the beer. Kettle fining removes substances that precipitate out of solution during wort cooling after boiling. Recent improvements in equipment have involved excluding oxygen so as to avoid oxidation and negative

Chapter 4 Colloidal stability of beer

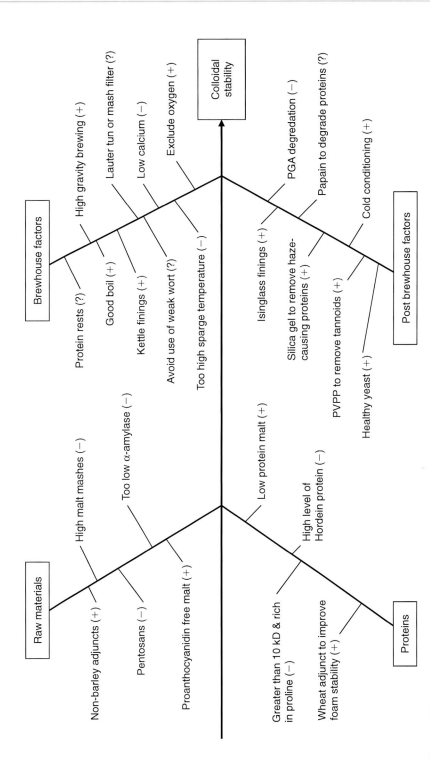

Figure 4.1
How raw materials and the brewing process affect the colloidal stability of beer.

effects on flavor. However, this also reduces oxidation of polyphenols which would normally polymerize to tannins and be removed with the trub. Sufficient calcium must be present in the wort, or there could be a build up of oxalate.

Apart from some loss of haze precursors in the cold break, little occurs during fermentation. However polyphenols become attached to yeast cell walls and are removed during yeast cropping. Yeast that is in poor condition can release polysaccharides, proteinases and nucleic acids into the beer.

Cold conditioning is important. Traditional brewing methods involved long maturation periods at 0–1°C (O'Rourke, 1994). Beer should be chilled as cold as possible without freezing to ensure precipitation of particulate matter. Work by Miedl and Bamforth (2004) has indicated that a short period of very low temperature storage prior to filtration can be as beneficial to colloidal stability as longer periods at less cold temperatures.

High gravity beers can be cooled to a lower temperature due to their higher alcohol content and this results in this type of beer being more stable, particularly on dilution to sales gravity. However, high gravity beers have been found to exhibit reduced foam stability. This has been shown to be due to lower levels of foam-causing polypeptides being extracted from high gravity mashes (Cooper et al., 1998).

If finings are used, it is desirable not to recover too much beer from the tank residue using centrifugation or filtration as there is the risk of returning haze precursors to the beer. The chosen stabilization technique such as silica gel to remove protein or Polyvinylpolypyrrolidine (PVPP) to remove polyphenols should be carried out at a suitable dosing rate. The benefit of using papain is debatable as it can damage foam. Using propylene glycol alginate (PGA) to restore foam stability should be carried out with care as it can degrade and cause haze.

Filtration should be carried out at 0–2°C after cold conditioning to allow particulates to settle. It is important to not allow the beer to warm up at this stage or chill haze particles could disassociate and release haze precursors (Chapon, 1994). Beer should be kept cold after filtration and oxygen and metal ions must be excluded. All equipment and containers must be clean and the carbonation gas should be of the highest purity.

Chill haze

This type of haze forms when beer is cooled to below 0°C, but will disappear if the beer is warmed. Its composition and formation are identical to those of permanent haze except that it is reversible (Power and Ryder, 1989–1991), this will be discussed later. However, it is this haze that must be attended to by brewers as it will develop into permanent haze if not treated (Bamforth, 1999).

Protein in beer

Beer contains ~300–800 mg/l proteinaceous material. Most of this is in the form of polypeptides that range in size from 5 to 100 kD. As true proteins are considered to be over 17 kD, beer protein is in the form of proteins and polypeptides although generally referred to as protein (Bishop, 1975).

Beer polypeptides originate primarily from barley, with possibly some originating from cereal adjuncts and hops. Proteins are subjected to change during malting and are extracted during mashing. Much protein is broken down during malting by the activity of proteolytic enzymes into polypeptides and free amino nitrogen (FAN) or single amino acids. It is possible that this continues during mashing, although temperatures are too high to permit extensive proteolytic activity. During boiling, much protein is removed from wort as trub (hot break) and during wort cooling (cold break).

Some breweries employ an additional mashing stage before the main mash. This is referred to as a "protein rest" and takes place at around 45°C which is optimal for proteolytic enzymes, rather than the ~65°C temperature of the main mash which is optimal for amylolytic enzymes. The reason for this is to allow more proteolysis to create more FAN, this is important if a high level of adjunct is being used which could dilute the amount of FAN available. Lewis et al. (1992) state that the FAN level of wort is determined during malting and the FAN level of malt is essentially the FAN level of wort. Nevertheless, some breakdown of polypeptides does seem to occur during such rests. The type of mashing system used will influence this stage.

Measuring protein in beer

There are various methods available for measuring proteins in beer, these have been reviewed by Hii and Herwig (1982) and their advantages and disadvantages described. The Kjeldahl method is widely used in brewing but actually measures amino nitrogen so will also measure beer amino acids as well as true polypeptides, the Lowry assay tends to measure non-protein substances as well, the Biuret method is not sensitive to beer proteins, absorption of ultraviolet light at 280 nm can be affected by interference from hop components and gel filtration is time-consuming. Hii and Herwig (1982) recommend the use of the Bradford assay using the dye Coomassie brilliant blue, with beer polypeptides as they are over 2 kD, however it is known that this dye is insensitive to haze-causing proteins as it does not react with glutamic acid or proline (Siebert, 1997). Haze formation can be detected by centrifuging defoamed beer and resuspending the resulting pellet in supernatant. The absorbance of this solution is measured at 400 nm against a supernatant blank, this provides a measurement of light adsorbing particles (LAPs) (Walters et al., 1996).

Yang and Siebert (2001) have reported a detection method for haze-active protein using the dye bromopyrogallol red. This dye binds to proline-rich polypeptides in a way similar to polyphenols. The authors report that the test is accurate, but samples must be pretreated to remove polyphenols before assaying as these interfere with the dye binding.

A more recent review of potential methods has been carried out by Siebert and Lynn (2005). Their findings confirmed that adsorption at 280 nm was unsatisfactory due to interference from polyphenols. A method using bicinchoninic acid, a derivation of the Biuret method, was also rejected for the same reason. The authors found that the Bradford assay was free from polyphenol interference, but was suitable only for non-haze proteins due to the dye's inability to bind to hordein-derived polypeptides. The sensitive protein assay, based on the addition of tannic acid, was found to be influenced slightly by polyphenols, but was still able to produce usable results when used to measure levels of haze-causing protein.

In order to study beer proteins in detail, it is necessary to isolate them from beer. Various methods exist for extraction of different fractions. Total beer protein can be isolated by precipitation with ammonium sulfate followed by centrifugation, dialysis and freeze-drying (Deutscher, 1990). Precipitation using ethanol can also be used. Veneri et al. (2006) have described a novel purification method for beer protein. This makes use of sodium dodecyl sulfate (SDS) to complex proteins followed by isolation using potassium ions. This method is much faster, and produces results equivalent to existing purification methods such as precipitation with ammonium sulfate. This method should be of use in future studies of beer proteins.

Foam protein can be extracted by foaming beer with gas and collecting the foam which is then re-liquified. Various designs of "foam towers" have been devised for this purpose. Haze proteins can be isolated by centrifuging aged beer or by treating beer with silica gel followed by removal of the protein from the silica with ammonia.

Investigations into beer proteins generally follow four approaches. The first is to identify the molecular size of polypeptides by running samples on gels using SDS-PAGE (sodium dodecyl sulfate-polyacrylamide gel electrophoresis). The second is to measure the amino acid composition of polypepdides using acid hydrolysis and HPLC (high pressure liquid chromatography). The third is to measure the hydrophobicity of polypeptides by isolating the hydrophobic types on sepharose columns; this method is related to foam-forming proteins. The fourth makes use of the ELISA method (Enzyme Linked Immunoadsorbent Assay) with antibodies raised against isolated beer polypeptides, this method can be used to measure the amount of a particular component at various stages of the brewing process. More recent techniques such as LC-MS (liquid chromatography-mass spectroscopy) have yet to be used successfully on beer proteins.

Beer polypeptides and their functions

It has been known for many years that beer protein is responsible both for colloidal and foam stability. There have been many attempts to link these two attributes to polypeptides of particular size. This however, has been complicated by the restricted range of polypeptides in beer and reliable means for detecting them. Consequently, a number of authors have claimed different functions for various polypeptides and the resulting literature is often confusing and contradictory. One thing that is certain is that only a small amount of protein is involved in beer haze, 2 mg/l of protein is sufficient to cause a haze of 1 EBC (Chapon, 1994).

Proteins in beer generally originate from malted barley. Most adjuncts such as maize and rice grits contain little protein and brewing syrups are almost pure sugar. Other adjuncts such as unmalted barley are used at low levels and are thus unlikely to add significant amounts of protein. An exception is wheat, either malted or unmalted.

Wheat has been used as an adjunct for many years, particularly due to its ability to promote foam stability. Wheat should not be used at too high a proportion however, as it contains high levels of pentosans which can lead to filtration problems (Bamforth, 1999). Wheat may also contribute haze precursors to beer. This is not a problem for all brewers though, in Belgium and parts of Germany haze is actually desirable!

Delvaux et al. (2000) analyzed the hazes from a range of commercial and pilot brewed Belgian wheat beers. Proteins and polyphenols were detected as expected, but also carbohydrate in significant amounts (in one case 95% of the haze material). Most of this consisted of glucose from starch, but residual levels of arabinose, xylose, galactose and mannose were found, the last originating from yeast. Calcium was found at significant levels, most probably originating from oxalate.

Traditional Belgian white beers are brewed using up to 40% unmalted wheat which was assumed to be responsible for haze. However, work by Delvaux et al. (2001) has shown that while wheat contains haze-active polypeptides (gliadins from the breakdown of gluten), most are lost during wort boiling and fermentation. Work by Delvaux et al. (2003) showed that beer haze increased if 5% wheat was used, but at levels above this, haze was steadily reduced, until at 40% there was hardly any. The authors suggested that while low concentrations of gluten form hazes, those formed at higher levels form precipitates with polyphenols which do not persist into the finished beer. Further work by Delvaux et al. (2004) showed that a successful way of *increasing* beer haze was to use malted wheat as the malting process causes increased proteolysis leading to an increase of haze-active polypeptides in beer.

Depraetere et al. (2004) brewed beers with 40% and 20% unmalted wheat and found that the 40% beers did not exhibit cloudiness while the 20% beers did. The use of wheat affected foam too, but not in a straightforward way, when used with normal brewing malt, the wheat caused a loss of foam stability. Stability was only increased if an over-modified malt was used. The authors reported that the effects of wheat on foam and haze depend largely on the malt used.

Much more attention has been paid to foam-forming proteins than to haze-causing proteins. However, some of the foam work published made use of unsuitable extraction methods which call the results into question. Work related to size identification will be reviewed first, followed by that involving amino acid composition, then that by hydrophobicity.

Identifying polypeptides by size

Early research identifying beer polypeptides focused on foam and this has been reviewed by Bamforth (1985). The first polypeptide to be identified was antigen 1 which is the most prominent polypeptide in beer. Using immunological methods it was found to be present in beer foam and haze and to have originated from barley albumin. It survives malting and mashing with only minor chemical modification and has a molecular weight of around 40 kD.

Work by Hejgaard (1977) showed that antigen 1 was immunologically identical to a barley protein known as protein Z which is associated with β-amylase. Later work by Hejgaard and Kaersgaard (1983) involved purification of antigen 1 from beer and analysis of its properties. The beer protein had an almost identical amino acid composition to that of the barley protein (except for less lysine) but had the same molecular weight. The protein was not glycosylated in barley but was in beer and was present at levels of 22–170 mg/l depending on beer type. The authors stated that it was probably involved in both foam and haze formation.

More wide-ranging work was conducted by Asano and Hashimoto (1980). They reported extracting foam protein from beer, however this was extracted using ammonium sulfate precipitation which is not selective for foam protein, but will remove all beer protein. From their precipitate they identified three groups of polypeptides, a high molecular weight (MW) group containing polypeptides of 1000 kD, 400 kD and 90 kD, a medium MW group containing polypeptides of 36–40 kD and a low MW group of polypeptides of 10–15 kD. Analysis of these groups showed that the high MW group was only 21% protein and contained 64% carbohydrate. The medium and low MW groups contained 17% and 12% carbohydrate respectively, consisting of mainly glucose, xylose and arabinose. The authors measured the amino acid composition of the extracts and reported that they originated from barley albumin or globulin due to the similarity to these barley proteins. In addition they reported that 60% of the high MW, 55% of the medium MW and 15% of the low MW polypeptides existed as glycoproteins.

Asano and Hashimoto (1980) raised antibodies to their polypeptides and used them to investigate various stages of the malting and brewing process. They reported that foaming proteins were produced at the start of germination, but the level fell towards the end. If a protein rest was incorporated at the beginning of mashing, more could be released, but there was loss during mashing and boiling. Only around half the foam proteins in wort survived into beer.

Asano et al. (1982) later turned their attention to haze-forming proteins in beer. These were extracted by filtration of aged beer. Using various fractionation methods the authors identified four haze-causing polypeptides and estimated their sizes to be 40 kD, 19 kD, 16 kD and several fractions ranging from 1 to 10 kD. The protein content of these fractions was found to range from 65% to 76%. However this was measured by the Lowry method which is susceptible to interference from carbohydrate (Hii and Herwig, 1982). The samples contained carbohydrate as well, with the 1–10 kD, 16 kD and 19 kD fractions containing mostly glucose while the 40 kD fraction also contained arabinose and xylose. All of the fractions were rich in proline and glutamic acid. Immunological investigations showed that all the fractions came from hordein, but the 16 kD and 40 kD fractions could also have originated from albumin or globulin, this being in agreement with the earlier findings of Asano and Hashimoto (1980). However, the 16 kD polypeptide had a similar amino acid composition to the 10–15 kD polypeptide which had previously been related to foam.

As colloidal instability is presumed to be caused by haze proteins combining with polyphenols, Asano et al. (1982) combined their fractions with the polyphenol catechin. The 19 kD fraction produced the most haze, and this seemed to be directly related to the amount of proline. These authors tested other polypeptides that contained no proline and no haze was produced. As haze formation was inhibited by hydrogen bond acceptors, it was concluded that hydrogen bonding was taking place. The authors reviewed the possible reactions between proteins and polyphenols. Firstly hydrogen bonding between the oxygen atoms of peptide bonds and hydroxyl groups of polyphenols. Secondly, hydrophobic bonding between hydrophobic amino acids (proline, tryptophan, phenylalanine, tyrosine, leucine, isoleucine and valine) and the hydrophobic ring structure of polyphenol. Thirdly, ionic bonding between positive charged proteins such as the ε-amino groups of lysine and the negative hydroxyl groups of polyphenols. However, at beer pH the hydroxyl groups have no charge and ionic bonding cannot occur.

Because of the pyrrolidone ring of proline, proline-rich proteins have unfolded structures that permit the access of polyphenols. Also the pyrrolidone ring of proline cannot form intra and inter molecular hydrogen bonds with the oxygen atoms of peptide bonds, thus these free oxygen atoms can form hydrogen bonds with the hydroxyl groups of polyphenols. Also proline is hydrophobic and can participate in hydrophobic bonds with polyphenols. These two interactions are possibly responsible for chill haze formation.

Slack and Bamforth (1983) measured beer polypeptides by hydrophobic interaction chromatography to investigate how hydrophobicity related to foam stability. They found that the most stable foams were composed of hydrophobic polypeptides larger than 55 kD. They suggested that polypeptides can uncoil their structures during foaming to reveal more hydrophobic regions which results in a more stable foam.

Dale and Young (1988) measured beer polypeptides and found a large number of fractions ranging in size from 15 to 44 kD but gave no exact sizes. They also reported that some wheat adjunct polypeptides could survive the brewing process and were present in the final beer. Yokoi and Tsugita (1988) extracted polypeptides from evaporated beer. Using electrophoresis they identified three polypeptides of 40 kD, 10 kD and 8 kD as well as a faint band at 12 kD. Their bands were not sharp, but when analyzed, the upper and lower portions of the bands were found to be identical. The authors suggested this was because their polypeptides were glycoproteins. Barley proteins were measured and bands were found at 55 kD, 40–50 kD, 15–12 kD, 11 kD and 10 kD. In wort only the 40 kD and 10–15 kD bands were seen. In beer only the 40 kD and 10 kD were left, with the addition of a band at 8 kD which they suggested was a degradation product of the 10 kD polypeptide as the amino acid profiles were similar. Yokoi et al. (1989) identified the 40 kD polypeptide as being involved in foam due to its hydrophobicity. A 10 kD polypeptide was also isolated but had no foaming ability.

Outtrup (1989) extracted beer polypeptides on the basis of hydrophobicity and investigated their interactions with polyphenols. The polypeptides which were most hydrophobic, did not react with polyphenols indicating that haze-causing polypeptides are not hydrophobic. The least hydrophobic group had a size range of 14–19 kD and contained higher levels of proline and glutamic acid than the middle group of more hydrophobic polypeptides which had a wider size range of 12–40 kD. This indicated that haze polypeptides are concentrated in a size range of 14–19 kD. Previous work (Outtrup et al., 1987) showed that polyphenols gave haze when reacted with pure proline, and these above results support this. However, the author reported that the amount of proline was not directly linked to haze and that the sequence of the polypeptide was also important.

Hollemans and Tonies (1989) attempted to identify foam proteins by removing them from beer in order to gauge their effects. The authors identified three foam polypeptides with sizes of 90 kD (or larger), 40 kD and 15 kD (or smaller). These were isolated and antibodies were raised against them which were used to remove these polypeptides from beer. The authors were unable to remove the ~90 kD polypeptide. Beer with the 40 kD polypeptide removed only lost around 10% of its original foam stability. With the 15 kD polypeptide removed, the foam loss was between 0 and 33% depending on which of their results are considered. Removal of the 40 kD polypeptide resulted in a ~20% reduction in total beer protein, but removal of the 15 kD polypeptide gave no measurable protein reduction. This may have been due to this polypeptide being heavily glycosylated.

Mohan et al. (1992) produced foam using a foaming tower and analyzed it using electrophoresis and immunological methods. In beer they identified three polypeptides with molecular sizes of 66 kD, 40 kD and 20 kD. However, in foam they identified over 20 polypeptides. These included protein Z (40 kD) as well as polypeptides of 36 kD, 20 kD and 14 kD. Faint bands were also observed in the 60–65 kD and 80–100 kD ranges and it was suggested that these could be of yeast origin. They stated that any of these components could be involved in foam stability.

Kano and Kamimura (1993) extracted beer protein and identified two polypeptides with the approximate sizes of 40 kD and 20 kD. They investigated links with foam stability and found that the 40 kD polypeptide to be more related than the 20 kD although neither of their graphs is particularly convincing. This result was not in agreement with Hollemans and Tonies (1989). Kano and Kamimura (1993) tested their beers with various types of silica and found that very little protein (4–12%) was removed. No correlation was found between the amount of protein removed and physical stability. These authors also suggested that wet milling could improve foam stability. It was found that this milling procedure gave increased levels of both polypeptide types in wort and beer. This was possibly due to inhibition of proteolytic enzymes.

Sorensen et al. (1993) used a foam tower to extract foam protein from beer. Gel filtration of this protein resulted in three fractions, a high molecular weight fraction (90% carbohydrate, 10% protein) a low molecular weight fraction (90% protein, 10% carbohydrate) and a third fraction of amino acids and other small molecules. Only the low molecular weight fraction produced a foam. However, when the high molecular weight fraction was added to the low molecular weight fraction, this improved the foam stability. This fraction contained protein Z (40 kD). The low molecular weight fraction was found to contain polypeptides in the 6–18 kD range. It also contained a polypeptide of 9.7 kD as well as fragments originating from hordein. Using amino acid analysis and Western Blotting this polypeptide was identified as being LTP1 (Lipid Transfer Protein 1).

When LTP1 was purified from the hordein fragments it gave good foam when foamed in water or beer, but with a poor half-life. However, in the presence of the fragments, the foam produced was of good quality and had a good half-life. LTP1 was previously identified in barley where it originates from the aleurone. It has a molecular weight of 9.694 kD and consists of 91 amino acids including eight cysteine residues. It had not been connected to foam stability previously, possibly due to its small size. Its amino acid sequence was found to be homologous to amylase and protease inhibitors and was named PAPI (probable amylase/protease inhibitor). Later it was found to be homologous to lipid transfer proteins and was found to be able to transfer phospholipids from liposomes to mitochondria in potatoes *in vitro*. Because of this it was renamed LTP1.

It has been suggested that LTP1 is modified during malting and brewing which renders it more foam active. Addition of beer LTP1 to beer gave greater foam enhancement than the addition of barley LTP1. Sorensen et al. (1993) proposed that the low molecular weight fraction including LTP1 and the hordein fragments gave foam potential and the high molecular weight fraction including the 40 kD protein (protein Z) gave foam half-life/stability.

Further work by the same group continued to investigate the differences between the two types of LTP1 (Bech et al., 1995). Mass spectroscopy of barley LTP1 gave a mass of 9.663 kD and SDS-PAGE produced a sharp band on gels, while LTP1 extracted from beer gave a mass of 9.600–9.990 kD and a fuzzy gel band. NMR (nuclear magnetic resonance) analysis showed that the three dimensional structure of barley LTP1 was destroyed during brewing. Antibodies were used against the two types of LTP1 to follow LTP1 through the brewing process. No loss or conversion of LTP1 was found during malting. During mashing, LTP1 was extracted almost immediately. The conversion from barley to a foam-active form was found to occur during boiling where the barley form decreased and the foam form increased. All the barley form appeared to be converted to the foam form with no apparent loss.

Yokoi et al. (1994) presented further work on hydrophobic proteins; they extracted four beer fractions based on their hydrophobicity. Their most foam-active fraction contained the 40 kD protein which they linked to foam (but so did two of their less foam-active fractions). The amino acid profile of the most foam-active fraction indicated high levels of valine, leucine and isoleucine which are hydrophobic.

Onishi and Proudlove (1994) investigated foam-causing polypeptides in relation to hydrophobicity rather than size. They isolated fractions of increasing hydrophobicity, but when subjected to SDS-PAGE there were no recognizable differences between the samples. These polypeptides were found to originate from hordein. It was suggested that peptides "open up" their structure to reveal more hydrophobic sites thus aiding foaming. Bamforth (1995) showed there to be no correlation between total beer protein and foam stability, but that there was a correlation between hydrophobic protein and foam stability. Protein was measured using Coomassie blue, which although unsuitable for haze polypeptides (Hii and Herwig, 1982) is suitable for foam polypeptides (Chandley, 1994).

Lusk et al. (1995) investigated the role of specific polypeptide sizes on foam in addition to melanoidins and polysaccharides. Using a foam tower they produced foam and isolated three polypeptides with molecular sizes of 10 kD, 12 kD and 40 kD. The 40 kD polypeptide consisted of protein Z, the 10 kD was LTP1 and the 12 kD had no known function. When added to an artificial "beer" solution, the 10 kD polypeptide gave good foam stability. Using mass spectrometry they determined the size of LTP1 to be 9.975 kD. In addition, they sequenced the barley form of LTP1 and found it to have 91 amino acids and a molecular weight of 9.965 kD. The addition of protein Z gave some foam stability, but the 12 kD polypeptide had no effect. The addition of melanodins and polysaccharides (which were not characterized) gave some foam stability.

Curioni et al. (1995) further investigated the 40 kD polypeptide. Their results showed that it is actually composed of two polypeptides that are genetically different but immunologically related, known as antigen 1a and antigen 1b. They are derived from two barley albumins called protein Z4 and Z7 which are also present in wheat. Their structures are similar to plant serine protease inhibitors or serpins. It was also found that they originate from the barley endosperm and one (1a) is glycosylated and the other is not.

Ishibashi et al. (1996) developed an ELISA method which they claimed was able to detect foam and haze polypeptides. Foam protein was extracted by ethanol precipitation (although this would remove all beer protein) and haze protein by filtering old beer, and raised antibodies against them. The authors reported that foam protein was 41 kD and haze protein was 40 kD. It was mentioned that there were more polypeptides involved in haze, but their sizes were not given. Convincing correlations between foam protein and foam adhesion and haze protein and shelf-life were presented. They followed the brewing process with their antibodies and found a loss of proteins throughout, with about 10% of the proteins remaining in beer as compared to the levels at mashing. They reported around 50 mg/l foam protein in beer which would be expected, but around 130 mg/l haze protein in beer which seems high.

Douma et al. (1997) collected samples from dispensed foam and isolated several polypeptides from the various fractions. A 40 kD polypeptide was identified plus a range between 8 and 18 kD with major bands at 10 kD, 12 kD and 18 kD as well as a large band at 100 kD. They reported that both the 40 kD and the 8–18 kD range polypeptides were important for foam stability as they accumulated in beer foam.

Sheehan and Skerritt (1997) identified beer polypeptides originating from hordein and related them to their precursors in barley. These originate from hordein storage proteins which are designated B, C and D according to molecular size. Barley endosperm proteins consist of 80% B, 10–20% C and 5% D. Some of these are monomers and others are aggregated. They developed antibodies from barley hordeins, and found that the monomeric form was mostly C hordein with some B and consisted of polypeptides in the 30–45 kD range, and the aggregated fraction was mostly B size ranges 30–45 kD and 50–60 kD. In wort, the antibodies which bound to barley hordeins identified a polypeptide at 40 kD (the authors linked this to protein Z, but this is known to originate from albumin) as well as a group ranging from 29 to 8.6 kD which probably resulted from breakdown of hordein rich polypeptides as none of these sizes were found in barley. This group included a doublet at 29 kD and a single band at 23 kD. Hordein polypeptides with molecular weights greater than 40 kD were lost during mashing. Mashing with an extra 30 minute rest at 45°C followed by 68°C for 15 minutes produced wort containing 252 mg/l FAN instead of 230 mg/l without the extra rest. In beer, the antibodies identified polypeptides between 29 and 51 kD although the 45 kD fragment had disappeared. The doublet at 29 kD was detected although in a reduced amount, the 23 kD polypeptide was also found. This showed that some polypeptides are stable while others are affected by proteolysis.

Sheehan et al. (2000) treated beers with five silica types. The authors found this treatment had little effect on the total protein level, protein Z or LTP1, but there were great reductions in the levels of hordein derived polypeptides. No molecular sizes were given for these polypeptides.

Kakui et al. (1999) continued the ELISA work of Ishibashi et al. (1996) on foam proteins. These were extracted with ammonium sulfate precipitation (instead of ethanol precipitation as before, but still not specific) from two beers of 100% barley and 100% malted wheat. They reported that barley foam protein is 41 kD and that of wheat is 10 kD. They found that their antibodies did not cross-react indicating that their proteins are different. They found good correlation between wheat foam protein and foam stability. They followed the brewing process with regard to barley and wheat foam protein and reported the usual loss of protein throughout. In finished beers, the barley beer contained 51 mg/l and the wheat beer 355 mg/l foam protein. These values were 10% and 5% respectively of the amounts of protein at mash off. This indicated that wheat protein is less stable, but there is much more of it present.

Evans and Hejgaard (1999) further investigated the origins of LTP1. Two forms are expressed in barley, LTP1 and LTP2. LTP1 originates in the aleurone in early germination, they found the size to be 9.690 kD by mass spectroscopy, but also found another form with a size of 9.983 kD. LTP2 has not been found in beer. Both LTP1 and protein Z are affected by kilning, but survive into beer as they are able to inhibit proteases. Further work by Evans et al. (1999) showed that protein Z (the Z4 part) was related to foam stability along with beta glucan and arabinoxylan, but that the Z7 part and LTP1 were not related to foam.

Vaag et al. (1999) further investigated foam proteins. They isolated a polypeptide of 17 kD. This was conducted by ammonium sulfate precipitation, but was also extracted from foam produced in a foam tower. It was identified as a barley protein and had antibodies raised against it. The foam from the tower also contained LTP1 but not protein Z. Only the foam-active form of LTP1 was found in the foam. The authors removed protein Z from beer and found it had little effect on foam stability (except when an undermodified malt was used where its absence caused a decrease in stability). The 17 kD polypeptide like LTP1 was found to be rich in cysteine and contained disulfide bridges. The native form of the 17 kD polypeptide was converted to the foam-active form by mashing and boiling.

Jegou et al. (2000) further investigated the changes to LTP1 during brewing that make it more foam active. They found two isomers of LTP1 in barley. LTP1 with a weight of 9.689 kD and LTP1b with a weight of 9.983 kD, which is possibly the same as reported by Evans and Hejgaard (1999). This might also be the same as the 9.6–9.9 kD form reported by Bech et al. (1995) and the 9.975 kD form reported by Lusk et al. (1995). Both these isoforms are modified during brewing involving disulfide bond reduction, hydrolysis and glycosylation. This could explain why LTP1 is involved in foaming.

Leiper et al. (2003a,b) investigated beer proteins and how they were affected by treatment with silica gels. Three types of beer were analyzed: all malt, 30% maize adjunct and 30% syrup adjunct. Haze material was isolated by saturating silica by repeated exposure to fresh beer. Once the silicas could adsorb no more, the material was recovered. Total beer protein was recovered using ammonium sulfate precipitation and foam protein was isolated using a foaming tower. The samples were analyzed for chemical composition, molecular size and amino acid composition.

Molecular size analysis by SDS-PAGE showed a limited range of polypeptides. The total protein samples from the three beer types all had three main bands with sizes of 46.9 kD, 12.4 kD and 11.9 kD. The first two were the known proteins protein Z and LTP1 with their actual sizes of 40 kD and 9.7 kD increased by glycosylation. The haze samples contained small amounts of protein Z along with a group of polypeptides ranging from 15 to 32 kD with the most abundant being 16.5 kD and 30.7 kD. Foam protein samples were found to contain elevated levels of LTP1 confirming the results of Sorensen et al. (1993) that LTP1 is important for foam. Samples of beer that had had the foaming fraction removed were analyzed and this showed that protein Z was concentrated in these fractions indicating that it not involved with foam.

Chemical analysis of the samples showed that all beer proteins were heavily glycosylated. The protein content of total beer "protein" ranged from 40 to 60% depending on type, the remainder consisting of glucose, arabinose and xylose. Haze and foam proteins were found to be glycosylated with glucose. The amino acid composition of the fractions is described below. Thus, three distinct types of glycoprotein were found in beer. The first is those responsible for haze formation, ranging in size from 15 to 32 kD and accounting for 3–7% of the total. The second is those responsible for foam formation, particularly LTP1, accounting for ~25% of the total. The last, accounting for the remaining 70% of the total, including protein Z, appears to have no function related to beer stability.

Evans et al. (2003) extracted a 12 kD polypeptide which they showed to be involved with haze formation. The polypeptide was recovered from silica gel and analysis indicated that it originated from barley hordein. An antibody was raised against the polypeptide and this was used to detect the presence of the polypeptide throughout the brewing process. It was shown that malt varieties that lack this polypeptide produced beer with improved colloidal stability. This work was continued on a pilot scale by Robinson et al. (2004) with the same result.

Leisegang and Stahl (2005) investigated the conversion of LTP1 from barley to beer forms. While this process made the polypeptide more foam active, it also made it susceptible to degradation by proteinase A. This enzyme had no effect on protein Z and barley LTP1.

A method new to brewing research has been used by Hao et al. (2006). The authors extracted proteins from beer and foam using SDS-PAGE followed by

LC-MS (liquid chromatography-mass spectroscopy). Beer stabilized with silica gel was used and foam was extracted by foaming beer with nitrogen and freeze-drying the re-liquified foam. When run on gels, the beer and foam samples produced similar patterns, with significant bands at 40 and 7–17 kD. These bands were excised and digested with trypsin before LC-MS analysis.

This technique produces fragments of proteins, these are analyzed for their masses and these can be matched to known fragments stored in databases thus allowing identification of the original proteins. This is a well-established technique, although it would appear that this is the first report involving the analysis of beer. If carried out well, it is a reliable way of identifying proteins. To be reliable, sufficient fragments of each protein must be found to allow a confident identification. This did not occur in this paper. The authors claimed to have identified 30 proteins in beer and foam, however the evidence was far from convincing. Protein Z was identified in beer and foam, eight fragments matching the structure of protein Z were found, accounting for 82 of the 399 amino acids of that molecule, this is 20.6% of the total, thus giving a fairly safe-if hardly novel-result. In contrast, most of the other proteins were identified on the basis of single fragments, for example a ubiquitin/ribosomal protein was "identified" on the basis of one six-amino acid fragment (3.9% of the total amino acids in that protein). Thus, all of the results from this work must be viewed with caution. Many of the proteins identified were found to be too large or too small to fit into the 40 or 7–17 kD bands where they were found in the gels. This further highlights the unreliability of these results. This method shows great promise for the identification of beer proteins if it is carried out with greater care.

Identifying polypeptides by amino acid composition

The second part of this section will review the amino acid composition of the various beer polypeptides isolated. Data presented here comes from two sources, papers that dealt with amino acid composition only and also papers that dealt with polypeptide size that have already been discussed. When combined, these two sources provide a large amount of information. This has been summarized by being placed in tables containing related data.

Not all the amino acid data from the papers reviewed above have been included. Some of these were for isolated fractions that contained many different polypeptides, the combined composition of which are of little help in relating particular polypeptides to their potential functions. Also it must be remembered that all these samples have been hydrolyzed before analysis and this will change the levels of certain amino acids. Tryptophan is destroyed by hydrolysis and no data for this amino acid are included below. Hydrolysis results in asparagine and glutamine being converted to aspartic acid and glutamic acid, so the data below for aspartic acid and glutamic acid will include that which was asparagine and glutamine in the actual polypeptides. Lastly, some hydrolysis methods can destroy cysteine, so not all the cysteine values below may be accurate (shown in the tables as cysteic acid). Not all the data below will have been created using the same hydrolysis and detection methods, but comparisons can still be made.

A useful place to start is with the composition of all the polypeptides present in beer. Dale et al. (1989) analyzed total beer polypeptide composition in commercial beers and in their own beers consisting of 100% malt (three types), 30% syrup adjunct, 100% malted wheat and beers with various wheat adjuncts. The average data for the three all malt beers, the syrup adjunct beer, the average of four all wheat beers and a wheat adjunct beer are shown in Table 4.1. It can be seen that there is little difference between the profiles of the all malt beers and the syrup adjunct beer indicating that the syrup adjunct contributed no protein.

The all wheat beer profile was similar to the all malt profile with the exception of a much higher level of glutamic acid. The profile of the beer with 20% torrified wheat (a realistic inclusion rate for this adjunct) was again similar to the all malt beer, with only a slightly higher level of glutamic acid. This indicates that wheat contributes little in the way of distinctive protein to beer.

Dale et al. (1989) also measured more specific polypeptides by dividing their beers into three fractions containing polypeptides of >60 kD, 40–60 kD and 20–40 kD size. They found that the >60 kD fraction contained high levels of serine and threonine which suggested that yeast cell wall material was present, and in those beers brewed with wheat, that high molecular weight wheat polypeptides were also present.

The 40–60 kD fraction included material from albumin and globulin, as well as from wheat prolamines and glutelins in those beers brewed with wheat. The 20–40 kD fraction contained high levels of glutamic acid and proline, possibly from prolamines and glutelins. However, these data are of limited use here as the fractions could contain a wide range of polypeptides. For example, it is not clear if the 40 kD polypeptide protein Z, the major beer polypeptide is in the 20–40 kD or the 40–60 kD fraction, or both.

Amino acid composition of total, haze and foam proteins was measured by Leiper et al. (2003a,b). The published results have been recalculated to mol% and are shown in Table 4.1. The composition of total protein from all malt, 30% maize adjunct and 30% syrup adjunct beers was similar indicating that these adjuncts contribute little protein to beer. The composition of the proteins is similar to the results of Dale et al. (1989).

The total polypeptide data show that the principal amino acids are glutamic acid, proline, glycine, aspartic acid alanine and serine. As protein Z is the most abundant beer polypeptide, data on its composition are included in Table 4.1. This consists of the composition of antigen 1 as isolated and analyzed by Hejgaard and Kaersgaard (1983) (data recalculated to mol%), the "medium fraction" isolated by Asano and Hashimoto (1980) and the 40 kD polypeptide isolated by Yokoi and Tsugita (1988).

It can be seen that the polypeptide isolated by Yokoi and Tsugita (1988) is (as they reported) protein Z. However, the fraction isolated by Asano and Hashimoto (1980) does not match so well. The higher levels of glutamic acid and proline suggest contamination with haze polypeptides. The principal amino acids of protein Z are glutamic acid, leucine, alanine, serine, aspartic

Table 4.1
Amino acid composition of total beer polypeptides and protein Z fractions (mol%)

	Dale et al. (1989)		Leiper et al. (2003a)			Hejgaard and Kaersgaard (1983)	Asano and Hashimoto (1980)	Yokoi and Tsugita (1988)
	Average Total All Malt	Total syrup adjunct	Total all malt	Total 30% maize adjunct	Total 30% syrup adjunct	Antigen 1	40 kD	40 kD
Cysteic acid	1.3	2.2	9.8	9.4	9.0	–	–	–
Aspartic acid	8.7	9.0	5.9	5.7	5.7	8.9	6.6	8.4
Glutamic acid	21.1	19.4	17.5	18.3	18.5	12.3	14.2	12.5
Serine	6.0	6.2	7.4	7.4	6.8	9.7	7.5	10.6
Arginine	3.0	2.8	2.8	2.9	3.0	2.6	3.4	2.9
Threonine	4.4	4.2	3.7	4.0	3.4	5.1	3.9	5.6
Glycine	10.0	9.3	10.2	9.7	9.7	7.5	7.5	9.0
Alanine	6.8	6.7	7.3	6.7	6.9	10.3	8.6	10.1
Methionine	0.5	–	1.4	1.6	1.3	1.2	–	0.8
Proline	15.1	16.5	10.6	10.8	12.3	3.9	8.7	4.2
Valine	5.2	5.5	4.7	4.8	4.9	8.2	6.4	6.4
Isoleucine	3.3	3.2	3.1	3.2	3.2	4.4	3.4	4.0
Leucine	4.9	5.0	6.0	5.9	6.3	11.2	8.3	10.9
Phenylalanine	2.4	2.5	2.7	2.7	2.7	5.9	4.1	4.8
Lysine	3.1	3.0	2.9	2.8	2.8	4.7	3.3	2.9
Histidine	1.8	1.8	1.7	1.6	1.3	2.3	3.4	5.3
Tyrosine	2.2	2.3	2.3	2.5	2.2	1.8	1.1	1.6

acid, valine and glycine. This shows that although protein Z is the main beer polypeptide, its profile does not match that of total beer polypeptides.

The composition of polypeptides that are related to haze are presented in Table 4.2. This includes data collected by Djurtoft (1965). These results were obtained from the EBC Haze Group which involved analyzing haze from nine beer types produced in four breweries. Despite production differences, the compositions of the hazes collected were similar. For ease of comparison, an average value for the nine hazes is shown. It can be seen that the principal components are glutamic acid, proline, arginine and glycine.

Djurtoft (1965) compared the haze results to the amino acid composition of the barley protein fractions albumin, globulin, hordein and hordenin and stated that there was no correlation between any of them. This is certainly true for albumin and globulin, but the results for hordein and hordenin were reasonably close. The author compared the results to total barley composition and found a good match and concluded that haze polypeptides originated from all barley fractions.

Preaux et al. (1969) investigated the composition of material adsorbed by three adsorbents including silica gel. The material adsorbed by silica gel was similar to chill haze. The chill haze analyzed was not related to the composition of beer or foam. The composition of the chill haze and the composition of the material adsorbed by Stabifix® silica gel (Table 4.2), has been estimated from the histograms in the publication. It can be seen that the chill haze and the material adsorbed by the silica gel are very similar to that found by Djurtoft (1965).

Data showing the composition of barley hordein and albumin/globulin by Asano et al. (1982) is shown in Table 4.2. It can be seen that the main amino acids in hordein are glutamic acid and proline. It is due to this that hordein has been proposed as the source of haze polypeptides. The hazes discussed by Djurtoft (1965) and Preaux et al. (1969) have higher levels of arginine and glycine, but there still appears to be a connection. The high levels of aspartic acid, serine, alanine, leucine and valine eliminate albumin/globulin as a source of haze polypeptides.

The chill haze isolated by Asano et al. (1982) matched the composition of those of Djurtoft (1965) and Preaux et al. (1969) apart from having a lower level of arginine (the authors measured the composition of chill haze at a range of ages; the 0–3 day figure is given here). In the same paper Asano et al. (1982) isolated two polypeptides of 19 kD and 16 kD which they identified as being haze-causing. This can be seen as true, as their composition matches the chill haze values (again with arginine being lower).

Yokoi and Tsugita (1988) isolated beer polypeptides of 10 kD and 8 kD size, but did not propose any function. They contain low levels of proline and glutamic acid and this indicates that they are unlikely to be involved in haze.

Leiper et al. (2003a) isolated haze proteins from three different beer types using four silicas of differing structure. All the haze samples contained large amounts of glutamic acid and proline with only low levels of the other amino acids. It was found that the silicas bound to proline specifically. As there were only minor differences among the materials recovered by the four silicas from each beer, an average haze sample from each beer is included in Table 4.2. The three samples have similar compositions indicating that the differing production

Table 4.2
Amino acid composition of polypeptides related to haze (mol%)

	Asano et al. (1982)		Djurtoft (1965)	Preaux et al. (1969)		Asano et al. (1982)			Yokoi and Tsugita (1988)		Leiper et al. (2003a)		
	Hordein	Albumin and Globulin	Average haze values	Chill haze	Off stabifix	Chill haze	19 kD	16 kD	10 kD	8 kD	All malt haze	30% Maize haze	30% Syrup haze
Cysteic acid	0.6	1.5	2.5	–	2.5	–	0.9	1.9	–	–	3.9	2.7	2.9
Aspartic acid	1.2	9.3	4.2	4.5	4.0	5.4	3.0	8.5	12.8	11.4	3.2	2.2	2.5
Glutamic acid	29.1	9.6	16.5	19.0	19.0	14.3	20.9	14.3	12.8	13.9	31.7	35.1	34.0
Serine	3.3	6.0	4.1	4.0	4.0	4.7	3.4	5.9	8.5	7.6	3.9	4.3	4.5
Arginine	1.3	4.3	8.5	7.0	8.5	2.4	1.4	3.6	6.4	6.3	2.4	1.2	1.3
Threonine	1.4	3.5	2.7	3.0	2.5	2.6	2.0	4.1	4.3	3.8	2.4	1.9	1.8
Glycine	1.8	8.7	7.7	7.0	5.0	6.2	4.7	9.2	12.8	11.4	5.6	4.8	4.3
Alanine	2.1	8.6	3.3	3.0	3.0	2.7	3.1	7.4	6.4	7.6	3.8	2.4	2.6
Methionine	0.6	1.7	0.6	0.5	1.0	–	0.5	1.1	1.1	1.3	1.0	0.5	0.4
Proline	18.2	7.3	12.2	12.0	13	13.6	19.9	10.3	6.4	6.3	24.4	26.4	28.7
Valine	3.1	6.4	2.8	2.5	3.0	3.7	2.2	5.5	6.4	6.3	2.4	2.9	2.3
Isoleucine	2.8	3.2	1.9	2.0	2.5	1.9	2.3	2.8	5.3	5.1	2.8	2.4	2.7
Leucine	4.7	6.5	3.4	3.0	4.0	3.8	2.3	5.3	7.4	7.6	4.0	3.4	2.9
Phenylalanine	4.4	2.7	2.4	3.0	3.0	1.0	2.6	1.5	1.1	1.3	5.1	5.3	5.0
Lysine	0.3	3.6	2.7	2.5	3.0	1.9	1.3	2.5	2.1	2.5	1.2	0.7	0.8
Histidine	0.9	1.7	2.8	3.0	4.0	1.1	0.6	2.7	4.3	5.1	1.0	0.4	0.4
Tyrosine	1.6	2.7	2.4	2.5	2.5	1.5	2.0	1.7	2.1	2.5	1.2	3.4	2.9

methods had little effect. The composition is similar to the hordein sample analyzed by Asano et al. (1982).

The composition data related to foam polypeptides are presented in Table 4.3. The polypeptide most often related to foam is LTP1 and its composition presented by Sorensen et al. (1993) shows that the principal amino acids are aspartic acid, glycine, serine and cysteine.

Foam analyzed by Preaux et al. (1969) does not match the profile of LTP1, levels of cysteine, serine and glycine are lower and values for glutamic acid, arginine and proline are higher. If however, as Sorensen et al. (1993) proposed, protein Z and various hordein fragments were involved in foam, this would cause a difference in composition. Protein Z would reduce the levels of cysteine, aspartic acid and glycine, and increase the levels of glutamic acid and proline. Preaux et al. (1969) also analyzed a de-foamed fraction, but its composition was practically identical to that of their foam fraction.

Asano and Hashimoto (1980) isolated a polypeptide of 15 kD which they suggested was involved in foam. However, Asano et al. (1982) proposed that a polypeptide of 16 kD with identical amino acid composition was related to haze. These are undoubtedly the same, and as has been shown above, this polypeptide is probably involved in haze formation.

The two unidentified polypeptides isolated by Yokoi and Tsugita (1988) are included in Table 4.3, to see if they might be related to LTP1. Neither appeared to contain any cysteine, but this amino acid may have been lost during hydrolysis. Apart from having higher levels of glutamic acid, arginine and glycine, the composition of the 10 kD polypeptide is not dissimilar to that of LTP1 as suggested by its similar size. The higher levels of glutamic acid, arginine and glycine may have been caused by contamination with haze-causing polypeptides. Due to its similar composition, the 8 kD polypeptide may be, as suggested by Yokoi and Tsugita (1988), a degraded form of the 10 kD polypeptide.

Foam protein from all malt and 30% maize adjunct beers was analyzed by Leiper et al. (2003b). Unlike haze proteins, the foam proteins do not contain any amino acids in distinctive amounts. They contain low levels of proline, indicating that foam protein is unlikely to be affected by silica treatment. Foam from the all malt and 30% maize adjunct beer had similar compositions. The profiles were not a particularly close match to that of LTP1 as analyzed by Sorensen et al. (1993), probably due to the presence of polypeptides other than LTP1. The samples were more like the composition of the 15 kD sample analyzed by Asano and Hashimoto (1980).

Samples from de-foamed all malt and maize adjunct beers were also analyzed by Leiper et al. (2003b). In both samples, the amounts of glutamic acid and proline were much higher than found in the foam samples, indicating how haze polypeptides are concentrated in the non-foaming fractions.

Identifying polypeptides by hydrophobicity

Protein extraction based on hydrophobicity was used by Cooper et al. (1998a) to quantify foam-causing proteins throughout the brewing process. It had

Table 4.3
Amino acid composition of polypeptides related to foam (mol%)

	Sorensen et al. (1993)	Preaux et al. (1969)		Asano and Hashimoto (1980)	Yokoi and Tsugita (1988)		Leiper et al. (2003b)			
	Barley LTP1	Foam	De-foamed	15 kD	10 kD	8 kD	All malt foam	All Malt de-foamed	30% Maize foam	30% Maize de-foamed
Cysteic acid	8.8	2.0	1.5	1.9	–	–	12.7	8.2	12.9	7.6
Aspartic acid	16.5	5.0	4.5	8.5	12.8	11.4	7.8	4.3	6.8	4.1
Glutamic acid	6.6	12.0	12.0	14.3	12.8	13.9	11.2	20.0	10.6	21.1
Serine	8.8	3.5	3.5	5.9	8.5	7.6	8.2	6.5	8.5	7.0
Arginine	4.4	11.5	11.0	3.6	6.4	6.3	3.7	2.3	3.7	2.3
Threonine	3.3	2.5	2.5	4.1	4.3	3.8	3.3	3.6	3.8	2.8
Glycine	9.9	6.5	7.0	9.2	12.8	11.4	9.9	10.7	9.5	11.2
Alanine	4.4	6.0	7.5	7.4	6.4	7.6	7.5	7.5	7.7	6.8
Methionine	1.1	0.5	0.5	1.1	1.1	1.3	1.7	1.4	1.9	1.1
Proline	6.6	14.0	15.0	10.3	6.4	6.3	5.6	12.8	3.7	14.3
Valine	6.6	3.5	3.5	5.5	6.4	6.3	7.2	4.5	7.0	4.2
Isoleucine	6.6	2.0	2.0	2.8	5.3	5.1	4.8	3.0	4.9	3.0
Leucine	6.6	3.0	2.5	5.3	7.4	7.6	8.8	5.3	9.1	5.4
Phenylalanine	–	2.5	2.5	1.5	1.1	1.3	2.3	3.4	2.8	3.7
Lysine	4.4	5.0	5.0	2.5	2.1	2.5	3.2	2.7	3.7	2.3
Histidine	2.2	5.0	5.0	2.7	4.3	5.1	1.5	1.4	1.7	1.3
Tyrosine	3.3	2.0	2.0	1.7	2.1	2.5	0.6	2.4	2.0	1.8

been known for some time that high gravity beers exhibited poor foam stability. Cooper et al. (1998a) showed that this was due to poor extraction of hydrophobic polypeptides during mashing. This was proved by the detection of foam polypeptides in the wort extracted by re-mashing the spent grains from the high gravity brews. Using this liquid for subsequent mashes did not result in the extraction of more polypeptides indicating that there is a saturation point for this process (Cooper et al., 1998b). Further work (Cooper et al., 2000) showed that foam polypeptides were lost during fermentation due to the action of yeast proteinases (particularly proteinase A), the losses in high gravity brewed beers being greater due to the higher pitching rates used.

Work in this area was continued by Brey et al. (2002). It was found that the most important loss of foam polypeptides took place during the formation of hot and cold break, with much greater loss during high gravity brewing due to the larger amounts of break formed. Losses during fermentation due to proteinase activity were of importance, but only accounted for 7% of the foam protein lost during high gravity brewing. It was found that proteinase A does not actually degrade hydrophobic polypeptides but modifies them in some way which stops them being foam active. Fortunately, ~20% of hydrophobic polypeptides and 57% of LTP1 appear to be resistant to proteinase A (Brey et al., 2003). One way of reducing the loss of hydrophobic polypeptides in high gravity brewing was found. This involves the use of copper finings, this reduces the amount of polyphenol in hot wort, resulting in less precipitation of proteins by the remaining polyphenols. It was shown that this method resulted in higher levels of foam polypeptides surviving into the finished beer (Brey, 2003).

This area has also been investigated by He et al. (2006), the authors examined the levels of foam protein during storage and found that in unpasteurized beer, foam stability all but disappeared after 90 days. Analysis of proteinase A levels and levels of proteins by SDS-PAGE showed that proteins, particularly LTP1 (but not protein Z), were degraded during this period. This did not occur in pasteurized beer. This disputes the robustness of LTP1 described by Brey et al. (2002).

Summary of polypeptides

Reviewing the papers related to beer polypeptides shows that a limited range has been identified and that there is little consensus as to their functions. It is helpful to list the various sizes identified by the various authors (Table 4.4).

The amino acid data indicates that beer polypeptides have distinctive compositions. The polypeptides antigen 1/protein Z and LTP1 have been identified and linked to precursors in barley and survive malting and mashing. Polypeptides linked to haze appear to have high levels of glutamic acid, proline, arginine and glycine with low levels of the others and appear to originate from barley hordeins. LTP1 has been definitely linked to foam stability. Some authors have linked protein Z to foam stability too, but this is less certain.

Table 4.4
Summary of polypeptides isolated from beer by various authors

Molecular size (kD)	Author(s)	Comments
1000	Asano and Hashimoto (1980)	Only 21% protein, no function given
400	Asano and Hashimoto (1980)	Only 21% protein, no function given
100	Douma et al. (1997)	No function?
90	Asano and Hashimoto (1980)	Only 21% protein
	Hollemans and Tonies (1989)	Could not isolate
66	Mohan et al. (1992)	No function given
41	Ishibashi et al. (1996)	Barley foam
40	Hejgaard and Kaersgaard (1983)	Antigen 1/protein Z
	Ishibashi et al. (1996)	Barley haze
36	Mohan et al. (1992)	Foam?
29	Sheehan and Skerritt (1997)	No function given
23	Sheehan and Skerritt (1997)	No function given
20	Kano and Kamimura (1993)	Foam?
	Mohan et al. (1992)	Foam?
19	Asano et al. (1982)	Hordein origin and haze
18	Douma et al. (1997)	Foam?
17	Vaag et al. (1997)	Foam?
16	Asano et al. (1982)	Hordein, albumin or globulin
15	Hollemans and Tonies (1989)	Foam?
14	Mohan et al. (1992)	Foam?
12	Yokoi and Tsugita (1988)	Faint
	Lusk et al. (1995)	No function
	Evans et al. (2003)	Haze
10	Yokoi and Tsugita (1988)	No foaming ability
	Sorensen et al. (1993)	LTP1
8	Yokoi and Tsugita (1988)	Degradation of 10 kD

The range 14–19 kD linked to haze by Outtrup (1989) The range 6–19 kD linked to foam by Sorensen et al. (1993) The range 60–100 kD has been linked to yeast polypeptides by Mohan et al. (1992) The range 15–32 kD heavily glycosylated has been linked to haze by Leiper et al. (2003a)

Polyphenols

Polyphenols in beer originate from barley and hops. Polyphenols are ubiquitous in higher plants where they have been associated with signal and defensive roles (Doner et al., 1993). The structure of phenolics is based on phenol (mono-hydroxylated benzene), and the term "polyphenol" covers all molecules with two or more phenol rings (Bamforth, 1999).

Polyphenols have been identified as having at least two roles in beer. Firstly certain species are known to be involved in causing colloidal instability through binding with proteins, and secondly as being involved in flavor stability by protecting against oxidation (Whittle et al., 1999).

Beer contains around 100–300 mg/l polyphenol (McMurrough and O'Rourke, 1997). This is also referred to as tannin, although not all polyphenols have the ability to tan animal skins (Hough et al., 1982). Two types are present in beer, the first group are derivatives of hydroxybenzoic and hydroxycinnamic acids, the second group are the flavanols and their derivatives (Hough et al., 1982). Flavanols account for around 10% of total beer polyphenols, and this group includes the species related to colloidal instability.

Flavanoids (oligomers of flavanols) all have the same basic structure of two six carbon rings linked by a three carbon unit. They are often hydroxylated to varying degrees and these groups are sometimes glycosylated or methylated (Doner et al., 1993). Flavanols found in beer are catechin, epicatechin, gallocatechin and epigallocatechin, the structures of these are shown in Figure 4.2. These

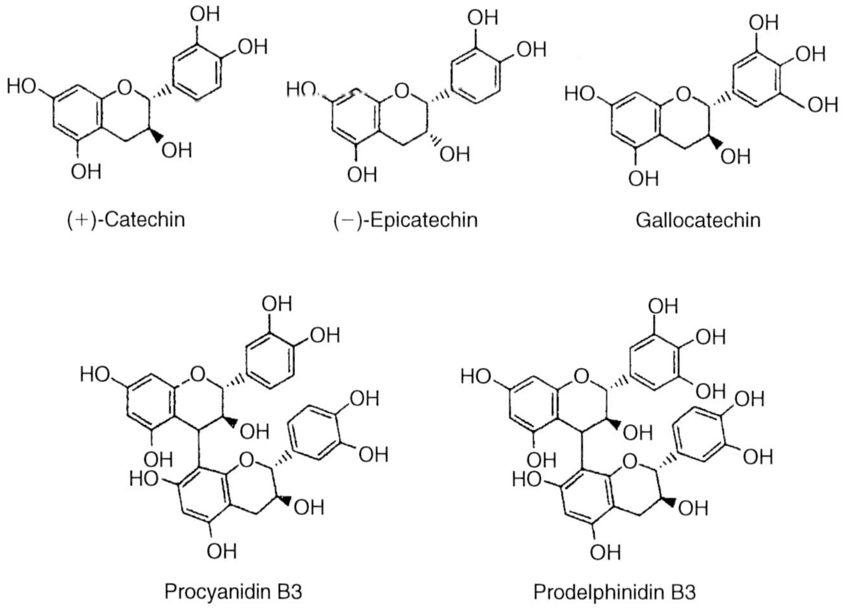

Figure 4.2
Structures of the main beer flavanol monomers and dimers (Siebert and Lynn, 1998).

can exist on their own as monomers, but are more commonly joined together to form flavanoids as dimers, trimers and larger polymers. These are classified as being "simple" or "complex" according to chromatographic mobility. Monomers, dimers and trimers are classed as simple flavanoids.

Methods of detecting polyphenols

Total polyphenols in beer can be measured by reaction with ferric ions in an alkaline solution to produce a red/brown color that is measured spectrophotometrically. Polyphenols can also be measured by the Folin-Ciocalteu method by reaction with a molybdotungstate reagent to give a blue color that is measured spectrophotometrically. Flavanoids can be measured spectrophotometrically following reaction with p-dimethylaminocinnamaldehyde. Haze-active polyphenols, or "tannoids," can be measured by precipitation with PVP (polyvinylpyrrolidone) to produce a haze that can be measured turbidimetrically (Chapon, 1994).

These methods have been reviewed by Siebert and Lynn (2006). The authors found that the Total Polyphenol and the Folin-Ciocalteu methods were able to measure both haze-active and non-haze–active polyphenols satisfactorily. The Folin method was slightly affected by protein while the Total method was not. The Tannoid method was able to detect haze-active polyphenols with very little interference from protein.

Individual polyphenols can be identified using gas liquid chromatography (GLC) and HPLC. Methods involving HPLC-MS have been developed more recently (McMurrough et al., 1993).

Polyphenols in beer

Barley polyphenols are subject to great change during the brewing process. Whittle et al. (1999) investigated flavanoids in barley. They found and tentatively identified 56 different types ranging from dimers to pentamers. Little change occurs during malting, the main changes taking place during mashing and boiling. Extraction of flavanoids is complex, with the highest concentration being recovered at the start of the run-off, however much is lost through precipitation and sweet wort contains only 23% of the flavanols in the malt (McMurrough et al., 1993).

It is possible that some polyphenols may originate from adjuncts. Polyphenols are normally found in cereal husks so adjuncts such as maize grits would be unlikely to be a source. However whole cereals such as wheat are possible sources. Hough et al. (1982) have reported that wheat contains no haze-active polyphenols.

During the boil, polyphenols will precipitate with protein. The larger polymers are so reactive that they are soon lost (Siebert and Lynn, 1998). During the boil, barley polyphenols are joined by those from hops. Hops contain more polyphenol by weight, but considerably less hops are used than malted barley.

Hops contain flavanoids in the form of glycosides (joined to sugars) and these survive into the beer. Polyphenols account for 2–4% of the dry weight of hops with aroma hops having more than bittering hops. Hop varieties with no polyphenols are available. Only 20% of hop polyphenols survive boiling compared to 70% of malt polyphenols (Bamforth, 1999).

There is debate as to whether hop polyphenols are important. Chapon (1994) considers that they are of importance, Delcour et al. (1985) states that barley and hop polyphenols are of equal importance, Srogl et al. (1997) reports they have little effect in boiling, but remain in beer to cause haze and that barley polyphenols are less extractable in a buffer at pH 5 (wort pH) than hop polyphenols. Mikyška et al. (2002) found that both malt and hop polyphenols contributed positively to beer flavor stability.

After boiling more polyphenols are lost in cold break, and later in cold conditioning, although little is lost during fermentation. The flavanoids in beer consist of monomers, dimers and a few trimers at a level of 15 mg/l (McMurrough and O'Rourke, 1997). These are mostly catechin, epicatechin and complexed flavanoids. Whittle et al. (1999) found 24 such species in beer and these included those responsible for haze formation.

Two dimers have been particularly associated with haze formation, procyanidin B3 (catechin-catechin) and prodelphinidin B3 (gallocatechin-catechin), the structures of these are shown in Figure 4.2. These are referred to as proanthocyanidins (also known to brewers as anthocyanogens) and originate from both barley and hops. These are present in beer at low levels and according to McMurrough and Baert (1994) these two dimers account for 32% of beer flavanoids, but only 3.3% of total beer polyphenols. This shows that a small fraction of total polyphenols are involved in instability. Monomers on their own do not appear to be involved in haze (Siebert, 1999), although catechin and epicatechin have been shown to produce haze in model systems (Siebert and Lynn, 1998).

The exact mechanism by which flavanoids bind to proteins and cause haze is uncertain. It has been proposed that the simple dimeric flavanols are too small to cause haze on their own and must polymerize into larger molecules (oligomers) before they are large enough to cause haze. It is thought that the flavanoids must first be oxidized before they can form polymers (McMurrough and O'Rourke, 1997). Oxidation can occur throughout the brewing process. This can be enzymatic during mashing, or non-enzymatic during boiling. The products can be either polymeric flavanoids or polyphenol–protein complexes. Oxygen radicals may be involved (Kaneda et al., 1990). Acetaldehyde has also been shown to react with polyphenols (Delcour and Dondeyne, 1982).

Protein binding is due to the number and position of the hydroxyl (OH) groups on the flavanoids' aromatic rings. Thus rings with only one OH group are almost inactive, those with two are more active, especially when they are adjacent (vicinal), rings with three groups are even more active (Siebert, 1997). In this way prodelphinidin B3 is more haze-active than procyanidin B3 as gallocatechin has three vicinal OH groups while catechin has two (Figure 4.2). Gallotannins, such as tannic acid, have eight or nine gallic acid units, each with three vicinal OH groups, connected to a glucose unit (Siebert and Lynn, 1998, 2000).

This has a strong ability to bind to proteins which is why it is used for fining in brewing. These OH groups mean that proanthocyanidins are highly susceptible to oxidation, and it is thought that the late oxidation of these surviving dimers creates the species that will bind to sensitive proteins to cause haze (McMurrough, 1995).

Chapon (1994) refers to haze-active polyphenols as "Tannoids." These are presumably similar to the dimeric proanthocyanidins described above, probably in their oxidized form, although not all tannoids are proanthocyanidins. Tannoids later convert to "Tannins" which are able to form haze, these presumably are similar to the flavanoid oligomers (McMurrough, 1995). Chapon (1994) reports tannoids being present in beer at 10–60 mg/l and that there are no tannins in beer. Chapon (1994) defines tannoids as polyphenols that precipitate with PVP and are thus haze-active.

Rehmanji et al. (1998) measured haze formation in untreated beer and found that the level of flavanoids and total polyphenols decreased while the level of tannoids and haze increased. However, McMurrough and O'Rourke (1997) found poor correlation between tannoids and beer stability due to Chapon's test being insensitive to smaller tannoids.

McMurrough and O'Rourke (1997) also removed and purified tannoids from beer and added them to beer stabilized with PVPP that is with the haze-causing polyphenols removed. This caused haze to form in proportion to the addition rate and ~60% of simple flavanoids added were converted to tannoids. However, more haze resulted than could have been caused by the amount added, so it would appear that haze formation depends on the oxidation of the simple flavanoids and the pre-existing levels of partly oxidized polyphenols (tannoids) in the beer. However Siebert and Lynn (1998) report that the degree of polymerization has more haze-causing effect than the number of hydroxyl groups.

There is disagreement about the other roles that polyphenols play in beer. For some time there has been a caveat about removing too much polyphenol. Polyphenols have been linked to mouthfeel, antioxidant properties against staling and nutritional value (Bamforth, 1999). More recently beer has been related to health as it contains a worthwhile amount of polyphenols which could act as antioxidants. This has been recognized in red wine for many years (Walker et al., 2001).

McMurrough et al. (1996) investigated the possible flavor stabilization role of flavanoid polyphenols. They removed oxidizable polyphenols (with PVPP) which increased the beer's shelf-life, but also lowered the beer's reducing capacity by around 20%. This however had no effect on the beer's flavor stability.

Siebert and Lynn (1997) states that polyphenols might be important for flavor and as antioxidants. It has been reported that stabilization by removal of polyphenols did not contribute to staling as there was enough reducing power remaining, the role of flavanoids in beer is minimal as they only act negatively by forming tannins and giving unpleasant astringent flavors.

Whittle et al. (1999) states that polyphenols are important as they influence the redox state during malting and brewing by removing free radicals. These are believed to be formed via the Fenton Cycle with the reaction of oxygen and transition metals (particularly iron and copper). Polyphenols prevent this as they are hydroxy and hydroperoxide scavengers.

Barleys with low levels of anthocyanogens have been produced and have been shown to be beneficial to colloidal stability, but have other properties which have discouraged their use. Although free of anthocyanogens, there are still other polyphenols present (Delcour et al., 1985). Erdal (1986) investigated brewing with proanthocyanidin free barley. Proanthocyanidins can be quite easily removed from barley as the plant does not require them for growth. The beer produced was more stable than that produced from normal malt, even after stabilization. Other findings included that the grains take up water faster during steeping, worts produced are darker, contain more soluble nitrogen and have 1–2% less fermentability. As barley polyphenols are found in the husk, it is possible to remove the husk with alkaline steeping (Dadic et al., 1976). Winter varieties of barley have more polyphenols than Spring varieties.

In conclusion, it would appear that the polyphenols responsible for causing haze consist of dimers of flavanols which oxidize and polymerize into tannins which can then bind with protein. These precursors account for a very small proportion of beer polyphenols, and their removal does not damage the desirable affects of other polyphenol species.

Haze-forming reactions between polypeptides and polyphenols

It has been known for many years that the most frequent cause of colloidal haze in beer is the combination of polypeptides and polyphenols. The polypeptide component consists of proline-rich fragments originating from malt hordein and the polyphenol component consists of proanthocyanidin dimers which undergo some form of activation before becoming haze-active. Both these constituents form a very small part of the total beer protein and polyphenol content. Free amino acids do not appear to bind to polyphenols (Siebert et al., 1996a). In forming colloidal haze, polypeptides and polyphenols combine to form soluble complexes which then grow to colloidal size at which point they are able to scatter light, these complexes are visible and will form a sediment (Siebert et al., 1996b).

Initially a chill haze will form in beer. This becomes visible when the beer is cold (below 0°C) but will disappear when the beer is warmed to room temperature. Chill haze is formed from polypeptides and polyphenols, but unlike permanent haze (which is insoluble) remains a soluble complex, binding being with non-covalent bonds (Siebert et al., 1996b). It will precipitate at low temperatures and forms particles of 0.1–1 µm diameter. This will re-dissolve quickly if the temperature rises. Chill haze can be removed at the filter before packaging as long as the beer is kept cold.

Permanent haze forms initially in the same manner as chill haze. However the soluble complexes soon convert to insoluble complexes with covalent bonds which will not dissolve if heated. The particle size is larger than chill haze, ranging from 1 to 10 µm diameter. The reactions are catalyzed by oxygen and metal ions and the amount of haze increases over time (O'Rourke, 1994).

The reaction is closely related to the amounts of haze-active polypeptide and polyphenol present (McMurrough et al., 1992).

Haze development has two stages. At first no visible haze forms, then the level increases in a linear pattern (Siebert, 1997). Chapon (1994) states that chill haze evolves into permanent haze, and proposed the following equilibrium, where P is protein and T is tannoid.

$$\underset{\text{soluble}}{P + T} \Leftrightarrow \underset{\text{soluble}}{P.T} \rightarrow \underset{\text{insoluble}}{P.T}$$

The haze precursors are on the left hand side, they can freely move into the loosely associated chill haze state in the center and back again, however if the chill haze converts to the permanent haze state on the right, there can be no return to a soluble state.

According to Siebert (1999) the two stages could be due to the protein-polyphenol complexes being initially too small to scatter light, or by there being some kind of reaction which has to take place with the precursors before haze can form. Siebert (1999) has proposed two pathways, firstly simple flavanoids polymerize to form tannins which then combine with protein and grow to produce haze. Secondly, already complex phenolics are oxidized, these active phenols then combine with protein and grow to produce haze. The second pathway is supported by the finding that labeled oxygen in beer headspace was found incorporated in beer haze (Siebert, 1997).

Siebert et al. (1996b) proposed a model of haze formation. Each polyphenol has two binding sites and each polypeptide has three binding sites. The greatest amount of haze will occur when the number of polyphenol binding sites is equal to the number of polypeptide binding sites. If there were more polypeptides than polyphenols, as is the case in beer (Siebert et al., 1996a), two polypeptides could be joined but it would be unlikely that any further binding could occur due to the shortage of polyphenols. If there was more polyphenol and polypeptide, all the polypeptide binding sites would be filled with polyphenols, but again there would be little chance of further binding due to the shortage of protein.

Siebert and Lynn (1997) report that it is certain that proline is involved in polypeptide and polyphenol binding. They incubated catechin with poly-glutamine and polyproline, but only the latter produced haze. Siebert et al. (1996) reported that proline's structure is important in binding as the closely related structure of polyhydroxyproline produced no haze when exposed to catechin. As more polyphenol binding sites on the protein are exposed by heating, this suggests that hydrophobic bonding is involved in polypeptide-polyphenol reactions.

Rehmanji et al. (1998) have produced a model of haze formation. It proposes that simple flavanoids do not produce haze when they bind to proteins as they are too small. However, oxidized flavanoid oligomers can produce chill haze and this converts to permanent haze when the flavanoids convert to fully oxidized tannoids. Collected haze from beer often contains high levels of carbohydrate, sometimes up to 80%. Stabilization treatments however, usually remove either the polypeptide or polyphenol fraction. This would indicate that the carbohydrate is not directly involved in haze formation, but is merely associated or co-precipitated (Siebert et al., 1996a).

Outtrup et al. (1987) experimented with both dimeric and trimeric proanthocyanidins and synthetic polyproline, they found that the trimerics reacted more strongly. Polyproline reacted strongly with all the proanthocyanidins tested. Reducing the proline content of the polypeptides lowered the level of precipitation and the resulting complexes were less stable. They suggested that the binding sites are the o-hydroxyl groups in the polyphenols and the nitrogen-carbonyl arrangement in proline regions of polypeptides. Binding increases the hydrophobicity of the complex as hydrogen bonds shield the hydrophilic sites on both reactants.

Stabilization treatments

Over the years many different types of beer stabilization treatments have been developed. The more successful of these have assisted greatly in prolonging colloidal stability as well has having the added benefit of shortening maturation times (McMurrough et al., 1997). A secondary benefit is that stabilization helps to reduce fouling of vessels and pipework with contaminating solids (Siebert et al., 1996). With a few exceptions, all the methods involve adding a material to the beer which binds to the haze precursor. The treatment-precursor complex is then removed from the beer by filtration or settlement.

Whatever the treatment material chosen it is important to use a suitable loading of the stabilizer as using too much is wasteful. The treatment must also be specific to whatever substance is being removed so as to avoid removing other potentially beneficial substances (Siebert and Lynn, 1997). Most treatments remove either protein or polyphenol haze precursors, or in some cases both. Whatever treatment is used, it is not normally necessary to remove all of the precursors, just enough to ensure stability.

It is important to keep the beer at a low temperature as any binding that takes place between the treatment surface and the haze-causing particle tends to be weak and will dissociate if the beer is allowed to warm up. Also, it is important to maintain as low a load as possible on the filters to extend their lives. The proper use of copper finings, settling in cold conditioning, concentration of filter aids (they will also remove haze precursors), dosing rate of stabilizers and oxygen levels will all optimize filtration.

Stabilizers have advanced from early attempts using wood chips, saw dust, aluminium dust and glues to more sophisticated products (Basarova, 1990). Most techniques currently employed either act on proteins or polyphenols, with few acting on polysaccharides or metal ions. The treatments currently in use are discussed in detail below.

Silica

Silica gels are adsorbents that stabilize beer by removing those proteins which are responsible for causing haze. Silica has a highly porous structure and proteins bind to and are held by the internal surface of the pores. The starting point in their production is the manufacture of sodium silicate. There are two methods available, the first involves fusing silica sand and soda ash in a furnace at a temperature of 1200–1500°C to produce a glass. The molten glass is dissolved in water to produce sodium silicate (waterglass). The other method involves fusing the sand and soda ash in an autoclave, referred to as the hydrothermal method.

The prepared sodium silicate ($SiO_2:Na_2O$) is reacted with a mineral acid such as sulfuric acid, this condenses the orthosilicic acids (Si–(OH)$_4$) to form silica (Mitchell, 1966). This takes place at a pH of 0 and forms a hydrosol-containing polysilicic acids. The particles formed fall within the 15–20 Angstrom range, this being affected by silica concentration, electrolyte level, pH and temperature (Fernyhough et al., 1994).

The hydrosol sets to form a gel, although the structure contains water (65% w/w) this is held within the highly porous structure and the material behaves as a dry non-dusty powder which can be milled to the required particle size. Alternatively, the hydrogel may be aged under controlled temperature conditions followed by washing and drying. This produces a xerogel with a low-water content (~5%), this dry dusty product can then be milled to the desired size. Milling xerogels results in a large amount of fine particles which must be removed or they will cause filtration problems (Fernyhough et al., 1994).

The ageing process is important as this changes the silica's structure, reducing surface area and increasing pore diameter. For silica, surface area, pore diameter and pore volume are crucial to protein adsorption. The magnitude of surface area does not necessarily indicate a silica's effectiveness as some of the area may lie inside pores that are too small for protein to enter.

Characterization of silica structure is generally made using nitrogen adsorption. This provides information on surface area, pore volume and pore size. In brewing applications, nitrogen is much smaller than protein but these values are conventionally used to compare silica physical structures. Pore size measured by nitrogen adsorption can however, usually be correlated with protein uptake. The protein adsorption capacity of a silica is determined by particle size, surface area, porosity and the number of free OH groups, but the specificity is influenced by the pore size. Silica gels have very large surface areas, up to around $800 \, m^2/g$. A diagram illustrating silica surface chemistry and pore structure is shown in Figure 4.3.

Another important property of silica is permeability or filterability. Silica can be used during cold conditioning and removed by settling, if so a dense hydrogel would be appropriate. If however the silica is to be removed from beer by filtration it is important that it can be removed easily. If silica has too many small particles these can fill in the gaps between the larger particles at the filter and cause blinding. However it has been found that silicas with lower permeability give better stability. Nock (1997) investigated this problem and concluded that the beneficial effect of reduced permeability was due to the trapping of haze particles in the filter bed that could otherwise have passed through a bed consisting of a high permeability silica.

The surface of silica exists as siloxane groups (SiOSi), but these "residual valences" are unstable and react with water so that the surface is covered

(a) Silanol groups at silica surface

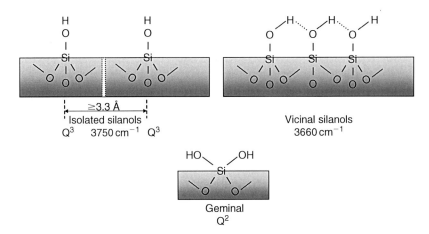

(b) Silica surface chemistry

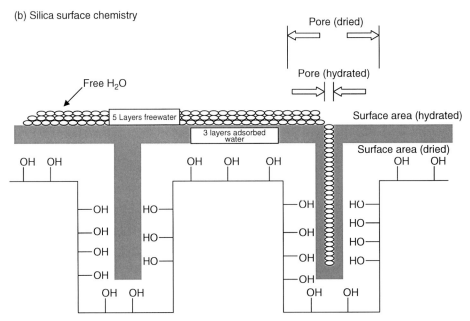

Figure 4.3
Silica structure (two parts)
(a) Silanol groups at silica surface
(b) Silica surface chemistry.

with silanol (SiOH) groups. The adsorption of proteins by silica is complex, hydrogen bonding with OH, NH and CO groups, ionic bonding with quaternary ammonium ions and hydrophobic bonding with hydrophobic regions of proteins are all involved, with hydrogen bonding being the most important (Fernyhough et al., 1994). Protein binding depends on pH. In beer which has a pH of 3.8–4.7, proteins which have isoelectric points (and are thus less soluble at this pH) between 3 and 5 can be adsorbed (Fernyhough et al., 1994). It would appear that proteins bind in a monolayer (Nock, 1997). Silicas tend to have far more available silanol groups than required for complete haze protein uptake and only 10–15% are occupied during stabilization (McKeown and Earl, 2000).

Siebert and Lynn (1997) found that silica could bind to polyproline but not to polyglutamine. This indicated that silica binds to proline in the same way as polyphenols that is via the silanol OH groups on silica instead of the aromatic ring OH groups on polyphenols. Kano and Kamimura (1993) found that one of the silicas they tested was able to remove polyphenols, but possibly only at the filter.

Leiper et al. (2003a) tested the haze removing properties of four types of silica xerogels. All were able to remove haze, but two types were particularly successful, presumably because their pore sizes suited the sizes of the polypeptides. Haze material was recovered by saturating the silicas by repeated exposure to fresh beer, the amount of haze-causing or sensitive protein taken up in each cycle was measured. The total number of sensitive protein units contained in each silica sample at the end of saturation was found to be incompatible with the dry weight of material recovered from the silicas. This indicated that haze "protein" was not pure protein.

Leiper et al. (2003a) showed that haze material was heavily glycosylated with glucose, with the protein content ranging from 39 to 69%. The amount of protein matched the amount of sensitive protein taken up by the silicas to a greater extent. However, when the amount of proline was measured in the samples, the figures gave a much better match. Thus silicas bind to haze material according to the amount of proline present.

The amount of carbohydrate present in some of the samples led the authors to suspect that silica might be binding to that material as well as protein. To test this, a protein-free "beer" was prepared using dextrin, ethanol and water with the pH buffered to 4.5. This beer was treated with silica, but no carbohydrate was removed. Thus silica was shown to bind to protein only.

Leiper et al. (2003b) investigated beer foam proteins. The foaming fraction the authors isolated from beer was found to be very highly glycosylated with glucose and to contain very little proline. This indicates that silica treatment will have very little effect on beer foam.

It has been suggested that having the surface silanol groups of silica covered with a layer of water reduces a silica's performance by reducing its surface area and pore volume (Matsuzawa and Nagashima, 1990). To avoid this, another type of gel has been produced which has around 10% water in its structure (close to the level found in xerogels) but the surface area and pore volume values are closer to that of a hydrogel. This gel still has a layer of surface bound water, but one water molecule is bound to two silanol groups, so there is in effect only half a layer of water present. This type of gel is referred to as vicinal hydrated gel (Matsuzawa and Nagashima, 1990).

In the brewery silica gels can be added to beer at various points in the process, and different silica types are suited to these applications. If silica is added during liquid transfer from the fermentation vessel to conditioning tank, high density hydrogels would be most suited to this purpose as they settle rapidly to form a compact sediment. This results in a contact time of more than 2 days which is actually longer than necessary (McKeown and Earl, 2000). Low permeability, but highly stabilizing silica xerogels are also suitable for this application as they are not removed by filtration, but are left on the tank bottom as the beer is drawn off (Guzman and Nock, 1996).

Silica can be added just prior to filtration where the contact time required is very short. The beer must be cold (less than 0°C), with many breweries now using high gravity brewing, this can be carried out even colder (due to higher levels of alcohol preventing freezing) resulting in greater stability. A range of silicas can be added depending on the shelf-life desired. Filtration grade silicas have been developed with permeabilities similar to that of filter aids. It is possible to use this silica type to replace some of the diatomaceous earth filter aid and thus extend the filter run (Guzman and Nock, 1997).

Wherever the silica is added it must first be slurried in cold deaerated water at a level of 10–25% w/w depending on the silica type. If it is to be used at the filter, it can be mixed with the filter aid. The slurry is added to the beer stream using a metering pump. The amount of silica required to ensure stability depends on the beer type, the desired shelf-life and if any other type of stabilization is being used. Application rates range from 20 to 120 g/hl depending on the silica type (McKeown and Earl, 2000).

Silica sol

Silica sols are manufactured in the way described above, but are not allowed to set and are used as aqueous solutions of silicic acid. Such solutions contain 15–50% w/w acid depending on type. When added to wort or beer, the sol cross-links to form a gel trapping any haze material. The gel is removed at filtration. Sols are mostly used for wort clarification after boiling, but can be used to stabilize beer too; they are marketed under the trade name Stabifix (Niemsch and Obermann, 2004).

Polyvinylpolypyrrolidine (PVPP)

PVPP is an adsorbent which binds to polyphenols. It selectively binds to those polyphenols responsible for chill and permanent haze without affecting beer quality. PVPP is a cross-linked polymer of vinylpyrrolidone, it is insoluble in beer and water, and must be kept well mixed in suspension to ensure efficiency. It has a high surface area due to its amorphous structure with values around 1.2 m^2/g (much lower than silica). PVPP has no apparent effect on haze-active polypeptides (McMurrough, 1995).

It is claimed that PVPP does not remove those polyphenols that are involved in resistance to staling (Gopal and Rehmanji, 2000). This is important as certain polyphenols play important roles in wort clarification, hot break formation, removal of oxygen radicals and beer flavor stability. If used at lower dosing rates, PVPP does seem to be selective in removing haze-active polyphenols only. Work by Mikyška et al. (2002) showed that beer stabilized with 50 g/hl PVPP retained good flavor stability.

There are two types available for beer treatment. The first is designed for single use, this is in the form of a micronized powder with a very large surface area. This is added to the beer stream, a contact time of 5–10 minutes being required, or it can be dosed into the cold conditioning tank. Around 10–40 g/hl should be used, less is required if it is being used in conjunction with silica gel.

The second type is designed to be regenerated. This form has a larger particle size to withstand repeated recycling. This means that it has a smaller surface area and will adsorb less polyphenol, however more can be added without extra cost as recycling is cost effective, generally 10–50 g/hl is used (Gopal and Rehmanji, 2000).

Both types of PVPP are slurried before use at around 8–12% w/vol and mixed for 15–60 minutes to allow hydration and swelling (Gopal and Rehmanji, 2000). Single use PVPP is dosed into the beer stream and removed by DE filtration. Regenerable PVPP is added to the beer after primary filtration and is removed at a separate filter. It is regenerated by washing with NaOH (1–2% w/vol) at 60–80°C for 15–30 minutes, then with dilute acid to restore the pH to 4. Regeneration losses as low as 0.5% have been claimed (Gopal and Rehmanji, 2000).

Regenerable PVPP is normally used as a sole treatment; however single use PVPP can be used in conjunction with other treatments. PVPP can be used with papain and tannic acid, but in this case the PVPP must be added later, or it will bind to the tannic acid before it has had time to operate. The most common combined treatment is with silica gel, and blended products are available (O'Reilly, 1994).

PVPP can be used for stabilization in the brewhouse. It is added to the kettle near the end of the boil where it binds to haze-active polyphenols. Blended products consisting of PVPP and kappa-carrageenan are available. Carrageenan is the main constituent of copper finings which are described later.

Rehmanji et al. (2002) reported a trial with the product Polyclar® Brewbrite, the authors claim that this product provides beer stability by removing tannoids from wort and also increases trub removal and shortens fermentation time. The beer produced showed good stability, but for a longer shelf-life, such beer would require further stabilization before packaging.

PVPP binds to polyphenols as it has a very similar structure to proline (Siebert and Lynn, 1997). Both have five member, saturated, nitrogen containing rings with amide bonds and no other functional groups. The structures of proline, polyproline and PVPP are shown in Figure 4.4. It is not certain whether PVPP binds to the same part of the polyphenol molecule to which polypeptides bind, or another part.

Binding to PVPP is very strong, only very aggressive chemical treatments can remove bound polyphenols (McMurrough and O'Rourke, 1997). It is thought that binding involves hydrogen bonding with phenolic OH groups, hydrophobic bonds and association between the aromatic ring of the polyphenol and the PVPP ring through the π-bonded complex, or a combination of these (Siebert and Lynn, 1998).

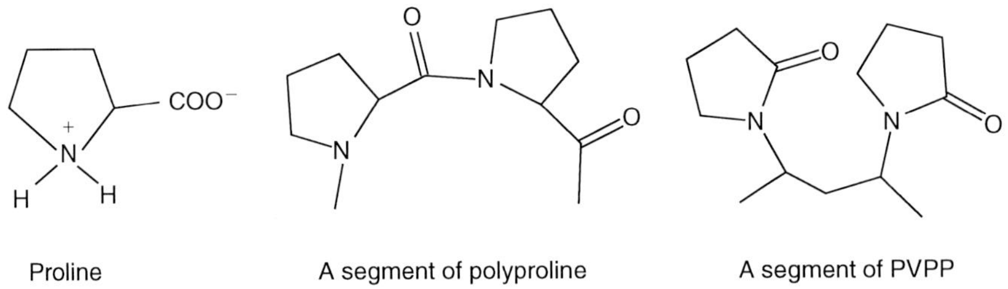

Figure 4.4
Similarities in structure between proline, polyproline and PVPP (partly Siebert and Lynn, 1998).

To obtain colloidal stability it has been found to be necessary to reduce flavanoid polyphenols and their oxidized products to less than 5 mg/l (McMurrough and O'Rourke, 1997). The removal of oxidizable polyphenols can increase beer shelf-life 20 times (McMurrough et al., 1996). At low dosage rates, PVPP preferentially removes highly hydroxylated polyphenols (Siebert and Lynn, 1998). Siebert and Lynn (1997) reported that PVPP can reduce the level of haze-active polyphenols to around half their level, but cannot reduce much further even at high loading. PVPP cannot operate when haze polypeptides are in excess due to the binding sites being blocked.

PVPP is known to remove prodelphinidin B3 and procyanidin B3, largely remove catechin and partly remove epicatechin (McMurrough, 1995). Whittle et al. (1999) reported that PVPP at 20 g/hl removed catechin, epicatechin, gallocatechin, epigallocatechin and proanthocyanidins.

Rehmanji et al. (2000) investigated the combined use of PVPP and silica xerogel with the combined product Polyclar Plus 730. This caused simultaneous removal of polypeptides, tannoids and tannoid precursors. The product had a large particle size which gave good filtration performance.

Lucilite TR

Lucilite TR is a recent addition to the beer stabilization arena as an alternative to PVPP. It is a PVP-modified silica gel consisting of amorphous silica coated with PVP which is strongly attached by hydrogen bonding. Unlike PVPP, it does not require to be pre-swelled. The surface chemistry of both these adsorbents and functional groups on the haze-forming tannoids assure selective adsorption of the haze-forming fractions.

PVPP and PVP-modified silica gel adsorption sites comprise the carbonyl groups of polyvinylpyrrolidone. These sites interact with the hydroxyl groups on the polyphenol molecule. The relative number of hydroxyl groups in the polyphenol molecule and the number of accessible adsorption sites on PVPP or PVP-modified silica gel determines selectivity by competitive adsorption. Thus polyphenols having high numbers of hydroxyls adsorb most readily and most strongly. Rehmanji et al. (1998) reported that PVPP is most effective at lower dosages because at these levels it preferentially adsorbs haze-active (tannoid) polyphenols. At higher levels, selectivity is poorer as non-haze–active polyphenols are adsorbed.

The PVP-modified silica gel has fewer adsorption sites than PVPP making it effective in adsorbing only the most haze-active polyphenols. Because PVPP and PVP-modified silica gel use the same polymer type to provide adsorption sites, it is not surprising that they adsorb essentially the same haze-active polyphenols. This product is used at similar dosage rates to PVPP, 5–40 g/hl, less if used with silica.

Due to its recent advent, there have been few publications covering Lucilite TR. Mitchell et al. (2005) reported a comparative study of the stabilizing properties of PVPP, silica xerogel and Lucilite TR, and claimed that PVPP is the more effective in removing polyphenol from beer.

Leiper et al. (2005) presented data comparing Lucilite TR with PVPP. This showed that both treatments were able to remove tannoid polyphenols from beer. PVPP was found to reduce the amount of total polyphenols by up to 25%. Lucilite TR did not have this effect, suggesting that it is more selective. The authors presented data from a range of brewery trials which showed that Lucilite TR was as effective as PVPP in reducing tannoid levels and ensuring long-term shelf-life of beer.

Isinglass finings

Isinglass consists of natural piscine collagen (98%) from tropical and subtropical fish swim bladders. The collagen is a triple helix bound with hydrogen bonds. It is rich in glycine, proline and hydroxyproline. It is a rod-shaped molecule of 300 kD molecular weight. The hydrogen bonds have to be broken by acid hydrolysis to produce soluble collagen. Finings work by accelerating particle settling by increasing their size and weight (AB Vickers Ltd., personal communication).

The prepared isinglass molecule is amphoteric, carrying both negative and positive charges. The overall charge is positive with some negative areas. In this way isinglass can bind to yeast cells (negatively charged) and proteins (positively charged). The principal bonds are ionic bonds to tannins, yeast cell walls and carrageenan, and hydrogen bonds to tannins, carrageenan and proteins. The optimum pH for isinglass is around pH 4.4, and thus is within the pH range of beer.

Various forms are available ranging from whole leaf to ready-to-use liquid. It must be kept cool or it will degrade to gelatin. It can be used by dosing in-line during transfer, or injecting into unitank systems after yeast removal and roused with CO_2, gentle stirring may help. The normal dosage rate is 2–6 g/hl (Taylor, 1988).

A 3–4000 hl tank of beer will be ready for filtration after 3–5 days, without treatment this would take weeks. Isinglass should not be present in finished beer. Beers treated with isinglass have been shown to exhibit improved head retention, it is thought that this is due to removal of head damaging lipid material (AB Vickers Ltd., personal communication).

The use of isinglass has been brought to wider attention in recent years due to concerns over food intolerance and allergenicity. Several countries have decided that the consumer should have more information about the substances present in foods and beverages and that potential allergens present in ingredients and process aids should be listed on food labels.

The EC enforced allergen labeling for food and drink from November 2005 and this included isinglass. The brewing industry considered this to be unfair as isinglass has been used for years without any intolerance problems being reported, and was successful in obtaining a temporary 2 year exemption from the regulations. It is now the responsibility of brewers and suppliers to prove that isinglass is not allergenic so that a permanent exemption can be obtained. Hopefully this situation will be resolved in the near future (Chlup et al., 2006).

Concern over the potential allergenicity of isinglass has stimulated interest in the development of methods for detecting isinglass in small-pack, kegged or cask-conditioned beers. Two methods have been published recently. Chlup et al. (2006) have developed a method where beer is hydrolyzed with acid followed by HPLC analysis. The objective being the detection of the amino acid hydroxyproline, this is a major component of isinglass, but is most unlikely to be found in other beer components. The presence of hydroxyproline thus indicates the presence of isinglass. Baxter et al. (2007) have developed an alternative method which concentrates any isinglass molecules from beer using polyclonal antibodies, these complexes are precipitated from the beer and are analyzed for hydroxyproline using GC-MS. Both authors found that levels of isinglass in packaged beers were very low, most being removed in the brewery or remaining in cask bottoms.

Copper or kettle finings

Copper finings are Carregeenan gels from the seaweed *Euchema cottonii*. They consist of a linear sulfated polysaccharide containing galactose and anhydrogalactose. Of the various forms, the kappa type is the most effective.

In the presence of calcium or potassium ions, copper finings will remove proteins, polypeptides and possibly polyphenols. This is caused by binding to the sulfate groups on the galactose units. They are added towards the end of the boil at 2–5 g/hl at a pH of above 5.2. Used at an optimum dose, copper finings give good wort clarity with a compact stable sediment (AB Vickers Ltd., personal communication). They do not help the clarity of hot wort, where they are solubilized but will improve the clarity of cold wort by helping to precipitate trub. High gravity worts require more finings than normal gravity worts. Filtered beers that have been previously fined with either carrageenan or isinglass have lower hazes than unfined beers. This is due to the finings either increasing the size of haze particles and removing them, or by removing the precursors.

Papain

Papain was first used as a stabilizing treatment in brewing in 1911 when it was used against "undesirable albuminoids" (Esnault, 1995) and was a popular treatment for many years. It consists of a mixture of proteinase enzymes obtained from the latex of the *Carica papaya* fruit and works by hydrolyzing beer proteins. However this activity is not specific to haze-causing proteins and is thus liable to damage foam stability. It is added on transfer to maturation or to maturation itself at around 2 g/hl. It might be possible to use a foam stabilizer to counteract the problem with foam.

Clarex

Brewers Clarex is a recent product from DSM food specialities. It is a proteinase preparation, but unlike papain it is specific to haze-causing polypeptides rich in proline. The product consists of a proline-specific proteinase derived from the fungus *Aspergillus niger*. Development of this product has been reported by Lopez and Edens (2005). The enzyme was found to be effective in model systems and was tested in pilot-scale trials at rates of 1.25–2.50 U/kg malt, being added to the fermentation vessel with the wort. The beers produced showed good haze stability following 6 months storage at room temperature, with haze levels below 1.5°EBC. The authors reported that the product had very little effect on foam stability. Beer produced had higher levels of polyphenols due to the absence of haze-active proteins which would otherwise bind to them, but this had the benefit of increasing the beers' reducing power.

Tannic acid

Tannic acid is a precipitative stabilizer extracted from gall nuts. With the chemical formula $C_{76}H_{52}O_{46}$ it has a structure similar to that of other polyphenols. It works by making the ratio of polyphenols to proteins more even (Siebert and Lynn, 1997). It is used as a specific precipitant of haze-active protein at a loading of 2–6 g/hl. As it forms a precipitate it is important to be careful when transferring from a vessel since haze-active material could be mixed with the beer. A centrifuge may be used to reduce the possibility of transfer.

The main type of tannic acid employed is gallotannin, marketed as "Brewtan," which can be used as a wort treatment following boiling as well as in maturation or added in line to the filter or centrifuge. It removes some metals (Fe, Al, Zn and Pb) and some polyphenols when they are bound to proteins (Mussche, 1994). It binds to amino acids with exposed SH groups, and it has been claimed to be effective against flavor staling and light instability (O'Rourke, 1994). It is not supposed to damage foam (Siebert and Lynn, 1997), but Ishibashi et al. (1996) found that tannic acid removed foam protein from beer as well as haze. Its use with PVPP has been reported (Esnault, 1995), and it may also act as an antioxidant.

Combined stabilization system (CSS)

This recent process is unusual as it does not involve the addition of any stabilizing substances to beer. It consists of a chamber filled with agarose adsorbent. Beer is passed through the vessel where haze-active proteins and polyphenols bind to the agarose. Not all the beer needs to pass through the chamber, just enough to ensure stability. As filtration proceeds, the agarose bed becomes more loaded with haze precursors, so the proportion of the beer passing through the chamber can be increased to ensure that enough material is removed. The agarose can be regenerated using salt and caustic. The manufacturers (Handmann Filtration, Biberach, Germany) claim that this can be done several hundreds of times without loss of performance (Jany and Katzke, 2002).

A CSS unit has been installed at a brewery in Australia. Initial trials have indicated that the system removes haze precursors, particularly the polyphenols prodelphinidin B3 and procyanidin B3, as well as some sensitive proteins without affecting foam stability (Taylor et al., 2006).

Other treatments

Some of these treatments are not currently used extensively, but have been included for completeness.

Bentonite is a protein adsorbent consisting of alumino silicate, and is also used to stabilize wine and fruit juice (Power and Ryder, 1989–1991). It has been found not to discriminate between foam and haze-causing proteins and thus damages foam stability (Siebert and Lynn, 1997). It also has the disadvantage of swelling when wet (Basarova, 1990) and can require 24 hours for complete hydration (Power and Ryder, 1989–1991).

Gelatin is commonly used in the wine and juice industries to remove polyphenols. Siebert and Lynn (1997) found that gelatin removed haze proteins but did not damage foam proteins, causing the unusual situation of one protein removing another, it probably involves sedimentation or cross-linking.

Nylon, perlon and polyamides have been used to remove polyphenols by adsorption. They are effective, but damage bitterness, foam and color (Basarova, 1990). Formaldehyde has been used as a precipitant of polyphenols, but is now forbidden in many countries.

Antioxidants such as ascorbic acid and glucose oxidase have been used to reduce free radicals (Basarova, 1990). The use of hydrogen peroxide or polyphenol oxidase has been proposed to oxidize polyphenols to aid their precipitation but this has not been commercialized (Power and Ryder, 1989–1991). The use of copper sulfate as a sulfur scavenger has been reported (Bamforth, 1999). The use of EDTA to remove ions has been reported (Basarova, 1990).

Carbonyl-bisulfite adducts are complexes of sulfite produced by yeast during fermentation with aldehydes. They inhibit free radical reactions in beer such as those involving isohumulones and polyphenols and are thus good for haze stability (Kaneda et al., 1994). Basarova (1990) reports the use of high pressure to speed up the precipitation of polyphenols and proteins before filtration.

Other non-biological hazes

These can result from substances other than protein or polyphenol originating from raw materials, yeast or process aids. This kind of haze usually develops before chill haze (Basarova, 1990). The discovery of this type of haze should alert the brewer to problems with yeast handling and beer filtration.

A common constituent of such haze is polysaccharide. Examples include glucans, mannan, starch and pentosan. Research by Wenn et al. (1989) on "invisible" or "pseudo" haze – detectable by haze meters but not by the human eye – found these to be composed of α-glucan. The presence of β-glucan in beer can cause filtration problems as well as haze formation. A solution is to use β-glucanase, but this is forbidden in some countries. Speers et al. (2003) have investigated this subject, the authors added β-glucan to wort and beer that had had their β-glucan removed. Haze was found to increase with increasing molecular weight and concentration of the added material. Shear stress, pH, maltose and ethanol also influenced haze formation. Haze caused by high molecular weight β-glucan could not be reduced by filtration or cold-conditioning.

Lewis and Poerwantaro (1991) investigated haze material from yeast cell walls. This was found to be composed of glucan, mannan and cell wall enzymes and was caused by agitation, the age of the yeast cells and the pH of the agitation medium.

The use of centrifuges in breweries has increased in recent years due to the incentive to reduce processing times. Centrifuges can be used to remove hot and cold break from wort and to remove yeast from green beer or from beer before filtration. Centrifugation of yeast causes stress to the cells, this area has been investigated by several authors.

Stoupis et al. (2003) found that yeast that had been subjected to vigorous agitation produced beer with poor foam stability and high levels of haze. It was found that the foam stability was damaged by the action of proteinases released from the yeast on foam polypeptides and the haze by release of mannan and β-glucan from the yeast cell walls.

This work has been continued by Chlup et al. (2007a,b) who investigated the effects on yeast of successive cycles of centrifugation using a disc stack centrifuge. Levels of mannan in the supernatant were found to increase with the number of cycles. Beers produced with centrifuged yeast showed lower foam stability than beer produced with uncentrifuged yeast.

Mannan can be detected by acid hydrolysis followed by HPLC analysis of mannose. This sugar is not found in brewing raw materials, so its presence in a sample indicates a yeast source and yeast-handling procedures should be reviewed if mannose is found in beer. This method is described by Stoupis et al. (2003).

An alternative method has been developed by Chlup et al. (2007a,b). Here beer is centrifuged to remove yeast and the supernatant analyzed for the presence of mannan by staining with the dye Concanavalin A. The dye is detected, and the amount of mannan quantified, using flow cytometry. This technique involves simultaneous measurement and analysis of multiple physical characteristics of single particles as they pass through a beam of light.

Starch can cause haze if there has been poor mash conversion in the brewhouse. Pentosans can originate from barley and wheat, mostly in the form of arabinoxylan which can cause filtration problems (Leclercq et al., 1999). Malcorps et al. (2001) have reported an invisible haze composed of glycogen which originated from yeast, although it is not clear why yeast would release glycogen in sufficient quantity to produce a haze.

Other than polysaccharides, other substances from raw materials that can cause haze include oxalate or "beer stone." Oxalate forms if malt oxalic acid is not precipitated in the brewhouse which will happen if calcium levels are low, it forms distinctive square crystals. Excess metals have also been found to cause haze (Basarova, 1990).

Large particles sometimes occur in beer, these include skins, flakes, strings and granules. They have been investigated by Walters et al. (1996) and found to consist of protein and dextrins composed of five and six carbon sugars. Skins are believed to form on the surface of collapsed foam bubbles, they then break apart to give strings or combine to form irregular masses. They suggest that the precursors for these particles are already present in the beer and are not formed by heating. Only a very small fraction of beer carbohydrates and proteins are involved in this type of haze. In contrast, Wenn et al. (1989) identified protein in beer that did produce large particles when heated.

Haze can occur if beer is frozen by accident (beer is generally assumed to freeze at below $-2°C$) and consists of β-glucan, oxalate and dextrins (Skinner et al., 1993). Tanaka and Sakuma (1999) reported that this type of haze cannot be predicted by normal methods. They used a freeze and thaw test and found it to be related to the concentration of β-glucan. This type of haze is also confusingly referred to as colloidal haze (Basarova, 1990).

The last type of non-biological haze is that resulting from process aids. Silica, from silicon oxide or silicates leached from stabilizers, filter aids or raw materials has been identified as a potential cause of haze (Basarova, 1990). Sharpe and Channon (1987) reported that can lubricants may cause haze, the active chemicals being butyl stearate and palmitate.

It has been reported that the foam stabilizer PGA can degrade if heated and cause haze (Outtrup, 1991). However, O'Reilly (1996) claimed that it does not contribute to haze. Parker (2007) investigated haze formation and found that haze formed in fresh beer due to pasteurization, but that this haze disappeared over time. Beers treated with PGA were especially susceptible to this phenomenon. Haze levels increased again after 6 months storage due to the normal formation of colloidal haze.

Testing the effectiveness of beer stabilization

It is important to know if the treatment(s) being used are providing the stability desired. A range of tests has been developed to assess the effectiveness of treatments. In general, there are two types of test available-forcing tests and precipitative tests. Forcing tests involve storing beer at elevated temperatures in an attempt to speed up the natural ageing process that would occur if the beer were stored at room temperature. Thus many months of storage can be shortened to a few days. Such tests are also known as accelerated ageing. Different brewing companies have their own preferred temperatures and incubation times with haze being measured before and after incubation. These procedures do not generally correlate with the results of actual shelf-life (Chapon, 1994), but are widely used as an indicator of beer stability or as a way of comparing different stabilizing products or dosing rates.

Precipitative tests can be used to measure the stability of beers before and after treatment. Normally a substance is added to a sample which causes a haze to form which is then measured using a turbidimeter. This can be carried out manually or by an automatic machine, the Tannometer, produced by Pfeuffer. As with forced ageing, these methods may not give totally accurate results, but are useful for comparative purposes.

Chapon (1993) developed the alcohol chilling test. This involves adding alcohol to beer while lowering the temperature to below 0°C. This forces out chill haze. The test depends on the starting alcohol of the beer, and a good correlation to sensitive protein and tannoid levels has been reported.

The sensitive protein assay measures the amount of haze-active protein. This involves precipitation with tannic acid, and measuring the increase in haze. The result is not a quantitative statement of a defined substance, but a good indicator of a beer's haze forming potential (Chapon, 1993).

The saturated ammonium sulfate precipitation limit (SASPL) is another method for measuring unstable protein. Saturated ammonium sulfate (SAS) solution is added to beer until the haze value begins to increase, as more stable beer has less unstable protein in it, stable beer will require more ammonium sulfate to cause a haze. The result is expressed in ml of SAS/100 ml of sample.

The assay for tannoids involves precipitation with PVP. It is supposed to measure the polyphenol precursors of the protein–polyphenol haze complexes referred to by Chapon (1994) as tannoids. This test does not give good correlation with stability as only 10% of polyphenols are precipitated by PVP as tannoids (McMurrough et al., 1997). The test is generally used together with sensitive protein measurement to give an indication of beer stability.

There have been various assessments of these methods which have attempted to link the results with actual haze values from stored beer. Siebert et al. (2005) used the SASPL, sensitive protein and Chapon tests in comparison with force-aged beer that had been stabilized with silica gel. The authors found that the Chapon alcohol chilling test and sensitive protein tests gave good correlation with the forced haze results while the SASPL test did not.

McCarthy et al. (2005) compared the sensitive proteins, tannoids, SASPL, Chapon, simple flavanols and the Schneider PT instrument on silica-stabilized beers in comparison to the actual haze of beers stored at up to 110 days. The authors found that only the Chapon alcohol chilling test accurately predicted stability.

Berg et al. (2007) reported that the SASPL test gave poor correlation to room temperature ageing. The authors proposed a simplification of the method where a single dose of saturated ammonium sulfate is added before incubation and haze reading, this removes the problem of deciding at which precise point the haze of the sample starts to rise.

Haze identification

As well as normal colloidal haze, other haze problems occur from time to time in breweries. It is desirable to identify these and there is a range of methods available. These tend to be time consuming, but they do help to identify the source of the problem and allow action to be taken.

Haze material tends to occur in very low amounts, therefore in order to obtain sufficient material to examine, the beer should be centrifuged and the pellet retained. Depending on the amount of material recovered, the type of test can then be selected.

The easiest way of identifying haze material is to examine it under a microscope. In this way such items as yeast, bacteria, oxalate, diatomaceaous earth, cellulose fibers and isinglass finings can be identified due to their distinctive shapes. Staining with a dye will aid this process, for example the stain methylene blue will stain dead yeast (live yeast do not stain with this dye), protein and carbohydrate material and silica. Thionin will stain carbohydrate material purple, orange G will stain protein yellow and iodine will stain starch purple or red (Buckee, 1989).

There is a wide range of chemical tests that can be carried out if microscopy is not conclusive. Methods for proteins, polyphenols, mannan and isinglass have been described above. Sugars can be identified using HPLC, metals by AAS (atomic adsorbance spectroscopy) and lipids and hop components by GC.

References

Asano, K. and Hashimoto, N. (1980) Isolation and characterization of foaming proteins of beer. *Journal of the American Society of Brewing Chemists*, 38, 129–137.

Asano, K, Shinagawa, K and Hasimoto, N (1982) Characterization of haze-forming proteins of beer and their roles in chill haze formation. *Journal of the American Society of Brewing Chemists*, 40, 147–154.

Bamforth, C. W. (1985) The foaming properties of beer. *Journal of the Institute of Brewing*, 91, 370–383.

Bamforth, C. W. (1995) Foam: Method, myth or magic?. *The Brewer*, 81, 396–399.

Bamforth, C. W. (1999) Beer haze. *Journal of the American Society of Brewing Chemists*, 57, 81–90.

Basarova, G. (1990) The structure-function relationship of polymeric sorbents for colloidal stabilization of beer. *Food Structure*, 9, 175–194.

Baxter, E. D., Cooper, D., Fisher, G. M. and Muller, R. E. (2007) Analysis of isinglass residues in beer. *Journal of the Institute of Brewing*, 113, 130–134.

Bech, L. M., Vaag, P., Heinemann, B. and Breddam, K. (1995) Throughout the brewing process barley lipid transfer protein 1 (LTP1) is transformed into a more foam-promoting form. *Proceedings of the 20th Congress of the European Brewery Convention, Helsinki*, 561–568.

Berg, K. A., Ding, L. L. and Patterson, R. E. (2007) The dangers of the SASPL test in chillproofing evaluation. *Technical Quarterly of the Master Brewers Association of the Americas*, 44, 29–31.

Bishop, L. R. (1975) Haze- and foam-forming substances in beer. *Journal of the Institute of Brewing*, 81, 444–449.

Brey, S. E. (2003) The effect of kettle finings on cold break formation and hydrophobic polypeptides in fermenting wort. *ICBD Research Newsletter* (Autumn), 5–7.

Brey, S. E., Bryce, J. H. and Stewart, G. G. (2002) The loss of hydrophobic polypeptides during fermentation and conditioning of high gravity and low gravity brewed beers. *Journal of the Institute of Brewing*, 108, 424–433.

Brey, S. E., deCosta, S., Rogers, P. J., Bryce, J. H., Morris, P. C., Mitchell, W. J. and Stewart, G. G. (2003) The effect of proteinase a on foam-active polypeptides during high and low gravity fermentation. *Journal of the Institute of Brewing*, 109, 194–202.

Buckee, G. K. (revised by Boden, J.A.F.) (1989) Identification of hazes, turbidities and sediments in beer. BRFI.

Chandley, P. (1994) An improved method for the measurement of foam polypeptides in beer. *BRFI Quarterly* (October), 26–27.

Chapon, L. (1993) Nephelometry as a method for studying the relations between polyphenols and proteins. *Journal of the Institute of Brewing*, 99, 49–56.

Chapon, L. (1994) The mechanics of beer stabilization. *Brewers' Guardian*, 123, 46–50.

Chlup, P. H., Leiper, K. A. and Stewart, G. G. (2006) A method of detection for residual isinglass in filtered and cask-conditioned beers. *Journal of the Institute of Brewing*, 112, 3–8.

Chlup, P. H., Bernard, D. and Stewart, G. G. (2007a) The disc stack centrifuge and its impact on yeast and beer quality. *Journal of the Institute of Brewing*, 65, 29–37.

Chlup, P. H., Conery, J. and Stewart, G. G. (2007b) Detection of mannan from *Saccharomyces cerevisiae* by flow cytometry. *Journal of the Institute of Brewing*, 65, 151–156.

Cooper, D. J., Stewart, G. G. and Bryce, J. H. (1998a) Some reasons why high gravity brewing has a negative effect on head retention. *Journal of the Institute of Brewing*, 104, 83–87.

Cooper, D. J., Stewart, G. G. and Bryce, J. H. (1998b) Hydrophobic polypeptide extraction during high gravity mashing – experimental approaches for its improvement. *Journal of the Institute of Brewing*, 104, 283–287.

Cooper, D. J., Stewart, G. G. and Bryce, J. H. (2000) Yeast proteolytic activity during high and low gravity wort fermentations and its effect on head retention. *Journal of the Institute of Brewing*, 106, 197–201.

Curin, J., Chladek, L., Skach, J. and Suran, J. (1989) The effect of the mechanical stimulations on the colloidal stability of beer. *Proceedings of the 22nd Congress of the European Brewery Convention, Zurich*, 601–608.

Curioni, A., Pressi, G., Furegon, L. and Peruffo, A. D. B. (1995) Major proteins of beer and their precursors in barley: Electrophoretic and immunological studies. *Journal of Agricultural and Food Chemistry*, 43, 2620–2626.

Dadic, M., vanGheluwe, J. E. A. and Valyi, Z. (1976) Alkaline steeping and the stability of beer. *Journal of the Institute of Brewing*, 82, 273–276.

Dale, C. J. and Young, T. W. (1988) Fractionation of high molecular weight polypeptides from beer using two dimensional gel electrophoresis. *Journal of the Institute of Brewing*, 94, 28–32.

Dale, C. J., Young, T. W. and Brewer, S. (1989) Amino acid analysis of beer polypeptides. *Journal of the Institute of Brewing*, 95, 89–97.

Dealing with persistent hazes in freshly filtered beer. *Ferment*, 1997, 10, 186–190.

Delcour, J. A. and Dondeyne, P. (1982) The reactions between polyphenols and aldehydes and the influence of aldehyde on haze formation in beer. *Journal of the Institute of Brewing*, 88, 234–243.

Delcour, J. A., Schoeters, E. W., Dondeyne, P. and Moermann, E. (1984) The intrinsic influence of catechins and proanthocyanidins on beer haze formation. *Journal of the Institute of Brewing*, 90, 381–384.

Delcour, J. A., Schoeters, M. M., Meysman, E. W., Dondeyne, P., Schrevens, E. L., Wijnhoven, J. and Moerman, E. (1985) Flavour and haze stability differences due to hop tannins in all-malt pilsner beers brewed with proanthocyanidin-free malts. *Journal of the Institute of Brewing*, 91, 88–92.

Delvaux, F., Delvaux, F. R. and Delcour, J. A. (2000) Characterisation of the colloidal haze in commercial and pilot scale Belgian white beers. *Journal of the Institute of Brewing*, 106, 221–227.

Delvaux, F., Gys, W., Michiels, J., Delvaux, F. R. and Delcour, J. A. (2001) Contribution of wheat and wheat protein fractions to the colloidal haze of wheat beers. *Journal of the American Society of Brewing Chemists*, 59, 135–140.

Delvaux, F., Depraetere, S. A., Delvaux, F. R. and Delcour, J. A. (2003) Ambiguous impact of wheat gluten proteins on the colloidal haze of wheat beers. *Journal of the American Society of Brewing Chemists*, 61, 63–68.

Delvaux, F., Combes, F. J. and Delvaux, F. R. (2004) The effect of wheat malting on the colloidal haze of white beers. *Technical Quarterly of the Master Brewers Association of the Americas*, 41, 27–32.

Depraetere, S. A., Delvaux, F., Coghe, S. and Delvaux, F. R. (2004) Wheat variety and barley malt properties: influence on haze intensity and foam stability of wheat beer. *Journal of the Institute of Brewing*, 110, 200–206.

Deutscher, M. P. (1990) *Guide to Protein Purification. Volume 182 of Methods in Enzymology*. Academic Press. pp. 290–293

Djurtoft, R. (1965) Composition of the protein and polypeptide fraction of EBC beer haze preparations. *Journal of the Institute of Brewing*, 71, 305–315.

Doner, L. W., Becard, G. and Irwin, P. L. (1993) Binding of flavonoids by polyvinylpolypyrrolidone. *Journal of Agricultural and Food Chemistry*, 41, 753–757.

Douma, A. C., Mocking-Bode, H. C. M., Kooijman, M., Stolzenbach, E., Orsel, R., Bekkers, A. C. A. P. A. and Angelino, A. S. A. G. F. (1997) Identification of foam stabilizing proteins under conditions of normal beer dispense and their biochemical and physiochemical properties. *Proceedings of the 26th Congress of the European Brewery Convention, Maastricht*, 671–679.

Erdal, K. (1986) Proanthocyanidin-free barley – malting and brewing. *Journal of the Institute of Brewing*, 92, 220–224.

Esnault, E. (1995) Beer stabilization with papain. *Brewers' Guardian*, 124, 47–49.

Evans, D. E. and Hejgaard, J. (1999) The impact of malt derived proteins on beer foam quality. Part I. The effect of germination and kilning on the level of Protein Z4, Protein Z7 and LTP1. *Journal of the Institute of Brewing*, 105, 159–169.

Evans, D. E., Sheehan, M. C. and Stewart, D. C. (1999) The impact of malt derived proteins on beer foam quality. Part II. The influence of malt foam-positive proteins and non-starch polysaccharides on beer foam quality. *Journal of the Institute of Brewing*, 105, 171–177.

Evans, D. E., Robinson, L. H., Sheehan, M. C., Tolhurst, R. L., Hill, A., Skerritt, J. S. and Barr, A. R. (2003) Application of immunological methods to differentiate between foam-positive and haze-active proteins originating from malt. *Journal of the American Society of Brewing Chemists*, 61, 55–62.

Fernyhough, R., McKeown, I. and McMurrough, I. (1994) Beer stabilization with silica gel. *Brewers' Guardian*, 123, 44–50.

Finings, *Brewing Room Book*, Pauls Malt, 1998–2000, 243–244.

Gopal, C. and Rehmanji, M. (2000). PVPP – The route to effective beer stabilisation. *Brewers' Guardian*, May, Supplement.

Guzman, J. and Nock, A. (1996) Improved beer stabilization using silica adsorbents. *Technical Quarterly of the Master Brewers Association of the Americas*, 33, 185–186.

Guzman, J. and Nock, A. (1997) Use of filtration grade silicas to provide combined filtration and beer stabilization benefits. *Technical Quarterly of the Master Brewers Association of the Americas*, 34, 272–273.

Hao, J., Li, Q., Dong, J., Yu, J., Gu, G., Fan, W. and Chen, J. (2006) Identification of the major proteins in beer by mass spectrometry following sodium dodecyl sulfate-polyacrylamide gel electrophoresis. *Journal of the American Society of Brewing Chemists*, 64, 166–174.

He, G-Q., Wang, Z-Y., Liu, Z-S., Chen, Q-H., Ruan, H. and Scwartz, P. B. (2006) Relationship of proteinase activity, foam proteins and head retention in unpasteurized beer. *Journal of the American Society of Brewing Chemists*, 64, 33–38.

Hejgaard, J. (1977) Origin of a dominant beer protein immunochemical identity with a β-amylase-associated protein from barley. *Journal of the Institute of Brewing*, 83, 94–96.

Hejgaard, J. and Kaersgaard, P. (1983) Purification and properties of the major antigenic beer protein of barley origin. *Journal of the Institute of Brewing*, 89, 402–410.

Hii, V. and Herwig, W. C. (1982) Determination of high molecular weight proteins in beer using Coomassie blue. *Journal of the American Society of Brewing Chemists*, 40, 46–50.

Hollemans, M. and Tonies, A. R. J. M. (1989) The role of specific proteins in beer foam. *Proceedings of the 22nd Congress of the European Brewery Convention, Zurich*, 561–568.

Hough, J. H., Briggs, D. E., Stevens, R. and Young, T. W. (1982) *Malting and Brewing Science vol. 2, Hopped Wort and Beer*, 2nd edition. Chapman & Hall.

Ishibashi, Y., Terano, Y., Fukui, N., Honbou, N., Kakui, T., Kawasaki, S. and Nakatani, K. (1996) Development of a new method for determining beer foam and haze proteins by using the immunochemical method ELISA. *Journal of the American Society of Brewing Chemists*, 54, 177–182.

Jany, A. and Katzke, M. (2002) CSS – A new beer stabilization process. *Technical Quarterly of the Master Brewers Association of the Americas*, 39, 96–98.

Jegou, S., Douliez, J., Molle, D., Boivin, P. and Marion, D. (2000) Purification and structural characterization of LTP1 polypeptides from beer. *Journal of Agricultural Food Chemistry*, 48, 5023–5029.

Kakui, T., Ishibashi, Y., Kunishige, Y., Isoe, A. and Nakatani, K. (1999) Application of enzyme – linked immunosorbent assay to quantitative evaluation of foam-active protein in wheat beer. *Journal of the American Society of Brewing Chemists*, 57, 151–154.

Kaneda, H., Kano, Y., Osawa, T., Kawakishi, S. and Kamimun, M. (1990) Effects of free radicals on haze formation in beer. *Journal of Agricultural and Food Chemistry*, 38, 1909–1912.

Kaneda, H., Osawa, T., Kawakishi, S., Munekata, M. and Koshino, S. (1994) Contribution of carbonyl-bisulfite adducts to beer stability. *Journal of Agricultural and Food Chemistry*, 42, 2428–2432.

Kano, Y. and Kamimura, M. (1993) Simple methods for determination of the molecular weight distribution of beer proteins and their application to foam and haze studies. *Journal of the American Society of Brewing Chemists*, 51, 21–28.

Leclercq, C., Dervilly, G., Saulnier, L., Dallies, N., Zimmerman, D. and Roue, C. (1999) Barley and malt pentosans: Structure and functionalities in the brewing industry. *Proceedings of the 27th Congress of the European Brewery Convention, Cannes*, 429–436.

Leiper, K. A., Stewart, G. G. and McKeown, I. P. (2003a) Beer polypeptides and silica gel. Part I. Polypeptides involved in haze formation. *Journal of the Institute of Brewing*, 109, 57–72.

Leiper, K. A., Stewart, G. G. and McKeown, I. P. (2003b) Beer polypeptides and silica gel. Part II. Polypeptides involved in foam formation. *Journal of the Institute of Brewing*, 109, 73–79.

Leiper, K. A., Stewart, G. G., McKeown, I. P., Nock, T. and Thompson, M. J. (2005) Optimising beer stabilisation by the selective removal of tannoids and sensitive proteins. *Journal of the Institute of Brewing*, 111, 118–127.

Leisegang, R. and Stahl, U. (2005) Degradation of a foam-promoting barley protein by a proteinase from brewing yeast. *Journal of the Institute of Brewing*, 111, 112–117.

Lewis, M. J. and Poerwantaro, W. M. (1991) Release of haze material from the cell walls of agitated yeast. *Journal of the American Society of Brewing Chemists*, 49, 43–46.

Lewis, M. J., Robertson, I. C. and Dankes, S. U. (1992) Proteolysis in the protein rest of mashing – An appraisal. *Technical Quarterly of the Master Brewers Association of the Americas*, 29, 117–121.

Lopez, M. and Edens, L. (2005) Effective prevention of chill-haze in beer using an acid proline-specific endoprotease from *Aspergillus niger*. *Journal of Agricultural and Food Chemistry*, 53, 7944–7949.

Lusk, L. T., Goldstein, H. and Ryder, D. (1995) Independent role of beer proteins, melanoidins and polysaccharides in foam formation. *Journal of the American Society of Brewing Chemists*, 53, 93–103.

Malcorps, P., Haselaars, P., Dupire, S. and Van den Eynde, E. (2001) Glycogen released by the yeast as a cause of unfilterable haze in the beer. *Technical Quarterly of the Master Brewers Association of the Americas*, 38, 95–98.

Matsuzawa, K. and Nagashima, Y. (1990) A new hydrated silica gel for stabilization of beer. *Technical Quarterly of the Master Brewers Association of the Americas*, 27, 62–72.

McCarthy, S. L., Melm, G. D. and Pringle, A. T. (2005) Comparison of rapid physical stability tests. *Journal of the American Society of Brewing Chemists*, 63, 69–72.

McKeown, I. and Earl, G. (2000) Lucilite Silica – The clear choice for stabilization. *Brewers' Guardian*, June, Supplement.

McMurrough, I. (1995) Colloidal stabilization of beer. *Ferment*, 8, 39–45.

McMurrough, I. and Baert, T. (1994) Identification of proanthocyanidins in beer and their direct measurement with a duel electrode electrochemical detector. *Journal of the Institute of Brewing*, 100, 409–416.

McMurrough, I. and O'Rourke, T. (1997) New insight into the mechanism of achieving colloidal stability. *Technical Quarterly of the Master Brewers Association of the Americas*, 34, 271–277.

McMurrough, I., Hennigan, G. P. and Loughrey, M. J. (1993) Contents of simple, polymeric and complexed flavanols in worts and beers and their relationship to haze formation. *Journal of the Institute of Brewing*, 89, 15–23.

McMurrough, I., Kelly, R., Byrne, J. and O'Brian, M. (1992) Effect of the removal of sensitive proteins and proanthocyanidins on the colloidal stability of lager beer. *Journal of the American Society of Brewing Chemists*, 50, 67–76.

McMurrough, I., Madigan, D., Kelly, R. J. and Smyth, M. R. (1996) The role of flavanoid polyphenols in beer stability. *Journal of the American Society of Brewing Chemists*, 54, 141–148.

McMurrough, I., Madigan, D. and Kelly, R. J. (1997) Evaluation of rapid colloidal stabilization with polyvinylpolypyrrolidone (PVPP). *Journal of the American Society of Brewing Chemists*, 55, 38–43.

Miedl, M. and Bamforth, C. W. (2004) The relative importance of temperature and time in the cold conditioning of beer. *Journal of the American Society of Brewing Chemists*, 62, 75–78.

Mikyška, A., Hrabák, M., Hašková, D. and Šrogl, J. (2002) The role of malt and hop polyphenols in beer quality, flavour and haze stability. *Journal of the Institute of Brewing*, 108, 78–85.

Mitchell, S. (1966) *The Surface Properties of Amorphous Silicas*. Joseph Crosfield & Sons Ltd.

Mitchell, A. E., Hong, Y.-J., May, J. C., Wright, C. A. and Bamforth, C. W. (2005) A comparison of polyvinylpolypyrrolidone (PVPP), silica xerogel and a polyvinylpyrrolidone (PVP)-Silica co-product for their ability to remove polyphenols from beer. *Journal of the Institute of Brewing*, 111, 20–25.

Mohan, S. B., Smith, L., Kemp, W. and Lyddiatt, A. (1992) An immunological analysis of beer foam. *Journal of the Institute of Brewing*, 98, 187–192.

Mussche, R. A. (1994) Beer stabilization with gallotannin. *Brewers' Guardian*, 123, 44–49.

Niemsche, K. and Oppermann, A. (2004) Clarification of wort and beer with silica sol. *The Brewer International*, 4 (6), 30–32.

Nock, A. (1997) Development of an improved silica-based beer stabilizer. *Technical Quarterly of the Master Brewers Association of the Americas*, 34, 174–179.

Onishi, A. and Proudlove, M. O. (1994) Isolation of beer foam polypeptides by hydrophobic interaction chromatography and their partial characterisation. *Journal of the Science of Food and Agriculture*, 65, 233–240.

Outtrup, H. (1989) Haze active peptides in beer. *Proceedings of the 22nd Congress of the European Brewery Convention, Zurich*, 609–616.

Outtrup, H. (1991) Foam stabilizers may contribute to beer haze. *Proceedings of the 23rd Congress of the European Brewery Convention, Lisbon*, 473–480.

Outtrup, H., Fogh, R. and Schaumburg, K. (1987) The interactions between proanthocyanidins and peptides. *Proceedings of the 21st Congress of the European Brewery Convention, Madrid*, 583–590.

O'Reilly, J. P. (1994) The use and function of PVPP in beer stabilization. *Brewers' Guardian*, 123, 32–36.

O'Reilly, J. P. (1996) The role of enhanced solubility PGA in beer head retention. *Brewers' Guardian*, 125, 22–24.

O'Rourke, T. (1994) The requirements of beer stabilization. *Brewers' Guardian*, 123, 30–33.

Parker, D. K. (2007) The study of haze formation in freshly packaged and stored beers. *Technical Quarterly of the Master Brewers Association of the Americas*, 44, 23–28.

Power, J. and Ryder, D. (1989–1991) Some principal points of colloidal stabilization. *Brewing Room Book*, Pauls Malts, pp. 55–65.

Preaux, G., Holemans, P., Van der Vurst, M. and Lontie, R. (1969) Amino acid composition of the beer haze components retained by inorganic adsorbents. *Journal of the Institute of Brewing*, 75, 42–49.

Rehmanji, M., Mola, A., Narayanan, K. and Ianniello, R. (1998) Polyclar (PVPP) for improving shelf life in laboratory treated lagers. *Technical Quarterly of the Master Brewers Association of the Americas*, 35, 95–100.

Rehmanji, M., Mola, A., Narayanan, K. and Gopal, C. (2000) Superior colloidal stabilization of beer by combined treatment with silica (xerogel) and PVPP, polyclar plus 730. *Technical Quarterly of the Master Brewers Association of the Americas*, 37, 113–118.

Rehmanji, M., Gopal, C. and Mola, A. (2002) A novel stabilization of beer with Polyclar® Brewbrite™. *Technical Quarterly of the Master Brewers Association of the Americas*, 39, 24–28.

Robinson, L. H., Evans, D. E., Kaukovirta-Norja, A., Vilpola, A., Aldred, P. and Home, S. (2004) The interaction between malt protein quality and brewing conditions and their impact on beer colloidal stability. *Technical Quarterly of the Master Brewers Association of the Americas*, 41, 353–362.

Sharpe, F. R. and Channon, P. J. (1987) Beer haze caused by can lid lubricant. *Proceedings of the 21st Congress of the European Brewery Convention, Madrid*, 9, 599–606.

Sheehan, M. C. and Skerritt, J. H. (1997) Identification and characterization of beer polypeptides derived from barley hordeins. *Journal of the Institute of Brewing*, 103, 297–306.

Sheehan, M., Halim, C. and Skerritt, J. (2000) Characterisation of the protein fractions removed from beer by silica gels with different physical properties. *Proceedings of the 26th Convention of The Institute of Brewing, Asia Pacific Section, Singapore*, 151–154.

Siebert, K. J. (1997) Beer clarity stability. *Proceedings of the 6th Convention of The Institute of Brewing, Commonwealth & South Africa Section, Durban*, 67–78.

Siebert, K. J. (1999) Effects of protein-polyphenol interactions on beverage haze, stabilization, and analysis. *Journal of Agricultural and Food Chemistry*, 47, 353–362.

Siebert, K. J. and Lynn, P. Y. (1997) Mechanisms of beer colloidal stabilization. *Journal of the American Society of Brewing Chemists*, 55, 73–78.

Siebert, K. J. and Lynn, P. Y. (1998) Comparison of polyphenol interactions with polyvinylpolypyrrolidone and haze-active protein. *Journal of the American Society of Brewing Chemists*, 56, 24–31.

Siebert, K. J. and Lynn, P. Y. (2000) Effect of protein-polyphenol ratio on the size of haze particles. *Journal of the American Society of Brewing Chemists*, 58, 117–123.

Siebert, K. J. and Lynn, P. Y. (2005) Comparison of methods for measuring protein in beer. *Journal of the American Society of Brewing Chemists*, 63, 163–170.

Siebert, K. J. and Lynn, P. Y. (2006) Comparison of methods for measuring polyphenols in beer. *Journal of the American Society of Brewing Chemists*, 64, 127–134.

Siebert, K. J., Carrasco, A. and Lynn, P. Y. (1996a) Formation of protein-polyphenol haze in beverages. *Journal of Agricultural and Food Chemistry*, 44, 1997–2005.

Siebert, K. J., Troukhanova, N. V. and Lynn, P. Y. (1996b) Nature of polyphenol-protein interactions. *Journal of Agricultural and Food Chemistry*, 44, 80–85.

Siebert, K. J., Lynn, P. Y., Clark, D. F. Jr and Hatfield, G. R. (2005) Comparison of methods for assessing colloidal stability of beer. *Technical Quarterly of the Master Brewers Association of the Americas*, 42, 7–12.

Skinner, K. E., Hardwick, B. C. and Saha, R. B. (1993) Characterization of frozen beer precipitates from single packages. *Journal of the American Society of Brewing Chemists*, 51, 58–63.

Slack, P. T. and Bamforth, C. W. (1983) The fractionation of polypeptides from barley and beer by hydrophobic interaction chromatography: The influence of their hydrophobicity on foam stability. *Journal of the Institute of Brewing*, 89, 397–401.

Soluble Silicates and Their Applications. Crosfield Ltd. May 1997.

Sorensen, S. B., Bech, L. M., Muldbjerg, M., Beenfeldt, T. and Breddam, K. (1993) Barley lipid transfer protein 1 is involved in beer foam formation. *Technical Quarterly of the Master Brewers Association of the Americas*, 30, 136–145.

Speers, R. A., Jin, Y-L., Paulson, A. T. and Stewart, R. J. (2003) Effects of β-glucan, shearing and environmental factors on the turbidity of wort and beer. *Journal of the Institute of Brewing*, 109, 236–244.

Srogl, J., Kosar, K., Mikyska, A. and Bousova, P. (1997) Changes occurring in polyphenol substances content during wort production. *Proceedings of the 26th Congress of the European Brewery Convention, Maastricht*, 9, 275–282.

Stoupis, T., Stewart, G. G. and Stafford, R. A. (2003) Hydrodynamic shear damage of brewer's yeast. *Journal of the American Society of Brewing Chemists*, 61, 219–225.

Tanaka, M. and Sakuma, S. (1999) Prediction of the formation of frozen beer precipitates. *Journal of the American Society of Brewing Chemists*, 57, 104–108.

Taylor, R. (1998–2000) Traditional clarification procedures with modern cost benefits. *Brewing Room Book*, Pauls Malts, pp. 78–80.

Taylor, B., Clem, A. and David, P. (2006) Use of the combined stabilisation system and its impact on beer composition. *Brauwelt International*, 24, 158–163.

The mechanism of colloidal instability in beer and its consequences for haze and flavour stability. *Ferment*, 1998, 11, 189–196.

Vaag, P., Bech, L. M., Cameron-Mills, V. and Svendsen, I. (1999) Characterization of a beer foam protein originating from barley. *Proceedings of the 27th Congress of the European Brewery Convention, Cannes*, 157–166.

Veneri, G., Zoccatelli, G., Mosconi, S., Pellegrina, C. D., Chignola, R. and Rizzi, C. (2006) A rapid method for the recovery, quantification and electrophoretic analysis of proteins from beer. *Journal of the Institute of Brewing*, 112, 25–27.

Walker, C. J., Bolshaw, L. and Chandra, S. (2001) Healthy drinks? – Beer and cider antioxidants. *Proceedings of the 28th Congress of the European Brewery Convention, Budapest*, 92–101.

Walters, M. T., Seefeld, R., Hawthorne, D. B. and Kavanagh, T. E. (1996) Composition and kinetics of particle formation in beer post packaging. *Journal of the American Society of Brewing Chemists*, 54, 57–61.

Wenn, R. V., Wheeler, R. E. and Webb, D. J. (1989) The prediction of high haze levels in freshly filtered lager beers responsible for the generation of "invisible" hazes and "bits" in fresh bottled lager. *Proceedings of the 22nd Congress of the European Brewery Convention, Zurich*, 9, 617–624.

Whittle, N., Eldridge, H., Bartley, J. and Organ, G. (1999) Identification of the polyphenols in barley and beer by HPLC-MS and HPLC-Electrochemical detection. *Journal of the Institute of Brewing*, 105, 89–99.

Yang, J-I. and Siebert, K. J. (2001) Development of a method for assessing haze-active protein in beer by dye-binding. *Journal of the American Society of Brewing Chemists*, 59, 172–182.

Yokoi, S. and Tsugita, A. (1988) Characterization of major proteins and peptides in beer. *Journal of the American Society of Brewing Chemists*, 46, 99–103.

Yokoi, S., Maeda, K., Xiao, R., Kameda, K. and Kamimura, M. (1989) Characterization of beer proteins responsible for the foam of beer. *Proceedings of the 22nd Congress of the European Brewery Convention, Zurich*, 9, 593–600.

Yokoi, S., Yamashita, K., Kunitake, N. and Koshino, S. (1994) Hydrophobic beer proteins and their function in beer foam. *Journal of the American Society of Brewing Chemists*, 52, 123–126.

5
Microbiological stability of beer

Anne E. Hill

In ancient civilizations, the process of brewing developed as a means of producing a nutritious beverage that was safer to drink than water. Microbiological stability may therefore be regarded as inherent in its very creation. Beer has a spread of properties that impede microbial development: low pH, high alcohol concentration, antiseptic action of hop acids, low nutrient level, low oxygen concentration and carbonation. Haze has also been implicated in hindering bacterial contamination (Thelen et al., 2006).

To date no known human pathogens have been found to survive in beer. However, brewers cannot be complacent. A variety of yeasts and bacteria are able to flourish in beer causing product deterioration. There is also a range of new products containing lower levels of alcohol (or none at all), and/or fruit juices, that are more susceptible to spoilage than traditional ales and lagers. In addition, production of non-pasteurized beer is continuing to increase. Run losses, product retrieval, and the prohibitive cost effect on brand integrity excite the need for premium quality control mechanisms. Control of microbial contaminants is also important in ensuring consistently uniform and high quality beer.

In this chapter, we examine the microorganisms most commonly encountered at each stage of the brewing process and discuss their effect on fermentation and final product. We also assess the impact of new production methods and new products on the flora observed. Finally, we survey the methods available for microbiological detection and discuss techniques employed to reduce microbial contamination.

Overview of microbial spoilage

In general the bacterial contaminants brewers face today are the same as those encountered two centuries ago. This is despite modern brewing techniques and new packaging methods. The names assigned to the various spoilage organisms have changed more frequently than the prevalence of different species!

There are literally millions of food spoilage microorganisms. However, in general those responsible for beer spoilage consist of a limited number of bacteria and a small number of "wild yeasts." Moulds are not regarded as beer spoilage organisms due to their need for oxygen to grow, however they can indirectly cause spoilage through growth on raw materials. A number of fungi cause gushing of packaged beer and/or produce toxins.

Beer-spoiling bacteria are characterized as microorganisms capable of multiplying in beer, resulting in product deterioration. Bacteria may be divided into Gram positive or Gram negative depending on the structure of their cell wall. Gram positive bacteria appear purple under the light microscope following appropriate Gram staining. The Gram stain is of particular relevance in brewery microbiology as the cell wall structure determines which bacteria are able to grow in hopped wort. Growth of the vast majority of Gram positive organisms is inhibited by hop bitters, whereas growth of Gram negative bacteria is unaffected. A teichoic acid glycosylation protein, essential for Gram positive cell wall formation, has been identified as highly specific to Gram positive beer spoilage strains (Fujii and Hayashi, 2004). Of the Gram positive bacteria, the most dangerous members, in terms of beer stability, are lactic acid bacteria of the genera *Lactobacillus* and *Pediococcus*. Indeed, Lactobacilli are the most common beer-spoilage bacteria, regardless of beer type (Thelen et al., 2006). Some members of the genera *Micrococcus* and *Staphylococcus* can survive in beers and cause spoilage. The Gram positive Bacilli are generally not a serious threat, although have been problematic in unhopped worts. The occurrence of Gram negative bacteria in breweries is regarded as undesirable. The most significant are acetic acid bacteria, *Zymomonas* and certain members of *Enterobacteriaceae*, for example *Rahnella* and *Hafnia*, and *Acidaminococcaceae*, for example *Pectinatus*, *Megasphaera*, *Selenomonas* and *Zymophilus*.

Wild yeasts are generally defined as those yeasts "not deliberately used and not under full control" (Gilliland, 1971); these include contaminant yeast in the pitching yeast culture and those from the air or other raw materials. The major types of wild yeast encountered in the brewery are *Debaromyces*, *Dekkera*, *Pichia*, *Hanseniaspora*, *Kluyveromyces*, *Torulaspora*, *Williopsis* and non-brewing strains of *Saccharomyces*.

Outline of the brewing process

The microbiological stability of the final product can be compromised from a very early point in its production, with spoilage organisms able to access the brewing process at every stage, from raw materials to dispense. Figure 5.1 outlines a typical brewing process. Below we analyze the main parts of the process, detailing the most frequent spoilage microorganisms encountered and the consequences of their presence.

Chapter 5 Microbiological stability of beer

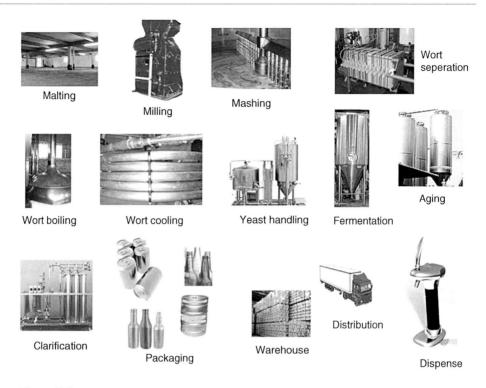

Figure 5.1
Outline of the brewing process.

Raw materials

Most potential contaminants of beer originate from raw materials and/or unclean brewing equipment. Brewing raw materials, such as malt, hops and occasionally brewing water, may be infected by microorganisms and these have to be killed during the brewing process to prevent wort and beer spoilage.

Barley and Malt

The microorganisms that are normally associated with different barleys are remarkably similar (Flannigan, 2003). In the field, they are commonly infected with fungi from the *Alternaria*, *Cladosporium*, *Epicoccum* and *Fusarium* genera, whereas in storage the most commonly encountered fungi are members of *Aspergillus*, *Eurotium* and *Penicillium*. Moulds and bacteria are less commonly isolated from barley.

The effects on brewing and beer of contamination of growing, stored or malted barley are varied. The best known effect of the microbiota of both barley and malt is that of reduced gas stability or gushing (spontaneous ejection of beer from its container). A variety of different fungi have been associated with gushing, the most notable of which are *Fusarium graminearum* and *F. moniliforme*. Small fungal proteins, hydrophobins, present in fungal cell walls have been isolated from strains of the genera *Fusarium*, *Nigrospora* and *Trichoderma* and shown to act as gushing factors in beer (Sarlin et al., 2005).

A second consequence of fungal infection of barley and malt is the potential for release of mycotoxins, compounds toxic to man or animals. Aflatoxin B1, ochratoxin A, zearalone, deoxynivalenol (DON) and fumosins B1 and B2 are mycotoxins that may be transmitted from contaminated grains into beer (Scott, 1996; Schwarz et al., 1995). In addition to the potential harm to humans, mycotoxins may effect fermentation due to their influence on yeast activity. There is also an apparent relationship between the ability of strains to produce the mycotoxin zearalone and gushing.

Inoculation of malt or barley (during steeping or in the field) with *Fusarium* results in increased wort nitrogen and formol nitrogen in wort and beer. Bacterial growth on barley can contribute nitrosamines (which are carcinogens) and cause flavor and filtration problems. The effect of mould growth on the raw material is principally on beer flavor; they may cause a range of off-tastes and odors.

Accumulating evidence has indicated that the evolving microbial flora of the malting process may influence the final malt quality and its ultimate performance in the brewhouse.

Steeping does reduce the levels of naturally occurring mycotoxins although growth of moulds during germination increases mycotoxin production.

Water

Breweries and good water have long had a close association, and water quality is generally taken for granted. Large quantities of water are used in production (water making up over 90% of the final product) as well as for cleaning, washing and sterilizing of equipment. Current consumption in the UK averages approximately 5.0 hl water used/hl beer produced, although there is large variation between breweries. Increasing costs of town supply have led to alternative, less expensive, sources of water being investigated, such as natural springs and boreholes. Water used for brewing must be fit for human consumption (potable). As such it must be free from contaminating organisms. However, what is fit to drink is not necessarily fit for brewing use.

Water for brewing is of course boiled during the process. From a microbiological point of view the main concern is the introduction of spoilage organisms from water introduced after fermentation, for example during dilution of beer following high gravity brewing or from vessels rinsed with contaminated water.

A variety of methods are available for water purification, and generally microorganisms may be removed very effectively. One of the most popular is carbon purification for removal of organic, and some inorganic, compounds. However, this method does not remove microorganisms. Indeed, it can be a source of contaminants. Membrane filtration is generally used for complete removal of bacteria, viruses, proteins, salts and ions. Chlorine dioxide can be applied to water systems to reduce or eliminate brewery spoilage organisms. At levels of approximately 0.2 ppm chlorine dioxide significantly reduces microbial count while causing no off-flavors or odors in the final beer (Dirksen, 2003).

For public supplies, water quality is rigorously tested. Bacteriological analysis of water is designed to detect recent faecal pollution, for example by farm animals or by sewage effluent. Such analysis ensures absence of coliform bacteria, which may be spread by contaminated water supplies. The most dangerous, *Salmonella/Shigella* spp. and *Vibrio cholerae*, occur irregularly and in small numbers in contaminated water, and it is not normal practice to culture these pathogens directly. For brewers, the most significant of the coliform bacteria are *Aerobacter aerogenes* which may be the cause of biologically unstable wort.

Algae and fungi from water supplies are also able to create problems within a brewery causing undesirable odors and taints, clogging filters and providing nutrients for bacterial growth. The wild yeast *Pichia* may be found in some supplies; *Pichia* is quite tolerant to anaerobic conditions, is able to spoil wort, and grows readily in unpasteurized finished beer.

Pitching yeast

The most common source of bacterial contamination in the brewery is probably from pitching yeast, which can transfer contaminants from fermentation to fermentation. Any microbial contamination of pitching yeast compromises product quality and taste and can have a significant effect on the final beer. Wild yeast can cause a variety of off-flavours and aromas, damage by diastatic and proteolytic activity, altered flocculation characteristics of the pitching yeast and may posses killer activity. Wild yeast include members of *Dekkera (Brettanomyces), Candida, Debaryomyces, Hansenula, Kloeckera, Pichia, Rhodotorula, Saccharomyces, Saccharomycodes, Torulaspora* and *Zygosaccharomyces* (Back, 1987). *Brettanomyces* may be used for fermentation and is commonly used in secondary fermentation, for example for Belgian style beers and lambics.

Enteric, acetic and lactic acid bacteria may also contaminate pitching yeast causing serious problems in downstream processes. For example, *Pediococcus damnosus* contamination of pitching yeast can cause extension of the fermentation time and accumulation of high levels of diacetyl. The Gram negative bacteria *Hafnia proteus* (formerly *Obesumbacterium proteus*) and *Rahnella aqualitis* (formerly *Enterobacter agglomerans*) have also been detected in pitching yeasts. These bacteria retard fermentation and may cause excessive diacetyl and dimethyl sulfide in final beer.

Acid washing has been shown to be effective in eliminating contaminating bacteria without adversely affecting the fermentation performance, and is common practice. However acid washing does not remove wild yeast.

Hops

The fourth brewing ingredient, hops, is known for its antiseptic properties. As described, the majority of Gram positive bacteria are inhibited by hops, although Gram negative bacteria are unaffected. The main cause of losses for hop growers are insect pests such as *Aphis humuli* (aphis blight) and *Tetranychus telarius* (red spider). However mould or mildew caused by the fungus *Podosphaera castagnei* is also problematic. Hop cultivars are generally bred to be resistant to the commonest pests and diseases, and any that do infect crops may be tempered by the use of pesticides or fungicides.

Whole hops are dried following harvesting. This process reduces the chances of subsequent microbial contamination, and for brewers who do use whole hops no beer spoilage effects attributable to infected hops have been reported. Similarly, no beer spoilage organisms have been reported to have been introduced by other herbs or plant-derived products used in brewing.

Sugars

Free flowing sugar, syrups or honey are commonly used adjuncts, generally added during wort boiling. Some are also added as non-fermentable sweeteners. Specially tailored syrups allow the production of better quality low alcohol beers. Sugars are in fact used to prevent microbial spoilage of preserves, jams, syrups and jellies, due to their effect on water activity (osmotic pressure), and problems from spoilage are rare. The main concern in brewing involves transfer of bacterial spores, principally from *Bacillus* sp., which can withstand heat treatment, including boiling, and may persist into the finished beer (although beer does not support the subsequent growth of these organisms).

Brewery surfaces

Microorganisms are all around us, and it is unrealistic to sterilize every surface within a brewery. The use of closed fermentation vessels and pipework does much to reduce the ingression of potential spoilage organisms. Similarly, the use of automated cleaning systems that can apply high pressure, high temperature or high power cleaning agents is very effective in reducing the levels of contaminants within brewing equipment.

Despite rigorous cleaning regimes, most industrial plants are inhabited by biofilms. Biofilms usually have no impact on the final product (Timke et al., 2005). However, if they are colonized by product-spoiling microorganisms beyond wort boiling, the risk that the final product will be contaminated increases significantly. Different surface materials used in the brewing process differ in their susceptibility to biofilm formation. Polytetrafluoroethylene (PTFE), nitrile rubber (NBR) and viton (a fluoroelastomer), for example, are generally less susceptible to biofilm formation than stainless steel. The microorganisms present also affect biofilm formation.

Wort

The nutrient-rich environment of wort and the availability of yeast growth factors make it an attractive habitat for microbial growth. The most common contaminants are enteric bacteria, acetic and lactic acid bacteria and wild yeast. Enterobacteria may grow during the initial stages of the brewing process causing unwanted off-flavors in the final product. They are indirect beer spoilage microbes. The effects of contamination include nitrate reduction, off-flavors, such as hydrogen sulfide and acetaldehyde, vinegar flavors, diacetyl, acidic flavors, phenolic flavors, fruity solvent off-flavors and haze.

Hafnia proteus is the best-known enterobacterial contaminant which has only been found in breweries. It can grow together with pitching yeast. Furthermore, it has been isolated from wort and early fermentation stages where it causes a parsnip – like odor (Koivula, 2006). Two further wort-spoilage symptoms are celery flavor from *Aerobacter* and phenolic odors from *Escherichia*.

After boiling, the wort is largely free from microbiological contamination. However, as described above, some microorganisms are able to form spores which can withstand heat treatment, including boiling, and persist into the finished beer.

Fermentation

Traditional fermentation

The most troublesome bacterial contaminants during fermentation are members of the lactic acid bacteria, such as *Lactobacillus* and *Pediococcus*, which cause diacetyl formation, lactic acid formation (Lactobacilli) and ropiness (Pedicocci). However, *Hafnia proteus* and *Rahnella aqualitis* can also tolerate ethanol concentrations greater than 5% v/v. They can therefore contaminate harvested pitching yeast and spread throughout the production plant. Spoilage may also arise from the presence of acetic acid bacteria (*Acetomonas* and *Acetobacter*), *Zymomonas* spp., and strict anaerobes (*Megasphaera* sp., *Pectinatus cerevisiophilus* and *Bacteroides serpens*). Early infection (days 1–2 of fermentation) by Enterobacteria can retard or accelerate fermentation (Koivula, 2006).

High gravity fermentation

Lactobacillus, *Pediococcus*, *Acetomonas*, *Acetobacter* and *Zymomonas* contaminants have been shown to survive levels of ethanol of 12–13% (Magnus et al., 1986) and therefore remain a risk during high gravity fermentation. *Hafnia*, *Rahnella*, *Citrobacter* and *Klebsiella*, known to exist through most stages of the fermentation of traditional gravity wort, have been found to be completely eliminated by elevated ethanol levels. Very high gravity fermentations (>22°P) therefore narrow the range of bacteria capable of spoiling the beer, thereby reducing the risk of bacterial spoilage problems. Most brewers do not modify microbiological testing regimes for high gravity brewing despite the lower risk.

Low calorie fermentations

Low calorie beers may be produced by either the use of amyloglucosidase to produce a wort which will ferment down to 8 or 9°P or lower, or by increasing wort fermentability via the mashing regime. Low or zero residual sugar and fairly high alcohol content leads to a fermentation product that is less prone to spoilage than traditional beers and as yet no specific microbiological issues in low calorie fermentations have been reported.

Non-alcoholic or low alcohol fermentations

Low alcohol beers may be produced by a variety of methods including distillation, dialysis, reverse-osmosis, ultrafiltration and pervaporation. However, the commonest method is by using wort that is not very fermentable and arresting fermentation at an early stage. A shorter fermentation time reduces the time available for any contaminants to "take hold," however the final product will be more susceptible with fermentable sugar present and no alcohol.

Other non-traditional fermentation methods include immobilized yeast and continuous fermentations, accelerated fermentation and high pressure fermentation. No new microbiological risks have been reported for such fermentations and testing regimes should follow those for traditional methods, although increased attention should be paid to continuous fermentations due to the requirement for complete closure if any contaminants are found.

Storage and finishing

Despite the anti-microbial properties of finished beer, a variety of bacteria are able to cause off-odors or flavors given access to storage or finishing tanks, most commonly members of the acetic and lactic acid bacteria (which cause vinegary off-flavors and odors, and may cause excessive gassing and strong head retention). The approach taken to maintaining beer quality post-fermentation is generally that of avoiding contamination with air or oxygen. This may be achieved by holding beer under a carbon dioxide atmosphere, using de-aerated water for chasing and holding the pipelines, and filling pipelines with de-aerated water when not in use. In addition to assisting in maintaining shelf life, such methods also help to reduce microbial contamination of finished beer while in storage or transit.

Packaging and packaged beer

Beer bottling and canning plants are complex machines with a lot of areas in which biofilms can develop (Timke et al., 2005). Beer-spoiling organisms can be harbored and protected from desiccation, heat and disinfection within biofilms, that is biofilms are considered a reservoir of beer spoilage organisms.

Ribosomal RNA studies of bottling plant biofilms have revealed a remarkably diverse habitat, despite intensive cleaning (Timke et al., 2005). Such diversity could enable the biofilm community to react to changes in detergent use or cleaning procedures. Although biofilm studies do not show evidence of strictly anaerobic bacteria or indeed important beer spoilers, there is a need to ensure that biofilms are not allowed to mature.

Finished beer is largely an unfavorable environment for microorganisms and few bacteria can grow in this medium. However, it is believed that even an initial contamination of only a few bacterial cells in a bottle can lead to beer spoilage. Beer-spoiling lactobacilli cause acidity and turbidity. Due to the reduced oxygen content of packaged beer, the importance of the strictly anaerobic beer-spoilage bacteria, *Pectinatus frisingensis*, *P. cerevisiiphilus* and *Megasphaera cerevisiae*, has recently increased (Juvonen and Suihko, 2006). These bacteria are secondary contaminants of beer that is not pasteurized after packaging, and they spoil the product by producing foul-smelling metabolic end-products, such as methyl mercaptan, dimethyl sulfide, and hydrogen sulfide, and causing turbidity.

Changes that can be detected in packaged beer which has been contaminated by microorganisms include turbidity or haze formation, over attenuation, continued fermentation, acidification (lactic, acetic), diacetyl (buttery or butterscotch flavor), liquid manure odor, sulfurs,for example rotten egg, cooked vegetable, phenolic aromas, fusel alcohols, that is propanol, isobutanol and ropiness.

Dispense

Beer dispensing systems in pubs and restaurants are prone to biofouling, and microbial contamination of draught beer is frequently encountered, as are cleanability problems with dispense equipment. The most common bacterial contaminants are acetic and lactic acid bacteria, both of which cause haze and surface films. More worryingly, coliforms such as *E. coli* have been isolated from beer dispensing systems. The aerobic environment also allows growth of wild yeasts which may be found as surface growth on components of a beer system that is exposed to the air such as faucets, keg couplers and drains.

A summary of the main beer spoilage organisms at each stage and their effect(s) is given in Table 5.1.

Detection

Knowledge of microorganisms found in the brewery environment and the control of microbial fouling are both essential in the prevention of microbial spoilage of beer. Low sample volume in relation to huge batch volume (typically 250 ml from >1000 hectoliter) and heterogeneity of the beer-spoiling bacteria makes detection of trace contaminants challenging. In addition, many of the "symptoms" of spoilage are identical to those of physical instability (see accompanying paper by K. Leiper).

Table 5.1
Principal causes of microbiological instability

Stage	Most common contaminant(s)	Effect
Raw materials		
Barley and malt	*Fusarium*	Gushing, mycotoxins, nitrosamines
Water	Coliforms	Unstable wort
Yeast	Wild yeasts	Off-flavor/odor, killer activity, super attenuation, altered flocculation
	Enteric, acetic and lactic bacteria	High diacetyl and dimethylsulfide
Wort	Enterobacteria	Decreased fermentation rate, off-flavor/odor, haze
Fermentation	Lactic acid bacteria	Retard/extend fermentation, off-flavor/odor
Storage and finishing	Acetic and lactic acid bacteria	Vinegary off-flavor/odor, excessive gassing, strong head retention
Packaging and packaged beer	Lactic acid bacteria, *Pectinatus*, *Megasphaera*	Off-flavor/odor, haze, rope
Dispense	Acetic and lactic acid bacteria	Surface film, haze
	Wild yeasts	Surface film, haze

Traditional methods of detection

Traditional methods of detection and identification are generally based on biochemical, morphological and physiological criteria (e.g. nutrient assimilation, microscopy and selective staining, respectively). Such methods tend to be time-consuming, often taking several days in order to first detect potential spoilage microorganisms. Further characterization (identification) can take several more days, once pure cultures have been isolated.

Generally, the first stage of detection involves sampling, either directly or through filtration, and direct microscopy. Samples are used to inoculate growth media and cultured for 2–3 days. Most commonly breweries use a combination of selective media, such as MRS (deMan, Rogosa and Sharpe), Raka-Ray and UBA (Universal Beer Agar). Cycloheximide (Actidione) may be used at a concentration of 0.004 g per liter, to suppress the growth of all brewery culture yeasts (it has no effect on bacterial growth). Anaerobic incubation is necessary to detect anaerobic bacteria such as *Megasphaera* and *Pectinatus*.

Traditional methods can yield inconclusive or incomplete results depending on the degree of characterization required and there is a growing need for more sophisticated methods. Further reasons to improve detection techniques in the brewing industry include:

- Increased consumer awareness on product quality
- Tightened government regulations
- Increased competition among brewers
- Increasing trend to avoid pasteurization of small-packed beer
- Technological advancements.

Rapid methods of detection

The basic premise of rapid methods is to significantly reduce the time required to establish the presence and/or nature of microorganisms found in particular samples of interest. Rapid microbiological detection systems should be faster than "traditional" methods, accurate, affordable, capable of being utilized for either low or high volume testing, easy to learn and user friendly. Fortunately for the brewing industry, research into such methods has been intense and a wide variety of new techniques, suitable for a range of budgets, are available. Here we examine rapid methods under three categories: physical, biochemical and molecular.

Physical methods

1. **Impedance/Conductance measurement:** Growth of microorganisms in culture media is usually measured by the increase in cell number. However, an alternative is to monitor changes in the chemical and ionic composition of the medium. If an alternating electric current is applied across a growth medium, the medium displays resistance to the flow of the current through it (impedence). Impedance is affected by the conductance (the ability of the medium to allow electricity to pass through it) and the capacitance (the ability to store electrical energy) of the medium. If the impedence or conductance of an actively growing culture is measured and the results plotted against time, a curve almost identical to the growth curve is produced. As such, impedimetric technology may be used as an alternative method to plating to measure microbial spoilage in breweries.
2. **Microcalorimetry:** As microorganisms grow they generate heat as a result of their metabolic reactions. The small fluctuations in temperature may be measured and recorded over time using a microcalorimeter. Microcalorimeters are instruments designed specially for microbial measurements that generally function by coupling heat to the production of an electric current. Heat production by microbial activity begins during exponential growth. As the culture reaches stationary phase the metabolic activity decreases accompanied by a decrease in heat generation. Microcalorimetry is often used in clinical applications to detect microorganisms, however it is rarely used in brewing.
3. **Turbidometry:** Spectrophotometry, or turbidometry, is commonly used in food and dairy industries and also in brewing. A number of automated methods are

currently available for detection of microbial contaminants such as *Lactobacillus casei*, *Rahnella aqualitis*, *Pediococcus damnosus* and *Saccharomyces diastaticus*, in pitching yeast. Such methods work within 2–4 days.

4. **Flow cytometry:** Flow cytometry is a modification of spectrophotometry that combines both microscopic and biochemical analyses. Cells are introduced into the center of a rapidly moving stream and then forced to flow at a uniform speed and in single file through an orifice of 50–100 µm in diameter. As they move through the orifice they are illuminated by a laser. The illuminating light is scattered by the cells and this scattering is characteristic of cell shape, size, viability, density and surface morphology. More advanced cytometry methods involve the use of fluorescent dyes to selectively stain different cells. In this way specific cells may be counted in a mixed population, for example ChemScan™. Fluorochromes may also be used to selectively label specific cellular components. When excited by the illuminating laser beam the fluorescence emitted can yield information on the expression of the component. Newer instruments are capable of counting well over 1000 cell per second and if "gating" a numerical boundary is applied, subpopulations can be differentiated from the cell population (dead/live). Also, a threshold applied to the gain will remove debris and noise, which is of particular of importance to brewers as wort and beer contain a lot of debris (Paul Chlup, Personal Communication). There are a number of disadvantages to using cytometry including slow process time and interference by small particulate debris. However its major advantage is sensitivity.

5. **Microcolony method:** This method uses microscopy to detect growing cells that have not yet reached visibly discernible colony forming units. Samples are filtered through a membrane and microcolonies are selectively stained and then examined under the microscope following approximately 24 hours of culture. The first stains used for the microcolony method included janus green, methylene blue and safranin for detection of yeasts. A method whereby both the membrane and the cells were stained was also developed for identification of bacteria. However, the early methods had the disadvantage that cells were killed using the stains and could therefore not be cultured for further examination. Nowadays a range of fluorescent dyes may be used. Such dyes are taken up by both yeast and bacterial cells and incorporated into the cell components, and may be visualized using fluorescence microscopy. Modern fluorescent dyes, or optical brighteners, may be used not only to identify yeasts (both brewing and wild) and bacteria (including member of the genera *Lactobacillus*, *Pediococcus*, *Bacillus*, *Hafnia* and *Escherichia*), but also differentiate between living and dead yeast cells. This technique is very useful for the rapid detection of very low numbers of microbes, but it does not give any indication of the organism's ability to spoil beer.

Biochemical methods

1. **Direct Epifluorescence Filter Technique (DEFT):** DEFT is an improved version of the microcolony method. Initially developed as a quality control method for counting bacteria in raw milk, DEFT combines filtration, using a polycarbonate

membrane, with epifluorescent microscopy to detect fluorochrome-stained cells within approximately 30 minutes. This method does not involve pregrowth, as required in the microcolony method, and may be used to detect single cells. The polycarbonate membrane used in DEFT has a very smooth and flat surface that is better suited for microscopy than the cellulose acetate filters used above. The most commonly employed dye for DEFT is acridine orange, a fluorescent dye that binds to nucleic acids. When the dye binds to single-stranded DNA or RNA, which are found in high numbers in growing cells, it stains orange. Double stranded DNA naturally fluoresces green, therefore actively growing cells may be distinguished from dead cells by the difference in color. Automated methods for DEFT analysis are currently available with throughput rates of over 150 samples per hour.

2. **ATP bioluminescence:** A number of commercially produced kits for detecting yeast and bacteria in wort or beer based on assaying ATP production by microbial metabolism, for example Microstar™, Bev-Trace™, Aqua-Trace™. The assay involves a two-step reaction employing the luciferin-luciferase enzyme reaction that is the basis of bioluminescence in fireflies. The light generated is measured using a luminometer. An enhanced kit is currently available that can detect a single yeast cell or 50 lactic acid bacteria (LAB) cells per sample (Nakakita et al., 2002). LAB produce a substantial amount of ATP in spoiled beer and detection could be used as a method of screening (Suzuki et al., 2005). ATP bioluminescence is now routinely used in surface swabbing, water analysis and beer analysis, and is considerably quicker and also comparable in cost to conventional plate-count techniques.

3. **Protein fingerprinting:** Expression of the microbial genome produces more than 2000 protein molecules. However, not all of the proteins are expressed at the same time. The cellular proteins may be divided into two different groups: the constitutively synthesized structural proteins, and a group of polypeptides, for example enzymes, that are either induced or repressed as a result of the environmental conditions. As such, under defined conditions, the complement of proteins in a cell is characteristic and may be used as a method of identification. In protein fingerprinting, polyacrylamide gel electrophoresis (PAGE) is used to separate cellular proteins based on size or on differences in ionic charge. Separation may be carried out in one dimension, based on size, or in two dimensions, based on both size and ionic charge (Figure 5.2).

    The pattern of proteins on the gel following separation is unique (and therefore referred to as the "protein fingerprint") and may be analyzed for relative similarities or differences to other strains. PAGE is still largely refined to research laboratories and is not generally used as a microbial identification method in breweries. This technique involves growth of cells followed by extraction of proteins. Subsequent electrophoretic separation and staining bring the total time to 1–2 days. In order to make comparisons between strains, the culture conditions must be absolutely exact.

4. **Immunoanalysis (ELISA):** The use of antibodies to detect contaminating microorganisms in breweries is growing in popularity due to its potential to identify microbes in a quantitative way. A range of assays has been designed employing polyclonal or monoclonal antibodies to differentiate microorganisms.

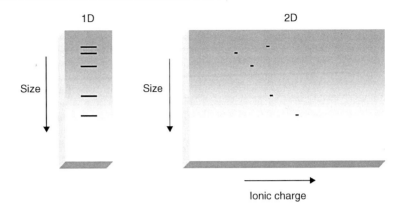

Figure 5.2
Polyacrylamide gel electrophoresis (PAGE).

A number of kits are commercially available for rapid colorimetric differentiation or identification of relatively abundant numbers of intact cells ($>10^4$). These kits involve the use of secondary antibodies to which enzymes such as alkaline phosphatase or horseradish peroxidase have been attached. This enzyme linked immunoassay (ELISA) combined with chemiluminescence detection can be used to detect microbial contaminants with high sensitivity and selectivity both during the brewing process and in the finished beer. A number of monoclonal antibodies are commercially available in kits for the detection of lactic acid bacteria such as *Lactobacilli* and *Pediococci*. Monoclonal antibodies for *Pectinatus* have been developed, but there are problems with cross-reactivity.

5. **Gas chromatography:** Metabolic end products, including volatile and non-volatile organic acids, may be analyzed qualitatively using gas chromatography. The organic acid profile generated is unique. The time taken is only a few hours, although further selection in recovery media may be needed to improve sensitivity. The advantage is that all important beer spoilage organisms may be detected, such as *Lactobacillus*, *Pediococcus*, *Pectinatus* and *Saccharomyces diastaticus* (Schisler et al., 1979).

Molecular methods

1. **Molecular probes (DNA/RNA):** Specific sequences of DNA can be detected by hybridization. Hybridization is the formation of a double-stranded nucleic acid (either DNA to DNA or DNA to RNA) between single-stranded nucleic acids from different sources. Short sequences of DNA generated either from a microbial cell or by chemical synthesis, may be labeled and then used to "probe" for cells that contains the complimentary sequence of DNA. Hybridizations may be carried out at different stringencies by altering conditions such as temperature or salt concentration. Under high stringency the probe will only hybridize to sequences that match it perfectly. Under low stringency the sequence of

nucleotides in the hybridized DNA may differ slightly from the order in the probe. DNA probes are sensitive and can detect 10^4–10^7 organisms per test sample. The advantage of DNA hybridization over protein analysis is that the detection of the target organism is not dependent on the products of gene expression, which can vary depending on the environmental conditions. DNA hybridizations may be performed using single colonies, cells on membrane filters or on nucleic acid purified from cells, digested with restriction endonucleases, and separated by agarose gel electrophoresis. A number of different DNA probes have been designed specifically for brewing spoilage organisms such as *Lactobacillus brevis*, *Pediococcus damnosus* and wild strains of *Saccharomyces cerevisiae* and are available in kit form, for example Vermicon identification technology (VIT). Recently kits have become available that are able to identify the most common beer-spoiling bacteria with no-cross-reactivity with naked DNA, dead bacteria or non-beer–spoiling bacteria, and with no influence from possible inhibitory components of the beer. Such tests take a maximum of 2 days, although some require pre-enrichment.

2. **Ribotyping/Riboprinting:** Ribotyping is a hybridization method identical to that described above. It is used for the comparison of similarities in ribosomal RNA (rRNA) gene sequence. Automated methods of ribotyping have been developed that allow identification of fastidious strains, such as *Lactobacillus lindneri*, and obligate anaerobes, such as *Pectinatus*, within 8 hours of isolating a pure culture (Barney et al., 2001; Amann et al., 2001). Recent innovations, known as FISH (fluorescent *in situ* hybridization), use fluorescent-labeled oligonucleotide probes that hybridize directly to the target region of nucleotides inside the bacteria, without extraction of the nucleotides. The results of FISH analysis reflect real-time physiological characterization and quantification of target bacteria in beer and are well suited to determine whether the spoilage bacteria would cause beer spoilage or not. Definition is greater than previous immunological, LPS and cell-surface protein analyses. For example, the detection limit for *Pectinatus* has been found to be approximately 1000 cells per 100 ml of beer (Yasuhara et al., 2001). FISH technology could potentially eliminate the need for culture-based detection systems traditionally adopted in microbiological quality control in breweries.

3. **Karyotyping:** Electrokaryotyping is a more recent method of characterizing and identifying yeast strains. It is an electrophoretic method that separates chromosomal DNA. It may be used to distinguish different strains of *Saccharomyces*, and therefore to detect non-brewing strains (Fologea et al., 2002).

4. **End-point PCR:** The development of the polymerase chain reaction in 1986 revolutionized molecular biology. This technique is used to amplify small amounts of nucleic acid by several orders of magnitude over the space of only a few hours. As the DNA sequence is unique to each cell, primers may be designed to complement DNA specifically for different strains of microorganisms. PCR primers have already been designed based on sequences found in *Lactobacillus*, *Megasphaera*, *Pectinatus*, *Leuconostoc* and *Saccharomyces diastaticus*. More recently, primers have been designed based on hop-resistance genes, such as *horc*, from Lactobacilli, which gives an indication of the spoilage capability of the strain (Iijima et al., 2006). In PCR technology, the design of the primers is important

because microorganisms that should be detected are not always monospecies (Tsuchiya et al., 1994). The total time taken from DNA extraction to PCR result may be less than 5 hours. In this way a microbial contaminant may be detected before the end of fermentation, in time to decide whether or not to collect the yeast for re-pitching. There may, however, be losses of target molecules during DNA extraction from brewing samples, therefore relatively high numbers of microbes are needed for detection. Inhibitive compounds in beer also decrease sensitivity levels.

5. **Real-time PCR (RT-PCR):** In RT-PCR, the products of each PCR cycle are quantified as they are produced, as opposed to quantifying reaction products at the end of 30 or more PCR cycles in End-Point PCR. The total time for RT-PCR detection is typically below 5 hours with limits of detection currently around 160–1600 cfu $100\,ml^{-1}$ for *Hafnia proteus* (*Obesumbacterium proteus*; Koivula et al., 2006). Cultivation methods are still more sensitive, being able to detect 1–5 cells in a beer sample. Pre-enrichment has been shown to be a suitable method to improve the sensitivity of PCR-based detection of beer spoilage bacteria, but this additional step increases the total detection time by 1–4 days.

6. **RAPD-PCR:** An alternative to using specific primers for PCR is to use a set of randomly designed primers in random amplified polymorphic DNA PCR (RAPD PCR). The product of RAPD PCR is a set of different sized DNA fragments that may be separated electrophoretically to give a characteristic pattern of DNA segments. This pattern is termed a "DNA fingerprint" because it may be genus or even species specific. A number of brewery microorganisms, such as *Pediococcus, Lactobacillus, Hafnia*, and a number of ale and lager yeasts have already been characterized by RAPD PCR and primers are commercially available. A mixture containing 600 primers has been used in reactions resulting in the identification of a highly specific region of DNA from beer-spoilage strains (Fujii et al., 2004; Fujii and Hayashi, 2004). The procedure requires only a small amount of crudely extracted DNA from a single colony isolate, however, the method is not useful for the characterization of mixtures of unknowns (Tompkins et al., 1996).

Improving microbiological stability

We have found that a wide variety of methods are available for detecting microorganisms. However, such methods do not predict the ability of a given contaminant to grow in and spoil beer. The simplest method of determining this is the forcing test, where a pure culture is re-inoculated into beer or beer enriched with concentrated nutrient medium. This test is unfortunately too time consuming for quality control purposes, and more recently attention has focused on developing databases of common physiological properties responsible for beer spoilage ability; detection of the genes responsible for determining the physiological properties using PCR or immunoassays is much faster than culture methods.

Unfortunately, both rapid detection and prediction methods are expensive. Realistically a pro-active response to microbiological instability would seem preferable to reactive testing, and research into methods of reducing microbial contamination remains strong. Generally, strict cleaning routines are employed to maintain a microbe-free plant. This does involve large amounts of cleaning agent and water, in addition to staffing.

A second major route to reducing microbial access is through use of closed vessels and pipework, limiting exposure to air. Filtered nitrogen is preferable to filtered air for any mixing of purging so that oxygen levels are kept to a minimum. Over the past decade, considerable progress has been made in packaging technology in breweries, which has led to a substantial decrease in the dissolved oxygen content of beer (Yasuhara et al., 2001). This trend has resulted in significant reduction of aerobic spoilage organisms, such as acetic acid bacteria. However, the frequency of detection of strictly anaerobic beer-spoiling bacteria, such as *Pectinatus* and *Megasphaera*, has increased dramatically.

A further method of minimizing bacterial growth is to ensure that the yeast starter culture is healthy and contains an adequate quantity of yeast cells. Bacterial growth is slowed under acidic, anaerobic conditions, and the quicker yeast can achieve this during fermentation the better the chance of limiting growth of contaminants.

For finished beer, there are two fundamental strategies to avoid microbial contamination: pasteurization or the application of preserving agents. Thermal pasteurization is the most common method used in beer treatment despite the disadvantage of high capital outlay and energy costs. Another aspect is the loss of quality, particularly taste, because oxidation processes run faster under increased temperatures. Problems also exist at filling and with heat-resistant microorganisms (Beveridge et al., 2004).

Over the past few years a number of non-thermal pasteurization methods have been investigated. These include the use of electric fields and hydrostatic pressure. The application of pulsed electric fields (PEF) as a non-thermal pasteurization method in food processing is of growing interest because of the inactivation of microorganisms and the maintenance of heat sensitive compounds and sensory properties (Evrendilek et al., 2004). One of the main advantages of this alternative processing technology is that the organoleptic and nutritional quality of foodstuffs sustains little or no degradation as a result of treatment. The application of PEF induces a relatively large transmembrane potential that can lead to electroporation of microbial cell walls.

The application of hydrostatic pressure ranging from 100 to 1000 MPa also allows the preservation of foods without altering food quality to the same extent as thermal treatments with a comparable preservation effect. Mild pressure treatments with little effect on cell viability have been shown to increase the permeability of the cytoplasmic membrane, and inactivate specific hop resistance mechanisms in *Lactobacillus plantarum* (Ganzle et al., 2001). Pressure treatment of both ale and lager beers at 600 Mpa for 5 minutes have been found to have no effect on pH, ethanol, bitterness or phenolics, and the beers displayed permanent haze similar to untreated beer. The microbiological stability of pressure treated beers has also been found to be comparable to heat-treated beer (Castellari et al., 2000). However the feasibility and cost of such treatments have not been defined.

An alternative to the above pasteurization methods is filtration. Advances in the understanding of the factors influencing cold beer filterability as well as the quality assurance procedures necessary for success have now made sterile filtration a viable alternative to flash pasteurization for the production of draft beer. The two most important considerations are the cost of filtration and the guarantee of the microbiological stability of the beer. Studies have shown that all beer-spoilage microbes are retained on either 1.2 μm or 0.22 μm filters. Membrane filtration may be applied at final packaging, bottle washing, blanketing/top pressure blending, line cleaning, carbonation, sparging, aseptic fill, bottle purging and tank charging.

The second strategy to reduce microbial contamination is the use of preserving agents, although this is not possible under German purity laws. A few manufacturers of low-alcohol and non-hopped beers have used preservatives such as sorbate or benzoate. Chitosan has been found to enhance the lag period of two strains of *Brettanomyces*. Growth rate of *Saccharomyces cerevisiae* is inversely proportional to chitosan concentration and in mixed fermentations *Brettanomyces* strains fail to grow whereas *Saccharomyces* is unaffected. Other naturally occurring "preservatives" have also been investigated, including zymocins, toxins produced by certain yeasts, which are lethal to sensitive yeast strains. Bacteriocins, produced by lactic acid bacteria, also have the potential to be used as antimicrobials (Vaughan et al., 2001).

Quality control

To summarize, there is a constant battle with microbes, and brewers have a range of tools available to detect and to limit access of undesirable microorganisms. Quality control recommendations vary in the strength of cleaning regimes and in types and number of testing procedures. However, the basic ingredients for good microbiological control may be regarded as good plant design, efficient plant maintenance/renewal, use of cleaning-in-place, effective detergents and sterilants, and strict microbiological monitoring.

The following locations are regarded as areas where any microorganism should either be completely absent or present in very low numbers.

1. Raw materials
2. Bright beer
3. Finished product
4. Strategic surfaces of process machinery, for example filler heads.

There are also three locations where brewing yeasts are present and where selective conditions that exclude the brewing yeast strain are required to detect contaminants: yeast slurry, fermentation and ageing.

For many companies, practical realities are far removed from desired sampling plans and detection methods, and the cost of imposing quality control systems is prohibitively high. However, awareness of threats, simple hygiene and "good practice" are often the most effective methods of managing microbiological risk.

References

Allen, F. (1994) The microbrewery laboratory manual – Part III: Wild yeast detection and remediation. *Brewing Techniques*, 2, 28–35.

Amann, R., Fuchs, B. M. and Behrens, S. (2001) The identification of microorganisms by fluorescence *in situ* hybridisation. *Current Opinion in Biotechnology*, 12, 231–236.

Back, W. (1987) Detection and identification of foreign yeast in brewing. *Brauwelt*, 127, 735–737.

Barney, M., Volgyi, A., Navarro, A. and Ryder, D. (2001) Riboprinting and 16S rRNA gene sequencing for identification of brewery Pediococcus isolates. *Applied and Environmental Microbiology*, 67, 553–560.

Beveridge, J. R., Wall, K., MacGregor, S. J., Anderson, J. G. and Rowan, N. J. (2004) Pulsed electric field inactivation of spoilage microorganisms in alcoholic beverages. *Proceedings of the Institute of Electrical & Electronics Engineers*, 92 (7).

Castellari, M., Arfelli, G., Riponi, C., Carpi, G. and Amati, A. (2000) High hydrostatic pressure treatments for beer stabilisation. *Journal of Food Science*, 65, 974–977.

Dirksen, J. (2003) Chlorine dioxide for the brewing industry. *Master Brewers Association of the Americas Technical Quarterly*, 40, 111–113.

Dowhanick, T. M. and Sobczak, J. (1994) ATP bioluminescence procedure for viability testing of potential beer spoilage microorganisms. *Journal of the American Society of Brewing Chemists Chem.*, 52, 19–23.

Evrendilek, G. A., Li, S., Dantzer, W. R. and Zhang, Q. H. (2004) Pulsed electric field processing of beer: Microbial, sensory, and quality analyses. *Journal of Food Science*, 69, 228–232.

Flannigan, B. (2003) The microbiota of barley and malt. In: *Brewing Microbiology* (F. G. Priest and I. Campbell eds). Kluwer Academic Press.

Fologea, D., Csutak, O., Vassu, Y., Gherasim, R., Stoica, I., Sasarman, E., Smarandache, D., Nohit, A. and Iftime, O. (2002) Field inversion gel electrophoresis of large DNA molecules extracted from a *Saccharomyces cerevisiae* strain. *Roum. Biotechnol. Lett.*, 7, 855–860.

Fuji and Hayashi (2004) Molecular cloning of genes from lactic acid bacteria based on RAPD-PCR. *Foods and Food Ingredients Journal of Japan*, 209 (9), 3.

Fujii, T., Nakashima, K. and Hayashi, N. (2005) Random amplified polymorphic DNA-PCR based cloning of markers to identify the beer-spoilage strains of *Lactobacillus brevis*, *Pediococcus damnosus*, *Lactobacillus collinoides* and *Lactobacillus coryniformis*. *Journal of Applied Microbiology*, 98, 1209–1220.

Ganzle, M. G., Ulmer, H. M. and Vogel, R. F. (2001) High pressure inactivation of *Lactobacillus plantarum* in a model beer system. *Journal of Food Science*, 66, 1174–1181.

Gilliland, R. B. (1971) Yeast Classification. *Journal of the Institute of Brewing*, 77, 276–284.

Iijima, K., Suzuki, K., Ozaki, K. and Yamashita, H. (2006) *horC* confers beer-spoilage ability on hop-sensitive *Lactobacillus brevis* ABBC45[cc]. *Journal of Applied Microbiology*, 100, 1282–1288.

Juvonen, R. and Satokari, R. (1999) Detection of spoilage bacteria in beer by polymerase chain reaction. *Journal of the American Society of Brewing Chemists*, 57, 99–103.

Juvonen, R. and Suihko, M. (2006) *Megasphaera paucivorans* sp. nov., *Megasphaera sueciensis* sp. nov and *Pectinatus haikarae* sp. nov., isolated from brewery samples, and emended description of the genus Pectinatus. *International Journal of Systematic and Evolutionary Microbiology*, 56, 695–702.

Koivula, T. T., Juvonen, R., Haikara, A. and Suihko, M. L. (2006) Characterisation of the brewery spoilage bacterium *Obesumbacterium proteus* by automated ribotyping and development of PCR methods for its biotype 1. *Journal of Applied Microbiology*, 100, 398–406.

Lee, S. Y., Moore, S. E. and Mabee, M. S. (1981) Selective-differential medium for isolation and differentiation of Pectinatus from other brewery microorganisms. *Applied and Environmental Microbiology*, 41, 386–387.

Magnus, C. A., Ingledew, W. M. and Casey, G. P. (1986) High-gravity brewing: Influence of high-ethanol beer on the viability of contaminating brewing bacteria. *Journal of the American Society of Brewing Chemists*, 44, 158–161.

Nakakita, Y., Takahashi, T., Sugiyama, H., Shigo, T. and Shinotsuka, K. (1998) Isolation of novel beer-spoilage bacteria from the brewery environment. *Journal of the American Society of Brewing Chemists*, 56, 114–117.

Nakakita, Y., Takahashi, T., Tsuchiya, Y., Watari, J. and Shinotsuka, K. (2002) A strategy for detection of all beer-spoilage bacteria. *Journal of the American Society of Brewing Chemists*, 60, 63–67.

Reid, G. C., Hwang, A. and Meisel, R. H. (1989) The sterile filtration and packaging of beer into polyethylene tetra phthalate containers. *Journal of the American Society of Brewing Chemists*, 48, 85–91.

Sarlin, T., Nakari-Setala, T., Linder, M., Penttila, M. and Haikara, A. (2005) Fungal hydrophobins as predictors of the gushing activity of malt. *Journal of the Institute of Brewing*, 111, 105–111.

Satokari, R., Juvonen, R., Mallison, K., von Wright, A. and Haikara, A. (1998) Detection of beer spoilage bacteria Megasphaera and Pectinatus by polymerase chain reaction and colorimetric microplate hybridisation. *International Journal of Food Microbiology*, 45, 119–127.

Schisler, D. O., Mabee, M. S. and Hahn, C. W. (1979) Rapid identification of important beer microorganisms using gas chromatography. *Journal of the American Society of Brewing Chemists*, 37, 69–77.

Schwarz, P. B., Casper, H. H. and Beattie, S. B. (1995) Fate and development of naturally occurring *Fusarium mycotoxins* during malting and brewing. *Journal of the American Society of Brewing Chemists*, 53, 121–127.

Scott, P. M. (1996) Mycotoxins transmitted into beer from contaminated grains during brewing. *Journal of the Association of Official Analytical Chemists International*, 79, 875–882.

Suzuki, K., Iijima, K., Ozaki, K. and Yamashita, H. (2005) Study on ATP production of lactic acid bacteria in beer and development of a rapid pre-screening method for beer spoilage bacteria. *Journal of the Institute of Brewing*, 111, 328–335.

Suzuki, K., Iijima, K., Sakamoto, K., Sami, M. and Yamashita, H. (2006) A review of hop resistance in beer spoilage lactic acid bacteria. *Journal of the Institute of Brewing*, 112, 173–191.

Thelen, K., Beimfohr, C. and Snaidr, J. (2004) VIT-Bier: The rapid and easy detection method for beer-spoiling bacteria. *Master Brewers Association of the Americas Technical Quarterly*, 41, 115–119.

Thelen, K., Beimfohr, C. and Snaidr, J. (2006) Evaluation study of the frequency of different beer-spoiling bacteria using the VIT analysis. *Master Brewers Association of the Americas Technical Quarterly*, 43, 31–35.

Timke, M., Wang-Lieu, N. Q., Altendorf, K. and Lipski, A. (2005) Fatty acid analysis and spoilage potential of biofilms from two breweries. *Journal of Applied Microbiology*, 99, 1108–1122.

Tompkins, T. A., Stewart, R., Savard, L., Russell, I. and Dowhanick, T. M. (1996) RAPD-PCR characterisation of brewery yeast and beer spoilage bacteria. *Journal of the American Society of Brewing Chemists*, 54, 91–96.

Tsuchiya, Y., Kano, Y. and Koshino, S. (1993) Detection of *Lactobacillus brevis* in beer using polymerase chain reaction technology. *Journal of the American Society of Brewing Chemists*, 50, 64–67.

Tsuchiya, Y., Kano, Y. and Koshino, S. (1994) Identification of lactic acid bacteria using temperature gradient gel electrophoresis for DNA fragments amplified by polymerase chain reaction. *Journal of the American Society of Brewing Chemists*, 52, 95–99.

Vaughan, A., Eijsink, V. G. H., O'Sullivan, T. F., O'Hanlon, K. O. and van Sinderen, D. (2001) An analysis of bacteriocins produced by lactic acid bacteria isolated from malted barley. *Journal of Applied Microbiology*, 91, 131–138.

Yamauchi, H., Yamamoto, H., Shibano, Y., Amaya, N. and Saeki, T. (1998) Rapid methods for detecting *Saccharomyces diastaticus*, a beer spoilage yeast, using the polymerase chain reaction. *Journal of the American Society of Brewing Chemists*, 56, 58–63.

Yasuhara, T., Yuuki, T. and Kagami, N. (2001) Novel quantitative method for detection of *Pectinatus* using rRNA targeted fluorescent probes. *Journal of the American Society of Brewing Chemists*, 59, 117–121.

6

Beer gushing

*Leif-Alexander Garbe,
Paul Schwarz and Alexander Ehmer*

Introduction

In the simplest terms beer gushing describes the sometimes violent over-foaming of beer that can occur when a bottle is opened (Figure 6.1). More specifically, Gjersten, (1967) stated that the

> phenomenon of gushing is characterized by the fact that immediately after opening, i.e., by removing the excess pressure above the beer, a very great number of fine bubbles are formed throughout the volume of beer and ascend very quickly, creating foam which flows out of the bottle or, in severe cases, actually spurts from the bottle.

Gushing was described in as early as the 1920s by Lüers (1924 as cited in Amaha and Kitabatake, 1981), Windisch (1923 as cited in Amaha and Kitabatake, 1981) and Vimpel (1922 cited in Gjertsen et al., 1963), and in the 1930s by Helm and Richardt (1938). It can occur in carbonated beverages such as beer, cider, fruit juice spritzers, lemonades and sparkling wines (Schumacher, 2002; Zepf, 1998). As gushing is by far more strongly observed by the consumer than many simple flavor defects, the brewery risks not only financial damage, but also a serious loss of public image. Gushing has been attributed to a number of causes, with some beers showing gushing immediately after filling, while in others this tendency emerges only after several weeks (Weideneder, 1992). Likewise the gushing potential can decrease after some months. Since there is no gushing test that is absolutely reliable, this phenomenon presents an incalculable risk to the brewer.

Considerable, albeit sporadic, research efforts have been devoted to the subject since the 1920s. The phenomenon of gushing was last extensively reviewed by Amaha and Kitabatake (1981). Since this time, several short or partial reviews have been presented by Casey (1996), Dilly (1988); Draeger (1996a), Linemann (1996), Schwarz (2003), Simon (1998), Wershofen (2004), Winkelmann (2004) and Zarnkow and Back (2001). The current review will focus on the literature since 1990, which in large part has been associated with mycroflora-related gushing.

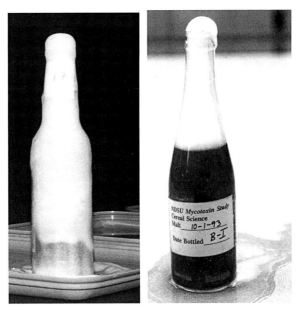

Figure 6.1
Gushing observed in two samples of beer which were spiked with extracts of FHB infected barley as per the test described by Vaag et al. (1993). The bottle on the left is shown immediately after removal of the crown, while the sample on the right is shown following the subsidence of gushing.

Terminology

Gushing in beer has been attributed to a number of factors including the malt, hop extracts, metal ions, calcium oxalate, rough interior bottle surfaces and bottle washing detergents. Over the years various terms have been used to in an attempt to classify or describe different types of gushing. Vogel (1949) divided gushing into three categories; mishandled beer, wild beer and gushing beer. Mishandled beer simply describes beer that has been subjected to extensive agitation or rough treatment immediately prior to, or during opening. Wild beer is over-carbonated beer or beer with an excessive air content. The term gushing was used to encompass cases which did not fall into the categories of mishandled or wild beer.

Munekata and coworkers (as cited by Amaha et al., 1978) described occurrences of gushing observed in Japan as falling into winter- and summer-types. Summer-type gushing occurred only when bottled beer was shaken or stored above 25°C. Summer-type gushing was believed to be the same as that frequently observed in Europe, and was attributed to the malt. Winter-type gushing occurred only when the beer was shaken or stored near 0°C, and was ascribed to polypeptides derived from beer proteins through the action of chill-proofing enzymes like papain. The winter- and summer-type terminology was not widely accepted, and is little used today.

Perhaps the most useful and utilized definition of gushing was by Gjertsen et al. (1963). They suggested that gushing be divided into two classes: primary and secondary gushing. Primary gushing was defined as occurring periodically and being related to the malt. Secondary gushing was defined as being caused by production faults during processing or incorrect treatment of the beer. Casey (1996) later advocated for the use of this terminology, and further indicated that secondary gushing was caused by the presence of colloidal or solid particles, while primary gushing results from the use of mold infected malt. As such, primary gushing will likely impact the entire production from a given lot of infected malt. The occurrence of secondary gushing, however, might be expected to be more sporadic within a lot, as in the case of particulates resulting from filter-breakthrough.

Physical background of gushing

Packaged beer usually contains 4–5 g CO_2/l (Moll, 1991). This translates to approximately two volumes. As such beer is supersaturated with CO_2, and over-foaming or overflow upon opening would not seem surprising. However, under normal conditions, when the package is opened and the excess pressure released, the CO_2 slowly comes out of solution, and forms the bubbles typically associated with beer. In the case of gushing the release of CO_2 occurs very rapidly with opening, resulting in loss of beer from the bottle. Gushing has often been associated with large numbers of minute or micro-bubbles, and the excessive and uncontrolled release of CO_2.

A bubble is stretched by the free movement of molecules inside the bubble. The requirement for bubble formation is the existence of several fast moving molecules at the same time and at the same place. According to Wilt (1986) and Draeger (1996a,b) the theoretical probability of the new bubble formation in a freshly opened beer bottle is low. However, in practice numerous bubbles are formed upon opening. Schumacher (2002) stated that physically dissolved CO_2 is present as 0.1–0.2% carbonic acid and their corresponding dissociation products in water. Approximately 99.8–99.9% of the CO_2 is dissolved as gas molecules and small gas bubbles, respectively. These are covered with a water hydrate coat, which is stabilized by hydrogen bonds. Similar interaction effects take place with other polar beer ingredients, such as ethanol. Other authors also postulate that micro-bubbles serve as nuclei for further bubbles (Draeger, 1996a,b; Franke et al., 1995; Gardner, 1973; Guggenberger and Kleber, 1963; Yount et al., 1984).

The work (W) required to form a bubble was calculated by Krause (1936).

$$W = 4/3 \pi \sigma r^2$$

where
σ = surface tension and
r = radius of the bubble.

Therefore the necessary work to form a bubble directly depends on the surface tension of surrounding liquid. Reducing the surface tension leads to more micro-bubbles. After the pressure is released with the opening of the bottle, the micro-bubbles expand. This leads to an imbalance between the strongly CO_2 oversaturated liquid and the CO_2 in the head area of the bottle. Carbon dioxide gas molecules diffuse from the oversaturated liquid into the micro-bubbles, increase, ascend to the surface and produce visible foam. The bubbles may divide, and a small part remains as renewed blister germ. If enough bubbles are present, the liquid is dragged along and gushing occurs (Draeger, 1996a,b; Fischer et al., 1997; Fischer, 2000; Schumacher, 2002; Zepf, 1998). In other words, with lower surface tension, the bubbles are more unstable and the gushing risk is higher. As an example, this explains why tensides act as gushing factors.

The temperature dependence of gushing was investigated by Helm and Richardt (1938), who reported the lower the temperature during storage and the higher the temperature when opening the beer bottle, the greater the amount of gushing. Casey (1996) found that test beers from *Fusarium* infected malt did not exhibit gushing unless they were equilibrated to room temperature, or were agitated, prior to opening. Gardner (1973) stated that micro-bubbles can form as the result of agitation. Casey (1996) also reported that gushing could temporarily be eliminated by re-pausterization, with the increased pressure during treatment presumably resulting in transient solubilization of the micro-bubbles. The flavor stability of beer is also extremely influenced by storage temperature and cold storage of beer results in less stale flavor. The solubility of CO_2 in water or beer is dependant on the temperature, and at low temperature CO_2 is better dissolved in the beverage.

Primary gushing

Barley or malt as a cause of gushing has long been recognized. Helm and Richardt (1938) reported that epidemics of gushing in North America, South America and Europe were associated with specific barley varieties and crop years, although the specific cause was only later identified as mycoflora from the grain (Gjersten, 1967). The periodic nature of these primary gushing problems has contributed to a sporadic nature of research efforts on the subject, and the thus somewhat interrupted gains in knowledge on the cause and mechanism. However, over the past 15 years there has been a considerable amount of research on the subject, with a consequent advancement in knowledge. This is largely due to recent and widespread outbreaks of fusarium head blight (FHB) that have occurred in many of the world's barley production regions (Steffenson, 2003), and the resultant scrutiny that has been applied to the pathogenic species of *Fusarium*.

Mycoflora

A strong empirical connection between grain microflora and gushing was first made in the USA in 1950 when it was observed that malt from a specific barley variety, that was severely attacked by net blotch (*Pyrenophora teres*), produced gushing in experimental brews (Gjersten, 1967). Problems were more severe when the barley was subject to weathering conditions during maturation and harvest. Prentice and Sloey (1960) and Sloey and Prentice (1962) later definitely proved the relationship between grain fungi and gushing. In their work a total of 29 bacterial, 10 yeast and 58 fungal isolates were applied to barley during the malting process. The impact upon malt quality parameters and beer gas stability was then determined. Beers prepared from malts treated with several *Fusarium* sp. produced pronounced gushing, with losses of 14–44% of beer volume upon opening. The importance of *Fusarium* sp. in gushing was confirmed by Gjersten (1967), who postulated that gushing resulted from an interaction between the growing mycelium and barley during malting. Several other fungal genera have also been identified as causing gushing. These include species of *Aspergillus* (Gyllang et al., 1977; Prentice and Sloey, 1960), *Alternaria*, *Stemphylium*, *Penicillium*, *Rhizopus* (Amaha et al., 1973), *Nigrospora* (Amaha et al., 1973; Sarlin et al., 2005b) and *Trichoderma* (Sarlin et al., 2005b) although *Fusarium* species seem to be most problematic in commercial practice.

Different species of *Fusarium* may be associated with FHB, a disease of both wheat and barley. The specific species, and the strains of individual species, depend upon both geographic location and climate. In the USA, Canada, China and southern and eastern Europe, *F. graminearum* predominates, while *F. culmorum* is more important in northern Europe (Gale, 2003). Other species encountered in association with FHB infected grain may include, but are not limited to *F. avenaceum*, *F. poae*, *F. sporotrichiodes*, *F.equiseti* and *F. tricintum*.

The etiology of FHB has been extensively reviewed elsewhere (McMullen et al., 1997), but a basic understanding is relevant to this discussion of gushing. As implied by the name, FHB is a disease which impacts the head or inflorescence of the grain. The pathogen exists as a saprophyte on crop residues, and ascospores or conidia are wind blown or rain splashed to the developing head. Prolonged periods of high humidity and warm temperatures (24–29°C) favor FHB, and in barley infection may occur at any time from heading until harvest (Prom et al., 1999). With probable dependence upon environmental factors and the timing of infection, the pathogen can be limited to the surface of the grain, or it may be found internally. Aside from the propensity to cause gushing, there are several impacts upon grain quality and suitability for malting and brewing. In wheat, a pronounced reduction in grain size and weight is often observed, but with barley this often may not be the case (Schwarz et al., 2006). The pathogen produces cell wall degrading enzymes and proteases which can alter quality parameters (Schwarz et al., 2001, 2002), but most detrimental is the production of mycotoxins. The *Fusarium* species are associated with the production of tricothecene mycotoxins and zearalenone. Deoxynivalenol (DON) is commonly associated with *F. graminearum* (Salas et al., 1999), and *F. culmourm* strains may produce DON or nivalenol (NIV) (Desjardins, 2006). DON present on the malt will largely be extracted into the beer (Schwarz, 1995), and as a consequence, malting barley produced in regions impacted by FHB, is generally screened for DON or other tricothecenes. In many cases the mycotoxins present on the barley may be significantly reduced or eliminated during steeping (Schwarz, 2003). However, in some cases the amount of mycotoxin may actually increase during malting, presumably through growth of the *Fusarium*, or perhaps by the liberation of bound mycotoxins.

Schwarz et al. (1997) reported that the compounds which cause gushing are present on FHB infected barley or wheat prior to malting. This was based on the observation that extracts prepared from infected barley were able to induce gushing, when added to beer. Following steeping a significant decrease in gushing propensity was typically observed. This is not surprising as the gushing factor(s), by nature should be at least partially soluble. However, gushing propensity does not always disappear with the rinsing of barley, which suggests that at least in some cases, the factors are produced beneath the husk (Vaag et al., 1993; Vaag and Pederson, 1992). As observed for mycotoxins, additional gushing factor(s) may be produced during germination when there is growth of the *Fusarium*. Data illustrating the change in gushing during the malting of two FHB infected samples is shown in Figure 6.2. The variable changes in the gushing potential of barley during malting was confirmed by both Munar and Sebree (1997) and Sarlin et al. (2005a).

As previously stated, Gjersten (1967) speculated that the gushing factor(s) were produced as a result of interaction between the barley and fungal mycelium. This was based upon the observations that gushing was induced in resultant beers only when viable fungal mycelium was added to the steep water, and no gushing was induced when either the substrate from the culture or an extract of mycelium was added to the steep. Based upon these results Gjertsen speculated gushing might be due to an antimicrobial substance formed

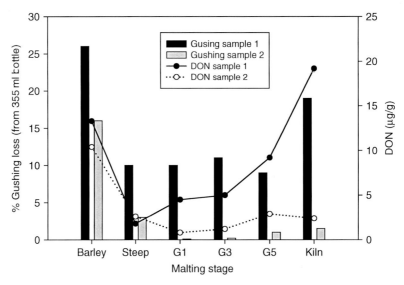

Figure 6.2
Change in the gushing potential of two FHB infected barley samples during malting. Gushing was measured by the Carlsberg test, and the barley samples were from North Dakota. G: indicates days of germination (freeze-dried green malts) (Schwarz, unpublished data).

by the barley in response to infection. However, Schwarz et al. (1997) later suggested that the gushing factor(s) are likely formed directly by the fungi, as cell-free extracts of *F. graminearum*, grown in peptone water, were found to induce gushing when added to beer in laboratory trials.

Also as observed for mycotoxin production, there seem to be significant differences between *Fusarium* species in terms of gushing propensity. Haikara (1983) evaluated the gushing propensity of two strains of *F. culmorum* and three of *F. avenaceum* over two crop years in Finland. The barley was inoculated during flowering, and *F. avenaceum* did not cause gushing, while within the *F. culmorum* samples, the tendency seemed to depend on the strain as well as weather conditions. Gushing was more pronounced with wetter weather during the growing season. Niessen et al. (1992) investigated the gushing potential of 160 samples of unmalted and malted wheat and barley from Germany. Most incidents of gushing were associated with *F. graminaerum*, and a subordinate role was played by *F. avenaceum* in barley and *Microdochium nivale* in wheat. A few cases of gushing associated with *Alternaria alternata*, in the absence of *F. graminearum*, were noted. Results were confirmed by inoculation studies, and different organisms appeared to have different threshold contamination levels for gushing. Results of a field inoculation study by Sarlin and coworkers (2005a) confirmed the gushing propensity of *F. graminearum*, *F. culmorum* and *F. poae*, with later being the least active. On a final note, while the above studies have all focused on field infected grain, Haikara (1980) has stated that gushing is not limited to field infection of barley or wheat, but also may occur as the result of poor storage conditions or malthouse sanitation.

Causal agents

Gushing inducing substances have been isolated from fungi by a number of groups, and the structure and characteristics of these compounds are described in Table 6.1. They are smaller proteins, with molecular weights up to 20kDa. The crucial criterion of a hydrophobic character is underlined, as a decrease of the surface tension of the beer leads to an intensified formation of micro-bubbles, and thus to gushing. Hydrophobic proteins formed by fungi should increase gushing, and Haikara et al. (1999), Linder et al. (2002, 2005) and Sarlin et al. (2005b) have all identified the fungal metabolites, hydrophobins, as so-called gushing factors. Hydrophobins are small extracellular proteins secreted by fungi.

Enzymes are also secreted into the grain by fungi, and it is possible that proteases can also play a role in gushing, since they can change the protein composition of the grain. An increased Kolbach Index can change the colloidal structure in the finished beer. This influences the foam stability, the solubility of gases and ultimately the formation of micro bubbles, which stand in close relationship with gushing.

Hydrophobins

Amaha et al. (1973) and Kitabatake and Amaha (1977), Kitabatabe (1978) and Kitabatake et al. (1980) first isolated gushing positive components with

Table 6.1
Characteristics of Gushing factors from different fungal and grain sources

Source	Chemical nature	Molecular-weight (kDa)	Concentration for Gushing (mg/l)	Reference
Nigrospora sp.	Hydrophobic polypeptide	16.5	0.05	Amaha et al. (1973)
				Kitabatake and Amaha (1977)
Stemphylium sp.	Glycoprotein (peptido-glucan)	n.d.	4	Amaha et al. (1973)
Fusarium graminearum	Hexapeptide	n.d.	0.4	Amaha et al. (1973)
Penicillium crysogenum	Cyclic tetrapeptide	n.d.	0.3	Kitabatake et al. (1980)
Malt	Hydrophobic peptide	10	0.5	Kitabatake (1978)
Trichoderma reesei	Hydrophobin I and II	7.5	0.003	Sarlin et al. (2005b)
Nigrospora sp.	Hydrophobin	8.5	0.03	
Fusarium poae	Hydrophobin	8.5	0.1	

n.d.: not determined

a hydrophobic character. Extensively characterized was a polypeptide isolated from *Nigrospora* sp. It had a molecular weight of 16.5 kDa, an iso-electric point of 4.0 and possessed strong surface activity. It contained 16 cysteine amino acids, which formed 8 intramolecular disulfide bridges. Methionine, histidine, thyrosine and tryptophan were not found in this protein. Current research data indicate hydrophobin dimers as the gushing positive compounds (Wessels, 1994, 1996, 1997; Wessels et al., 1991). They possess molecular weights of 7–20 kDa (Wessels et al., 1991) and play an important role in fungal growth. Hydrophobins occur in the cell walls and are involved with the formation of hydrophobic structures such as aerial hyphae, spores and fruiting bodies (Woesten 1999 1995, 1997). Hydrophobins can be released to the environment by fungi, resulting in a reduction of the surface tension of water, or changes in the hydrophobicity/hydrophilicity of surfaces Woesten and Wessels (1997). These fungal properties are useful for the fungi as it attaches to hydrophobic surfaces such as plant cuticle, or in the penetration of plant tissue (Talbot et al., 1996). Hydrophobins are the most active of all known surface active bio-molecules (Woesten and de Vocht, 2000). Characterized hydrophobins consist of approximately 100 (±25) amino acids. The hydrophobin molecule contains eight conserved cysteine residues, which form four intramolecular disulfide bridges. Thus the entire molecule is effectively accumulated in a very stable and compact structure. The aromatic amino acids tyrosine and tryptophan do not occur in hydrophobins. Likewise, methionine and histidine can be missing from some hydrophobins. Despite the hydrophobic character, different hydrophobins exhibit different amino acid sequences (Wessels, 1997; Wessels, 1994). A general amino acid sequence of a hydrophobin is as follows (Wessels, 1996):

$$X_{2-38}\text{-}C\text{-}X_{5-9}\text{-}C\text{-}C\text{-}X_{11-39}\text{-}C\text{-}X_{8-23}\text{-}C\text{-}X_{5-9}\text{-}C\text{-}C\text{-}X_{6-18}\text{-}C\text{-}X_{2-13}$$

where C = cysteine and X = other amino acids.

Hydrophobins can be divided into two groups (I and II) based on their physiochemical characteristics (De Vries et al., 1999; Wessels, 1996), and an overview of these two groups is presented in Table 6.2.

Hydrophobins independently form films between hydrophobic hydrophilic layers. Films formed by hydrophobins of the class I group are much more stable than those of class II (Woesten et al., 1993, 1994, 1995). Hydrophobins of the class I form rodlets with self-aggregation. The diameter of these rodlets is within the range of 5–12 nm (Bell-Pedersen et al., 1992). These structures play an important role with many morphogenesis processes, like sporulation and the development of fruiting bodies (Wessels, 1997). Hydrophobins of class II form highly arranged (regular) monolayer films (Paananen et al., 2003) and crystalline fibrils (Torkkeli et al., 2002). Secondary structures in monomer form are specified in detail in Table 6.3. The β-sheet structure is most abundant, and α-helical structure is involved in the formation of the rodlet form.

Table 6.2
Physiochemical properties of the hydrophobins I and II[1]

Hydrophobin	Surface activity ($mJ*m^{-2}$)	Hydrophilicity: hydrophilic sites (θ)	Hydrophobicity: hydrophobic sites (θ)	Lectin-activity	Rodlets
Class I					
SC3	32	36±3	117±8	Yes	Yes
SC4	35	48±3	115±3	Yes	Yes
ABH1	n.d.	63±8	113±4	n.d.	Yes
ABH3	37	59±5	117±3	n.d.	Yes
Class II					
Cryparin	32	22±2	>90	Yes	No
CFTH1	33	60±5	105±2	n.d.	No

[1]Data from Woesten and deVocht (2000).

Table 6.3
Secondary structure of hydrophobin SC3[1]

	α-Helix (%)	β-Sheet (%)	β-Turn (%)	Random coil (%)
Monomeric	23	41	16	20
Self-assembled on air–water interface	16	65	9	10

[1] Data from de Vocht et al. (1998).

Hydrophobins are also very interesting for medical and technical applications in respect of their physico-chemical properties. For example, coating of catheters with hydrophobins can prevent bacteria from adhering (Wessels, 1997).

The use of malt with fungal contamination can lead to the presence of hydrophobins in finished beer (Niessen et al., 2006). By spontaneous self-aggregation, which is caused by hydrophobic interactions, hydrophobins form amphiphatic layers on hydrophobic or hydrophilic surfaces (Woesten et al., 1994; Hakanpää et al., 2004; Rilva et al., 2003). This property leads to a stabilization of gas bubbles (CO_2) in liquids (Wessels, 1996). The reduction of the surface tension leads also to the formation of bubbles, since the reduction of the boundary layer leads to an energy gain. If a bubble exhibits a larger diameter than the critical bubble diameter, it continues to increase its size in an oversaturated solution when the bottle pressure decreases. This can lead to a spontaneous release of the CO_2 and thus to gushing. On the one hand the hydrophobins can induce gushing by their strong surface activity, but on the other, the rodlet layers of hydrophobins with its typical roughness of the surface can act as condensation germs for the release of CO_2 bubbles (Hippeli and Elstner, 2002). Both can likewise induce gushing.

The concentration of hydrophobins required to induce gushing as shown in Table 6.1, has been reported to fall between 0.05 and 0.5 mg/kg. In a recent study, Sarlin et al. (2005b) reported that hydrophobins from *F. poae* could induce beer gushing at concentrations as low as 0.15 mg/l. The type I and II hydrophobins from *Trichoderma reesei* however, were much more active, inducing gushing at concentrations as low as 0.003 mg/l. Zapf et al. (2006) reported the cloning of type I and II hydrophobins from *F. culmorum* into brewers yeast. These transgenic yeasts were used for experimental fermentations, and gushing potential of the resultant beers was evaluated. Beers prepared from yeast expressing the type II hydrophobin were subject to gushing, while those prepared from yeasts expressing the type I did not gush. This indicates different roles of the two classes of hydrophobins in stabilizing or destabilizing foam.

Hippeli and Elstner (2002) discussed the formation of non-specific lipid transfer proteins (ns-LTP) in response to fungal infection, and speculated that they might be responsible for gushing. However Zapf et al. (2006) cloned a LTP into yeast and found no gushing in resultant beers.

Determination of gushing risk

A variety of methods have been investigated for determining the gushing propensity of malt and beer. These range from laboratory brewing trials, which require a week or more for completion, to simple predictive tests. Predictive tests do not measure gushing, but rather a factor that is correlated with gushing, and are most likely useful for the reduction of problematic samples during grain procurement. Predictive tests, however, are unlikely to completely eliminate risk, and the determination of gushing potential with a high degree of accuracy requires specific laboratory tests. The precision of laboratory tests will undoubtedly only improve as the factors and mechanisms which cause gushing are better defined.

Correlative factors

As the contamination of the barley and malt with fungi, in particular with *Fusarium*, can induce gushing, inspection of incoming raw materials for these organisms seems a natural means to reduce gushing risk. The logic is that, the greater the amount of fungi or fungal metabolite present, the greater the risk of gushing. Infestation of the barley or malt with fungi can be quantified by growth on selective media, but these methods are time consuming, and not amenable to high throughput screening. In addition, the identification of organisms to the genus and species generally requires considerable experience and skill. The value of plate counts in determining gushing risk appears debatable. Haikara (1980) evaluated 21 commercial malts and found that beers produced from samples with >50% contaminated kernels caused gushing in 8 of 12 cases, while those prepared from samples with 0–48% contaminated kernels did not gush in 8 of 9 cases. Sarlin et al. (2005a) later showed that gushing was positively correlated with CFU (on Czapek Dox iprodione dichloran agar) on barley that had been field inoculated with *Fusarium* sp. Malt samples displaying in excess of 10^3 CFU frequently induced gushing. In contrast, Schwarz et al. (1996) reported that the percentage of barley kernels infected with *Fusarium* was poorly correlated with gushing propensity of the resultant malts ($r = 0.30$). This study involved 55 naturally infected samples that displayed 0–100% kernels infected. Plate counts suffer from the additional deficiency fact that only viable *Fusarium* is measured and its viability tends to decline in storage (Beattie et al., 1997). As such plating could underestimate or even fail to detect *Fusarium* damaged samples.

Schildbach (1987) indicated that the measurement of grain microflora following harvest was not practical, but conversely, since gushing is more problematic with cool and damp conditions at harvest, the damaged rain-soaked barley can be often visually recognized. This damaged grain can be easily eliminated from the processing chain. In fact many maltsters and brewers stipulate clauses for *weathered* or *blighted grain* (or similar) in contracts or purchase documents. *Fusarium* species can form napthoquinone pigments that contribute a red color to mycelium on grain or in culture (Desjardins, 2006), and Müller (1995) proposed counting red (and black) grains as an indicator of the microbiological condition of a batch. However, as pigment formation is variable in *Fusarium*, and pigments are also formed by other yeasts and fungi (e.g. *Rhodotorula* sp.), this method provides only a rough orientation (Bellmer, 1995; Narzsis et al., 1990; Niessen et al., 1991). Donhauser et al. (1991) in fact reported that the visual determination of red kernels did not provide prediction of gushing at the required level of assurance.

Several faster techniques for the determination and quantification of *Fusarium* are available, but only limited work has been conducted on their utility in predicting gushing. Vaag and Pederson (1992), and Vaag (1991) developed a method for the measurement of *Fusarium* antigens on barley and malt, and conducted some preliminary work on the utility of the method for the prediction of gushing. The advantage of an immunological method over standard plate counts is that antigen levels remain constant during grain storage while *Fusarium* viability may decline. More recently, Hill et al. (2006) reported an ELISA (Enzyme Linked Immuno Sorbent Assay) method that can be used for determination of *Fusarium* in barley. Recent years have also seen the development of numerous PCR-based methods for the detection of *Fusarium* on wheat and barley (Niessen, 1993, 1997, 2002; Niessen and Vogel, 1997; Sarlin et al., 2004, 2006; Schilling et al., 1996; Schnerr et al., 2002; Turner et al., 1999; Waalwijk et al., 2004; Xu et al., 2003; Yli Mattila et al., 2004).

Mycotoxin testing is perhaps the most widely used predictive test in current commercial practice. Testing is widespread in regions that are impacted by FHB, because of legal, food safety and quality concerns. The reduction of gushing risk is an indirect benefit of this screening. While mycotoxins, do not cause gushing, their production is roughly correlated with *Fusarium* biomass (Garbe et al., 2007; Hill et al., 2006; Niessen, 2002; Schwarz et al., 1996), and then to some degree, gushing propensity. In most cases the determination of the mycotoxin, DON is performed. Literature reports on the relationship between DON and the gushing potential of malt have reached various conclusions. According to Sarlin et al. (2005b) no correlation can be found between DON content of a malt sample and the gushing potential. However, Schwarz et al. (1996) determined a moderately strong correlation between the malt DON content and gushing potential ($r = 0.77$). Some of this difference can be explained by samples. In the study by Sarlin, samples ($n = 44$) were primarily from Finland, were thought to be mostly infected with *F. avenceum*, and all had less than 1.5 µg/g DON. In contrast, samples ($n = 50$) in the study by Schwarz were from the Midwestern USA, were primarily infected with *F. graminearum*, and ranged from 0 to 13.5 µg/g DON in the malt. Other investigators (Garbe et al., 2007; Haikara, 1983; Munar and Sebree, 1997) also have shown a clear relationship between mycotoxins and gushing potential. Niessen et al. (1993), in a survey of 196 German beers, found that DON was significantly higher in gushing beers than non-gushing beers. Approximately 70% of gushing beers were found to have pronounced DON levels. Finally a malt sample with increased DON content has to be rejected with respect to product security and consumer health. In the European Union maximum levels of DON in grain have been regulated since July 2006 (http://eur-lex.europa.eu/LexUriServ/site/en/oj/2005/l_143/l_14320050607en00030008.pdf).

Schwarz et al. (1996) also reported a moderately strong correlation between ergosterol content in the malt and the gushing potential of the beers ($r = 0.74$). Ergosterol is a component of the cytoplasmic membrane of fungi and yeasts, and the determination is by reverse phase HPLC (Jambunathan et al., 1991; Peacock, 1989; Rodriguez and Parks, 2002). However, the method is difficult and time-consuming, and this not really amenable for screening of large numbers of samples. Somewhat similar to the use of ergosterol as a correlative factor is the measurement of extra cellular polysaccharides (EPS) secreted by fungi. Schwabe et al. (1994) developed an EPS latex agglutination test to determine the barley. The EPS on grain is correlated with DON in malt, and the test is utilized to some degree by maltsters in Germany.

Hydrophobins

As previously discussed, recent research results (Linder, et al. 2002; Niessen et al., 2006; Sarlin et al., 2005b; Zapf et al., 2006) have suggested that hydrophobins, secreted by various fungal species may be responsible for gushing in beer. Hydrophobins in malt, wort and beer can be determined according to the method of Sarlin et al. (2005b). The hydrophobin concentration is determined by means of a competitive ELISA test (ELISA-enzyme-coupled immune adsorption regulation), which has been patented (Haikara et al., 1999; Haikara et al., 2006). This procedure utilizes polyclonal antibodies which were raised against the hydrophobin from *F. poae*. Other antibodies are not used with this test, since *F. poae* was held responsible for the occurrence of gushing. On the contrary, Zepf (1998) found that *F. graminerum*, *F. culmorum*, *F. avenaceum* and *F. sambucinum* showed very strong gushing, while *F. poae* and *F. tricinctum* did not cause gushing. The findings of Zepf (1998) were confirmed by Laible and Geiger, 2003. Sarlin et al. (2005b) indicated that a critical concentration of hydrophobin was required for gushing. They found that gushing risk increased as hydrophobins levels increased above 250 µg/g malt, and samples with 500 µg/g malt repeatedly induced gushing in trials.

However, there are also still some questions as to whether the content of hydrophobins in malt is a clear indicator for gushing propensity in the derived beer (Garbe et al., 2007). Beer is a complex mixture and gushing promoters and gushing inhibitors are in a balance, which can influenced by the yeast during fermentation.

Laboratory gushing tests

Laboratory gushing tests are those that utilize beer, wort, or another matrix to evaluate the actual gushing propensity of a sample. Casey (1996) has previously reviewed a number of laboratory tests, and the current review will largely focus on developments in the past 10 years. Despite the fact that various tests have been developed (Donhauser et al., 1990; Mitteleuropäischen Brautechnischen Analysenkommission (MEBAK), 2006; Vaag et al., 1993), there is currently no gushing test that supplies highly reproducible and reliable results. Furthermore, these tests are very labor and time intensive and again are not really amenable for screening a large number of samples.

Carlsberg test

One of the most widely utilized tests to date is the so-called Carlsberg test, described by Vaag and coworkers in 1993. In this test, the investigation of the gushing potential is performed in a beer matrix. A defined quantity (100 g) of malt is extracted with water (400 ml) in a blender, and then concentrated (200 ml) by boiling. A portion of this malt extract (50 ml) is added to a bottle of commercial beer, the bottle is pasteurized, and then agitated (rocked) under defined conditions for 3 days. The bottle is opened and the quantity of escaped beer is weighed. Standard deviations associated with the test are quite high, especially with samples displaying low gushing tendency. In addition, the type of beer utilized has an impact on the results. As a result a number of workers have endeavored to make improvements.

Modified Carlsberg tests

Radau et al. (1995) reported the development of a modified Carlsberg test (MCT), to correct problems that were seen with the original method. A large portion of the variability in the original test was associated with the fact that composition of the beer matrix was not standardized. In the MCT artificially carbonated water (7.0 g/l CO_2) was used instead of beer. Use of coarse malt grist was also adopted as it was found to yield more gushing than fine grist. The malt was extracted with cold water and then thermally concentrated. A portion (50 ml) of the concentrated extract is added to the carbonated mineral water and agitated for 3 days. The bottles are opened and gushing is ranked as follows: loss of 0–5 g is no gushing, loss of 5–50 g is possible gushing (50%), and loss of >50 g is indicative of gushing in beer (92%). The MCT was accepted by MEBAK as a forecast of malt induced gushing (MEBAK, 2006). However, in the experience of many brewers the significance of the MCT is low, as the risk of false negative results tends to be high (Garbe et al., 2007).

Gushing problems in Germany during 2005 led to a re-investigation of the MCT, and the development of further modifications. The proposed test is referred to as the doubly modified Carlsberg test or M^2CT (Garbe et al., 2007). Modifications for the M^2CT included the use of fine malt grist for the preparation of Congress worts. The remainder of the test was as for the MCT, including the use of the carbonated water matrix. The use of finely ground malt and then the Congress wort matrix, yielded a much greater gushing propensity than cold water extracts as utilized in the MCT. However, inter-laboratory tests of this additional modification are still needed.

Control of gushing

As brewers have continued to face malt mycoflora-related problems, including gushing, research efforts have been devoted to control measures. Several of these control measures were previously covered in the review by Schwarz (2003). Treatment application is typically to either the raw grain or in malt production, and it should be stressed that most of these measures are not specifically targeted towards gushing, but rather at controlling the growth of *Fusarium* or other problematic fungi during malting, or reducing mycotoxin contamination.

The simplest and most widely practiced means to control mold associated problems like gushing is to avoid the purchase and utilization of contaminated grain. Maltster or brewer imposed specifications for weather damaged, blighted or moldy grain, as well as for mycotoxins like DON, probably have the outcome of eliminating many problematic samples. However, as previously discussed, the relationships between these correlative factors and gushing are moderate at best. Complete elimination, thus may not be achieved. Also strict avoidance is not always practical, and severe economic costs can be associated with avoidance. Growers generally receive a discounted price for contaminated grain, and processors incur additional costs associated with testing and with importation of grain from other regions. Grain supply limitations may also compel the utilization of contaminated grain during some crop years, or within some regions. The ultimate solution to the problem will be development of FHB resistant cultivars of barley and wheat, and control of the disease through cultural practices and fungicides. Considerable work worldwide is currently being devoted to these goals (Leonard and Bushnell, 2003).

Treatment of barley or malt

Physical methods

The cleaning or sizing of grain prior to malting offers an opportunity for the reduction of *Fusarium* or mold contamination. Research has shown that cleaning operations can be effective for the removal of the shriveled *Fusarium* damaged kernels (FDK) in wheat (Dexter and Nowicki, 2003). Cleaning operations with barley, however, have met with mixed results (Schwarz, 2003). When FHB infection of barley occurs later during kernel development reduction in kernel size may be minimal, and overall, kernel size reduction in barley does not appear to be as pronounced as for wheat. Schwarz et al. (2006) has shown that sieving (screening) is impractical for reducing DON levels in many samples of FHB infected barley. DON is often present on plump kernels as well as on thins.

Kottapalli et al. (2003) evaluated the use of hot water treatment and electron beam irradiation for reducing *Fusarium* infection in malting barley. They found that electron beam irradiation of barley at doses of 6–8 kGy reduced *Fusarium* infection, with little impact upon germination. Hot water treatment at either 45 or 50°C for 15 min resulted in reductions in *Fusarium* infection from 32 to 1–2%, with again, only a very slight reduction in germination. Kottapalli (unpublished) later found that these hot water treatments could be applied in pilot-malting with minimal to no impact on malt quality. While the hot water treatment of barley always reduced the number of kernels contaminated with *Fusarium*, and in some cases DON, the *Fusarium* was never completely eliminated from the resultant malts.

Chemical treatments

The use of formaldehyde in the steep to overcome mold/dormancy problems was reported by Gjersten (1967), and Haikara (1980) confirmed that 1000 mg/kg barley in the first steep suppresses the growth of *Fusarium*. However, its use in commercial malting practice is not suitable. Kottapalli et al. (2005) showed that treatment of barley with hydrogen peroxide decreased *Fusarium* by 50 to 98% within 5 min of exposure, and had no effect on germination. The use of gaseous ozone as a grain fumigant was evaluated by both Allen et al. (2003) and Kottapalli et al. (2005). Allen evaluated the general fungicidal efficacy of ozone on barley without respect to specific genera. A 96% inactivation rate was achieved for fungal spores and mixtures of spores and mycelium after 5 min of ozonation at 0.16 and 0.10 mg ozone/g barley/min. Inactivation of fungi continued in silos, as long as ozone gas was retained in the storage atmosphere, and efficacy also increased with water activity (a_w) and temperature of the barley. Kottapalli et al. (2005) showed that 11 and 26 mg/g barley of gaseous ozone applied for up to 60 min reduced kernels infected with *Fusarium* by 24–36%. Some problems with seed germination were detected under certain conditions.

Biological treatments

Biological control, as an approach for the control of *Fusarium* and other fungi has received considerable attention since the mid-1990s. Cultures of biocontrol organisms are applied either in the steeping or germination phase of malting, and rely upon the competitive or antimicrobial properties of the culture organisms. Part of the attraction of this approach is the reliance on natural grain or brewery organisms. Boivin and Malanda (1997) first reported the use of the yeast *Geotrichum candidum*, and Haikara et al. (1993) the use of lactic acid bacteria. The use of lactic acid bacteria in malting and brewing was reviewed by Lowe and Arendt (2004). Starter cultures of *Lactobacillus plantarum* and *Pedicoccus pentosaceus* have been extensively investigated by researchers in the Biotechnology section of the Technical Research Center of Finland (VTT) (Haikara and Laitila, 1995; Laitila et al., 2002, 2006). Application of the cultures or cultures plus media to the steep has been shown to significantly reduce the counts of *Fusarium* sp. as well as other fungi and bacteria. Specific results have, however, varied somewhat with individual barley samples, undoubtedly reflecting differences in the microflora from different environments or crop year. Application of the cultures and culture media has been found to be most effective during steeping. The application of lactic acid cultures to barley in the field has also been evaluated and was found to reduce both *Fusarium* counts on the grain, and the gushing tendency of the beer (Reinikainen et al., 1997). As the *in-vitro* application of cell-free extracts of two *L. plantarum* strains were effective against several species of *Fusarium* the effect is assumed to be fungistatic.

In similar experiments, Boivin and Malanda (1997) utilized starter cultures of the yeast *Geotrichum candidum* in industrial-scale malting of infected barley. Application of the culture to the steep was found to completely eliminate *Fusarium* and DON on the steeped barley. *Geotrichum* is apparently able to grow faster and capture more of the available resources, thus limiting *Fusarium* growth. Both Boivin and Malanda (1997) and Haikara and Laitila (1995), Laitila et al. (2006) found that the application of starter cultures had numerous beneficial effects on malting and brewing quality. The malting process use of *G. candidum* has been patented (Boivin and Malanda, 1999).

Treatment of beer

A number of studies have evaluated treatments to reduce gushing in problematic beer. However, both the application and efficacy of post-production treatments are probably limited. Problems are frequently not detected until the beer is actually packaged and in the marketplace. Gjersten (1967) stated that the most practical measure is to blend gushing beer with non-gushing beer, which would have the effect of diluting gushing active components below a critical concentration.

Enzymes

Aastrup et al. (1995, 1996) reported that gushing caused by the *Fusarium* gushing-inducing factor could be reduced by treatment with commercial bacterial enzyme preparations. The action was attributed to proteolytic activity in the enzyme preparations, and their activity against a proteinaceous gushing factor. Enzyme treatment was proposed as an anti-gushing treatment in the brewery when precautionary measures have failed to eliminate problematic samples. Goh et al. (1980) patented a process for reducing gushing that involved use of an acid protease and a pepsin inhibitor from *Streptomyces*. This process was directed at reduction of gushing that was associated with chill-proofing enzymes. Burger and Bach (2001a,b) patented the use of beta-glucanase for the reduction or prevention of gushing, but there was no reference to primary gushing associated with mycoflora.

Adsorbents

As previously reviewed by Amaha and Kitabatake (1981), several researchers have reported the use adsorbents to be somewhat effective in reducing the gushing potential of problematic beer. Compounds evaluated included charcoal, Fuller's Earth, Tansul, kaolin, activated alumina and nylon. Musial and Schwarz (unpublished) treated beer prepared from a gushing positive *Fusarium* infected malt with either Polyclar-10, Polyclar SB-100 or silica xerogel Lucilite XL. Each of the chill-proofing agents appeared to somewhat reduce the amount of gushing observed, but differences were not statistically significant. The lack of statistical difference was attributed to the inherent high variability of the gushing method (Carlsberg test).

Secondary gushing

In general, secondary beer gushing relates to the introduction of nucleation sites for the dissolved CO_2. It has been attributed to a number of factors including, calcium oxalate, metal ions, tensides, isomerized hop extracts, the crown cork, filter aids, and the inner surface of the bottle (Amaha and Kitabatake, 1981;

Carrington et al., 1972; Franke, 1995; Gardner et al., 1973; Kieninger, 1976; Outtrup, 1980; Sandra et al., 1973; Weideneder, 1992; Zepf, 1998). This topic was extensively reviewed by Amaha and Kitabatake (1981), and there has been only a limited amount of research since this time (Wershofen, 2004; Winkelmann, 2004; Zarnkow and Back, 2001). The effect of secondary gushing can, in contrast to primary gushing, be influenced by changing brewing process parameters, such as Ca^{2+} and CO_2 concentrations (Wershofen, 2004).

Calcium oxalate

If the concentration of calcium oxalate in beer is too high, it will precipitate as crystals. These particles (seed crystals) then can induce the release of CO_2 resulting in gushing (Jacob, 1998; Madigan et al., 1994; Schur et al., 1980; Schildbach and Müller, 1980; Zepf and Geiger, 1999, 2000). The major source of calcium and oxalic acid is malt, brewery water, and to a small extent hops. The content of oxalate in barley malt ranges from 100 to 500 mg/kg, while wheat malts contain higher amounts. In barley malt, the amount of Ca^{2+} ranges from 100 to 200 mg/kg in barley malt, while around 100–120 mg/kg in wheat malt. In the brewery, the concentration of Ca^{2+} can mainly be influenced by the brewing water and by addition of $CaCl_2$. However, a large amount of Ca^{2+} is lost with the husks.

The final concentration of Ca^{2+} and oxalate leading to gushing in wort and beer has been reported as 60 mg/kg and 20–30 mg/kg, respectively. In pure water, the solubility product of calcium oxalate has been reported as $8 \times 10^{-9} mol^2/kg^2$, which is the product of the Ca^{2+} activity and oxalate activity (the activity of an ion is approximately the same as the ion concentration in an ideal diluted solution). Therefore about 15 mg/kg of calcium and oxalate leads to a precipitation of calcium oxalate crystals (Burger and Becker, 1999; Brenner, 1957). These differences can be explained by the different modes of solubilization and chelation of calcium and of oxalate. Dialysis of wort with a 10 kDa MWCO (molecular weight cut off) membrane for example eliminates only about 50% of the Ca^{2+} from wort indicating a large amount of Ca^{2+} is bound with proteins, etc. In addition, the enzymatic determination of oxalate leads to results that are 20–40% lower, than the results of the GC-EI-MS isotope dilution assay using 2-^{13}C-oxalate as isotope standard (Garbe, unpublished).

An aim of the brewing process, therefore should be to control Ca^{2+} concentration in order to insure that most calcium oxalate is precipitated prior to beer filtration. The process should therefore be influenced by adding Ca^{2+} before filtration, and then avoiding Ca^{2+} contamination after filtration (Schur et al., 1980; Zepf and Geiger, 1999, 2000).

Metal ions

The contribution of metal ions to beer gushing was recognized many years ago. Guggenberger and Kleber reported in 1963 the gushing of distilled water (with dissolved gas) after addition of transition metal ions. Gushing risk was significantly correlated with the concentration of the ion and pH. The literature clearly indicates that Fe^{3+} ions can be responsible for gushing, and while other ions like Ni^{2+} and Co^{2+} have been discussed, their contribution seems to be less important (Kieninger, 1976; Weideneder, 1992; Zepf, 1998). The physiochemical mechanism of the metal ion induced gushing is unknown. Some models assume the formation of metal-carbonates (Guggenberger and Kleber, 1963). The formation of metal carbonates in the phase boundary between the beer and CO_2 stabilizes the bubble. Pressure relief leads to hydrolyzation of the metal carbonate its stabilizing effect. Weak support for this assumption is that the addition of EDTA inhibits the metal induced gushing.

Tensides

Nickel and iron ions, and many other substances in beer such as protein, also influence the surface tension. The lower the surface tension, the higher the gushing risk of a beer. The origin of tensides in beer is mainly bottle washing caustic which contains tensides (Dachs and Nitschke, 1977; Linemann, 1996; Weideneder, 1992). The optimization of bottle washing in current practice usually prevents tensides in beer. Nevertheless tensides must be taken in account when secondary gushing occurs (Gardner, 1972, 1973; Linemann, 1996).

Isomerized hop extracts

Isomerized hop extracts (iso-humulones) have been used since the beginning of 1970s, and some breweries recognized an increased gushing risk with the addition of these compounds. Subsequently, dehydrated humulunic acid was identified as strong gushing promoter (Carrington et al., 1972; Laws and McGuinness, 1972; Outtrup, 1980). Some isomerized hop products contained up to 5% of dehydrated humulunic acid (Amaha and Kitabatake, 1981), and some of its further derivatives exhibited an even stronger gushing promotion. Hydrogenated iso-alpha acids (i.e. tetrahydro and hexahydro iso-alpha acids) have also been reported to weakly induce or even support gushing (Carrington et al., 1972; Laws and McGuinness, 1972). However, polyunsaturated fatty acids, which also are constituents of the hop oil, show inhibitory effects on gushing (Gardner et al., 1973). In total, hop induced gushing seems to be of lesser importance in modern practice.

Bottle and bottle filling

The relation of a rough inner surface of a bottle and the spontaneous release of dissolved gas is an effect even known from everyday life. Therefore the preparation and cleaning of bottles can be a factor contributing to the gushing problem. Alkaline bottle washing dissolves the glass surface and changes its properties. The introduction of PET bottles with different inner coatings might change this situation. However, coatings should be checked with respect to gas permeability, volatile organic carbon permeates and migrants in addition to its surface properties.

The shape of bottles, their level of fill, as well as the crown cork do not directly cause gushing (Gardner, 1973). However, in praxis, some gushing problems have been occurred with respect to these factors. Detergents excreted from the crown cork compound, or damage of crown resulting in a release of iron ions, have been reported to lower surface tension of the beer.

Particles and ions derived from Kieselgur

Very small particles and ions derived from Kieselgur transferred into the finished beer can cause gushing (Kieninger, 1976). However, this basic physiochemical process is very complex and is not fully understood. In modern practice, many breweries have changed to membrane filtration. While, membrane filtration processes bears new problems, it should eliminate the risk of particle induced gushing.

Summary

Secondary gushing would seem to be the least problematic form of gushing, as it can generally be handled with process changes. To avoid secondary gushing, it is important to maintain high concentrations of Ca^{2+} before final filtration and low concentrations thereafter. Avoidance of iron ions is also important not only for secondary gushing, but also with respective to flavor stability. Attention must also be paid to bottle washing operations, as tensides can induce gushing.

Primary gushing related to grain mycoflora remains the primary problem, and is not easily dealt with especially in years when weather conditions cause widespread contamination of the crop. This can cause extensive damage to an entire industry. The extent of gushing problems to some extent will be more problematic for beers using 100% malt, as the use of adjuncts has the effect of diluting the gushing factors contributed by the malt.

It has long been suspected that the fungal gushing factors were proteinaceous in nature, and the identification of fungal hydrophobins as a cause of gushing represents a major advance in knowledge of the problem. However, the mechanism of gushing is not yet fully understood, and it appears that multiple factors may be involved. As such, the simple screening of samples for hydrophobins may not be the definitive solution. Likewise, while the screening of grain for fungal contamination or mycotoxins can help to avoid some problematic samples, these techniques are unlikely to completely eliminate all gushing samples. Current laboratory tests for gushing are time-consuming and not fully reliable.

Elimination of primary gushing problems will come with development of resistant cereal varieties and/or the effective control of the microorganisms in the field and malthouse. Until this happens, additional research is required to improve our understanding of the gushing mechanism and associated factors, and to develop reliable laboratory tests that are amenable to the analysis of a large number of samples.

References

Aastrup, S., Legind-Hansen, P. and Nielsen, H. (1995) Enzymatischer Abbau der Gushing-Neigung im Bier. *Brauwelt*, 135 (28/29), 1385–1387.

Aastrup, S., Legind-Hansen, P. and Nielsen, H. (1996) Enzymatic reduction of gushing tendencies in beer. *Brauwelt International*, 14 (20), 136–137.

Allen, B., Wu, J. and Doan, H. (2003) Inactivation of fungi associated with barley grain by gaseous ozone. *Journal of Environmental Science Health. Part B: Pesticides, Food Contaminants, and Agricultural Wastes B*, 38 (5), 617–630.

Amaha, M. and Kitabatake, K. (1981) Gushing in beer. In: *Brewing Science* (J. R. A. Pollock ed.), Vol. 2, pp. 457–489. Academic Press: NY.

Amaha, M., Kitabatake, K., Nakagawa, A., Yoshida, J. and Harada, T. (1973) Gushing inducers produced by some mould strains. *Proceedings of the European Brewery Convention Congress*, 14, 381–398.

Amaha, M., Horiuchi, G. and Yabuuchi, S. (1978) Involvement of chill-proofing enzymes in the winter-type gushing of bottled beer. *MBAA Tech. Quart.*, 15 (1), 15–22.

Beattie, S., Schwarz, P. B., Horsley, R., Barr, J. and Casper, H. (1997) The effect of grain storage conditions on the viability of Fusarium and deoxynivalenol production in infested malting barley. *Journal of Food Protection*, 61 (1), 103–106.

Bellmer, H. G. (1995) Forschungsprojekt Gushing. *Brauwelt*, 24/25, 1167–1170.

Bell-Pedersen, D., Dunlap, J. C. and Loros, J. J. (1992) The *Neurospora* circadian clock-controlled gene, ccg–2, is allelic to eas and encodes a fungal hydrophobin required for formation of the conidial rodlet layer. *Genes and Development*, 6, 2382–2394.

Boivin, P. and Malanda, M. (1997) Improvement of malt quality and safety by adding starter culture during the malting process. *MBAA Technical Quarterly*, 34 (2), 96–101.

Boivin, P. and Malanda, M. B. (1999) Inoculation by *Geotrichum candidum* during malting of cereals or other plants. United States Patent US5955070 (date issued: 21 September).

Brenner, M. W. (1957) Gushing Beer II. Causes and some means of prevention. *Proceedings of the European Brewery Convention Congress*, 6, 349–362.

Burger, K. and Bach, H.-P. (2001a) Mixtures of enzymes containing an enzyme with beta-glucanase activity, to be used for decreasing or preventing gushing. European Patent EP1164184 (date issued: 19 December).

Burger, K. and Bach, H.-P. (2001b) Mixtures of enzymes containing one enzyme with beta-glucanase activity, their use to reduce or prevent gushing. European Patent EP1162259 (date issued: 12 December).

Burger, M. and Becker, K. (1949) Oxalate studies on beer. *Proceedings of the American Society of Brewing Chemists*, 102–115.

Carrington, R., Collett, C. R., Dunkin, I. R. and Halek, G. (1972) Gushing promoters and suppressants in beer and hops. *Journal of the Institute of Brewing*, 78, 243–254.

Casey, G. (1996) Primary versus secondary gushing and assay procedures used to assess malt/beer gushing potential. *MBAA Technical Quarterly*, 33 (4), 229–235.

Dachs, E., Nitschke, R. (1977) Fallstudie zum Problem Gughing Brauwelt, 117, 129–131.

Desjardins, A. E. (2006) Selected mycotoxigenic *Fusarium* species. *Fusarium Mycotoxins: Chemistry, Genetics and Biology*, pp. 145–194. American Phytopathological Society Press: St Paul, MN.

de Vocht, M. L., Scholtmeijer, K., van der Vegte, E. W., de Vries, O. M. H., Sonveaux, N., Woesten, H. A. B., Ruysschaert, J.-M., Hadziioannou, G., Wessels, J. G. H. and Robillard, G. T. (1998) *Biophysical. Journal*, 74, 2059–2068.

De Vries, O. M. H., Moore, S., Arntz, C., Wessels, J. G. H. and Tudzynski, P. (1999) Identification and characterization of a tri-partite hydrophobin from *Claviceps fusiformis*. A novel type of class II hydrophobin. *European Journal of Biochemistry*, 262, 377–385.

Dexter, J. E. and Nowicki, T. W. (2003) Safety assurance and quality assurance issues associated with Fusarium Head Blight in wheat. In: *Fusarium Head Blight of Wheat and Barley* (K. J. Leonard and W. R. Bushnell eds), pp. 420–460. American Phytopathological Society Press: St Paul, MN.

Dilly, P. G. (1988) Eine Bestandsaufnahme. *Brauwelt*, 45, 2062–2072.

Donhauser, S., Weideneder, A., Winnewisser, W. and Geiger, E. (1990) Test zur Ermittlung der Gushingneigung von Rohfrucht, Malz, Würze und Bier. *Brauwelt*, 32, 1317–1320.

Donhauser, S., Weideneder, A., Winnewisser, W. and Geiger, E. (1991) A test for determining the gushing propensity of raw grain, malt, wort and beer. *Brauwelt International*, 9 (4), 294–297.

Draeger, M. (1996a) Physical observations on the subject of gushing. *Brauwelt International*, 14 (4), 363–367.

Draeger, M. (1996b) Physikalische Ueberlegungen zum Thema Gushing. *Brauwelt*, 136 (6), 259–264.

Fischer, S. (2000) Bestimmung Gushingverursachender Partikel in Versuchslösungen und Bier mittels Photonenkorrelationsspektroskopie. Schlussbericht für das Forschungsvorhaben R. (Wissenschaftsförderung der Deutschen Brauwirtschaft. Berlin, Germany), 314, 9–1,

Fischer, S., Schwill-Miedaner, A., Illberg, V. and Sommer, K. (1997) Untersuchung von Einflussfaktoren des Gushingphänomens. *Brauwelt*, 137 (6), 210–214.

Franke, D., Pahl, M. and Vesting, M. (1995) *Theorie zur Existenz von Mikroblasen in Getränken Brauwelt*, 38/39, 1944–1949.

Gale, L. R. (2003) Population biology of *Fusarium* species causing head blight of grain crops. In: *Fusarium Head Blight of Wheat and Barley* (K. J. Leonard and W. R. Bushnell eds), pp. 120–143. American Phytopathological Society Press: St Paul, MN.

Garbe, L. A., Nagel, R., Rauschmann, M., Lamers, M., Ehmer, A. and Tressl, R. (2007) Correlation of Deoxynivalenol, Hydrophobins and Gushing 31. *International European Brewing Convention Congress Proceedings*, Venice, Italy, 06.-10. Mai 2007 only on CD-ROM, Fachverlag Hans Carl, 90268 Nürnberg, Germany.

Gardner, R. J. (1972) Surface viscosity and gushing. *Journal of the Institute of Brewing*, 78, 391–399.

Gardner, R. J. (1973) The mechanism of Gushing. A review. *Journal of the Institute of Brewing*, 79, 275–283.

Gardner, R. J., Laws, D. R. J. and McGuinness, J. D. (1973) The suppression of gushing by the use of the hop oil. *Journal of the Institute of Brewing*, 79, 209–211.

Gjersten, P. (1967) Gushing in beer: It's nature, cause and prevention. *Brewers Digest*, 42 (5), 80–84.

Gjertsen, P., Trolle, B. and Andersen, K. (1963) Weathered barley as a contributory cause of gushing in beer. *Proceedings of the European Brewery Convention Congress*, 9, 320–341.

Goh, H., Yabuuchi, S. and Amaha, M. (1980) Method of preventing gushing of packaged beer. United States Patent US4181742 (date issused: 1 Janaury).

Guggenberger, J. and Kleber, W. (1963) Über den Mechanismus des Wildwerdens von Bier. *Proceedings of the European Brewery Convention Congress*, 9, 299–319.

Gyllang, H., Sätmark, L. and Martinson, E. (1977) The influence of some fungi on malt quality. *Proceedings of the European Brewery Convention*, 16, 245–254.

Haikara, A. (1980) Gushing induced by fungi. European Brewery Convention Monograph VI; Relationship between malt and beer. European Brewery Convention Monograph IV, pp. 251–258.

Haikara, A. (1983) Malt and beer from barley artificially contaminated with *Fusarium* in the field. *Proceedings of the European Brewery Convention*, 19, 401–408.

Haikara, A. and Laitila, A. (1995) Influence of lactic acid star cultures on the quality of malt and beer. *Proceedings of the European Brewery Convention Congress*, 25, 249–256.

Haikara, A., Uljas, H. and Suurnäkki, (1993) Lactic starter cultures in malting: A novel solution to gushing problems. *Proceedings of the European Brewery Convention Congress*, 24, 163–172.

Haikara, A., Sarlin, T., Nakari-Setälä, T., and Pentillä, M. (1999) Method for determining a gushing factor for a beverage. PCT Patent WO 99/54725.

Haikara, A., Kleemola, T., Nakarai-Setälä, T. and Penttilä, M. (2006) Method for determining a gushing factor for a beverage. US Patent 7,041,464 B2 (date issued: 3 May).

Hakanpää, J., Paananen, A., Askolin, S., Nakari-Setälä, T., Parkkinen, T., Penttilä, M., Linder, M. B. and Rouvinen, J. (2004) Atomic resolution structure of the HFB II hydrophobin, a self-assembling amphiphile. *Journal of Biological Chemistry*, 279, 534–539.

Helm, E. and Richardt, O. C. (1938) Das Überschäumen (Wildwerden) des Bieres. *Wochenschrift für Brauerei*, 55, 89–94.

Hill, N. S., Schwarz, P. B., Dahleen, L. S., Neate, S. M., Glenn, A. E. and O'Donnell, K. (2006) ELISA analysis for *Fusarium* in barley: Development of methodology and field assessment. *Crop Science*, 46, 2636–2642.

Hippeli, S. and Elstner, E. F. (2002) Are hydrophobins and/or non-specific lipid transfer proteins responsible for gushing in beer? New hypotheses on the chemical nature of gushing inducing factors. *Zeitschrift. Für Naturforschung*, 57c, 1–8.

Jacob, F. (1998) Calcium oxalsäure: Technologische relevanz. *Brauwelt*, 138 (28/29), 1286–1287.

Jambunathan, , Kherdekar, M. S. and Pawan, V. (1991) Ergosterol concentration in mold-susceptible and mold-resistant sorghum at different stages of grain development and its relationship to flavan-4-ols. *Journal of Agricultural Food Chemistry*, 39 (10), 1866–1870.

Kieninger, H. (1976) Gushing des Flaschenbieres. Derzeitiger Forschungsstand. *Brauwelt*, 116, 1633–1636.

Kitabatake, K. (1978) A wort component responsible for gushing in beer. *Bulletin of Brewing Science (Tokyo)*, 24, 21–32.

Kitabatake, K. and Amaha, M. (1977) Effect of chemical modifications on the gushing-inducing activity of a hydrophobic protein produced by a *Nigrospora* sp. *Agricultural and Biological Chemistry*, 41 (6), 1011–1019.

Kitabatake, K., Fukushima, S., Kawasaki, I. and Amaha, M. (1980) Gushing-active peptides in beer produced by *Penicillium chrysogenum*. In: *Peptide Chemistry 1980* (H. Yonehara ed.), pp. 7–12. Protein Research Foundation: Osaka, Japan.

Kottapalli, B., Wolf-Hall, C. E., Schwarz, P., Schwarz, J. and Gillespie, J. (2003) Evaluation of hot water and electron beam irradiation for reducing *Fusarium* infection in malting barley. *Journal of Food Protection*, 66 (7), 1241–1246.

Kottapalli, B., Wolf-Hall, C. E. and Schwarz, P. (2005) Evaluation of gaseous ozone and hydrogen peroxide treatments for reducing *Fusarium* survival in malting barley. *Journal of Food Protection*, 68 (6), 1236–1240.

Krause, B. (1936) The stability of supersaturated carbonic acid solutions, especially that of beer. *Svenska Bryggareföreningens Manadsblad*, 5, 221–236.

Laible, B. and Geiger, B. (2003) Primary gushing and polar lipids. An important addition to gushing research. *Proceedings of the Congress of the European Brewery Convention*, 30, 915–922.

Laitila, A., Alakomi, H.-L., Rasca, L., Mattila-Sandholm, T. and Haikara, A. (2002) Antifungal activities of two Lactobacillus plantarum strains against moulds in vitro and in malting of barley. *Journal of Applied Microbiology*, 93, 566–576.

Laitila, A., Sweins, H., Vilpola, A., Kotaviita, E., Olkku, J., Home, S. and Haikara, A. (2006) *Lactobacillus plantarum* and *Pediococcus pentosaceus* starter cultures as a tool for microflora management in malting and for enhancement of malt processability. *Journal of Agricultural and Food Chemistry*, 54 (11), 3840–3851.

Laws, D. R. J. and McGuinness, J. D. (1972) Origin and estimation of the gushing potential of isomerized hop extracts. *Journal of the Institute of Brewing*, 78, 302–308.

Leonard, K. J. and Bushnell, W. R. (eds) (2003) *Fusarium Head Blight of Wheat and Barley*, p. 512. American Phytopathological Society Press: St Paul, MN.

Linder, M., Szilvay, G. R., Nakari-Setälä, T., Söderlund, H. and Penttilä, M. (2002) Surface adhesion of fusion proteins containing the hydrophobins HFBI and HFBII from *Trichoderma reesei*. *Protein Science*, 11, 2257–2266.

Linder, M. B., Szilvay, G. R., Nakari-Setälä, T. and Penttila, M. E. (2005) Hydrophobins: The protein-amphiphiles of filamentous fungi. *FEMS Microbiology Review*, 29, 877–896.

Linemann, A. (1996) Gushing: Spontanes Überschäumen von Flaschenbier. *Brauerei Forum*, 11 (14), 217–219.

Lowe, D. P. and Arendt, E. K. (2004) The use and effects of lactic acid bacteria in malting and brewing with their relationships to antifungal activity, mycotoxins and gushing: A review. *Journal of the Institute of Brewing*, 110 (3), 163–180.

Madigan, D. M., McMurrough, I. and Smyth, M. R. (1994) Determination of oxalate in beer and beer sediments using ion chromatography. *Journal of the American Society of Brewing Chemists*, 52 (3), 134–137.

McMullen, M., Jones, R. and Gallenberg, D. (1997) Scab of wheat and barley: A re-emerging disease of devastating impact. *Plant Disease*, 81 (12), 1340–1348.

Mitteleuropäischen Brautechnischen Analysenkommission (MEBAK) (2006) Brautechnische Analysenmethoden. Rohstoffe. Gerste. 1.2.8. Mykologischer Status, and Malz 3.14.21. Gushing. MEBAK, Freising, Germany.

Moll, M. (1991) Composition and properties of beer. *Beers and Coolers. Definition, Manufacture, and Composition*, pp. 270–345. Intercept Ltd.: Andover, UK.

Müller, C. (1995) Möglichkeiten zur Bewertung des Gesundheitszustandes von Braugerste und Malz. *Brauwelt*, 21, 1036–1054.

Munar, M. and Sebree, B. (1997) Gushing – A maltster's view. *Journal of the American Society of Brewing Chemists*, 55 (3), 119–122.

Narziss, L., Back, W., Reicheneder, E., Simon, A. and Grandl, R. (1990) Untersuchungen zum Gushing-Problem. *Monatsschrift fur Brauwissenschaft*, 43 (9), 296–305.

Niessen, L. M. (1993) Entwicklung und Anwendung Immunchemischer Verfahren zum Nachweis Wichtiger *Fusarium*-Toxine bei der Bierbereitung sowie Mycologische Untersuchungen im Zusammenhang mit dem Wildwerden (Gushing) von Bieren. Doctoral thesis. Technical University Munich. Program for Technical Microbiology and Brewing Technology 2.

Niessen, R. (2002) Weiterentwicklung von DNA-gestützten Verfahren zur schnellen Detektion von Fusarium – Kontaminationen und deren Anwendung bei der Qualitätskontrolle in Brauereien. Schlussbericht für das Forschungsvorhaben. (Wissenschaftsförderung der Deutschen Brauwirtschaft. Berlin, Germany, www.aif.de) AiF 12163/N.

Niessen, M. L. and Vogel, R. F. (1997) Moniotoring of trichothecene producing *Fusarium* species in brewing cereals using a group specific Polymerase Chain Reaction. *Proceedings of the European Brewery Convention*, 26, 61–67.

Niessen, L., Donhauser, S., Weideneder, A., Geiger, E. and Vogel, H. (1991) Möglichkeiten einer verbesserten visuellen Beurteilung des mikrobiologischen Status von Malzen. *Brauwelt*, 131 (37), 1556–1561.

Niessen, L., Donhauser, S., Weideneder, A., Geiger, E. and Vogel, H. (1992) Mycologische Untersuchungen an Cerealien und Malzen im Zusammenhang mit Wildwerden (Gushing) des Bieres. *Brauwelt*, 132 (16/17). 702, 704–706, 709–712, 714

Niessen, L., Böhm-Schraml, M., Vogel, H. and Donhauser, S. (1993) Deoxynivalenol in commercial beer-screening for the toxin with an indirect competitive ELISA. *Mycotoxin Research*, 9, 99–109.

Niessen, L., Hecht, D., Zapf, M., Theisen, S., Vogel, R. F., Elstner, E. and Hippeli, S. (2006) Zur Rolle oberflächenaktiver Proteine Entstehung des Gushings sowie zu den Möglichkeiten ihrer Beeinflussung. *Brauwelt*, 146 (19/20), 570–572.

Outtrup, H. (1980) The relation between the molecular structure and gushing potential of dehydrated humulunic acid. *Carlsberg Research Communication*, 45, 381–388.

Paananen, A., Vuorimaa, E., Torkkeli, M., Penttilä, M., Kauranen, M., Ikkala, O., Lemmetyinen, H., Serimaa, R. and Linder, M. (2003) Structural hierarchy in molecular films of two class II hydrophobins. *Biochemistry*, 42, 5253–5258.

Peacock, G. A. (1989) Separation of fungal sterols by normal-phase HPLC. *Journal of Chromatographic Science*, 469, 293–304.

Prentice, N. and Sloey, W. (1960) Studies on barley microflora of possible importance to malting and brewing quality. I. The treatment of barley during malting with selected microorganisms. *Proceedings of the American Society of Brewing Chemists*, 28–34.

Prom, L. K., Horsley, R. D., Steffenson, B. J. and Schwarz, P. B. (1999) Development of Fusarium Head Blight and accumulation of deoxynivalenol in barley sampled at different growth stages. *Journal of the American Society of Brewing Chemists*, 57 (2), 60–63.

Radau, B., Linemann, A., Krüger, E. (1995) Modifizierter Carlsberg-Test (MCT). *Brauerei-Forum 10*, 377–378.

Reinikainen, P., Peltola, P., Lampien, R., Haikara, A. and Olkku, J. (1997) Improving the quality of malting barley by employing microbial starter cultures in the field. *Proceedings of the European Brewery Convention Congress*, 27, 551–558.

Ritva, S., Torkkeli, M., Paananen, A., Linder, M., Kisko, K., Knaapila, M., Ikkala, O., Vuorimaa, E., Lemmetyinen, H. and Seeck, O. (2003) Self-assembled structures of hydrophobins HFBI and HFBII. *Journal of Applied Crystallography*, 36, 499–502.

Rodriguez, R. J. and Parks, W. (2002) HPLC of sterols: Yeast sterols. *Meth. Enzym.*, 111, 37–51.

Salas, B., Steffenson, B. J., Casper, H. H., Tacke, B., Prom, L. K., Fetch, T. G. Jr. and Schwarz, P. B. (1999) Fusarium species pathogenic to barley and their associated mycotoxins. *Plant Disease*, 83, 667–674.

Sarlin, T., Laitila, A., Pekkarinen, A. and Haikara, A. (2005a) Effects of three *Fusarium* species on the quality of barley and malt. *Journal of the American Society of Brewing Chemists*, 63 (2), 43–49.

Sarlin, T., Nakari-Setälä, T., Linder, M., Penttilä, M. and Haikara, A. (2005b) Fungal hydrophobins as predictors of the gushing activity of malt. *Journal of the Institute of Brewing*, 111 (2), 105–111.

Sarlin, T., Yli-Mattila, T., Jestoi, M., Rizzo, A., Paavanen-Huhtala, S. and Haikara, A. (2006) Real-time PCR for quantification of toxigenic *Fusarium* species in barley and malt. *European Journal of Plant Pathology*, 114 (4), 371–380.

Schildbach, R. (1987) Auswuchs der Gerste und Überschäumen des Bieres. *Brauwelt*, 127 (36), 1559–1567.

Schildbach, R. and Müller, J. (1980) Einflüsse der Technologie auf den Oxalatgehalt des Bieres und seiner Rohstoffe. *Brauwelt*, 120, 1648.

Schilling, A., Moeller, E. and Geiger, H. (1996) Polymerase chain reaction based assays for species-specific detection of *Fusarium culmorum, F. graminerum*, and *F. avenaceum*. *Phytopathology*, 86 (5), 515–522.

Schnerr, H., Vogel, R. F. and Niessen, L. (2002) Correlation between DNA of tricothecene producing *Fusarium* species and deoxynivalenol concentration in wheat species. *Letters in Applied Microbiology*, 35 (2), 121–125.

Schumacher, T. (2002) Gushing in Fruchtsaftschorlen. Ursachen und Gegenmaßnahmen. *Flüssiges Obst*, 5, 304–310.

Schur, F., Anderegg, P., Senften, H. and Pfenniger, H. (1980) Brautechnologische Bedeutung von Oxalat. *Brauerei Rundschau*, 91, 201–207.

Schwabe, M., Fenz, R., Engels, R., Krämer, J. and Rath, F. (1994) Nachweis von Fusarium auf Braugerste mit dem EPS-Latex-Agglutinationstest. *Monatsschrift fur Brauwissenschaft*, 5, 160–164.

Schwarz, P. B. (2003) Impact of head blight on the malting and brewing quality of barley. In: *Fusarium Head Blight of Wheat and Barley* (K. J. Leonard and W. R. Bushnell eds), pp. 395–419. American Phytopathological Society Press: St Paul, MN.

Schwarz, P. B., Casper, H. H. and Beattie, S. (1995) Fate and development of naturally occurring *Fusarium* mycotoxins during malting and brewing. *Journal of the American Society of Brewing Chemists*, 53 (3), 121–127.

Schwarz, P. B., Beattie, S. and Casper, H. H. (1996) Relationship between *Fusarium* infestation of barley and the gushing potential of malt. *Journal of the Institute of Brewing*, 102, 93–96.

Schwarz, P. B., Casper, H. H., Barr, J. and Musial, M. (1997) Impact of Fusarium head blight on the malting and brewing quality of barley. *Cereal Research Communication*, 25, 813–814.

Schwarz, P. B., Schwarz, J. G., Zhou, A., Prom, L. K. and Steffenson, B. J. (2001) Effect of *Fusarium graminearum* and *F. poae* infection on barley and malt quality. *Monatsschrift fur Brauwissenschaft*, 54 (3/4), 55–63.

Schwarz, P. B., Jones, B. L. and Steffenson, B. J. (2002) Enzymes associated with *Fusarium* infection of barley. *Journal of the American Society of Brewing Chemists*, 60 (3), 130–134.

Schwarz, P. B., Horsley, R. D., Steffenson, B. J. and Salas, B. (2006) Quality risks associated with the utilization of Fusarium Head Blight infected malting barley. *Journal of the American Society of Brewing Chemists*, 64 (1), 1–7.

Simon, R. (1998) Gushing: A never ending story. *Brauwelt*, 138 (27), 1244–1245.

Sloey and Prentice, N. (1962) Effects of *Fusarium* isolates applied during malting on the properties of malt. *Proceedings of the American Society of Brewing Chemists*, 25–29.

Steffenson, B. J. (2003) Fusarium Head Blight of barley: Impact, epidemics, management strategies for identifying and utilizing genetic resistance. In: *Fusarium Head Blight of Wheat and Barley* (K. J. Leonard and W. R. Bushnell eds), pp. 241–295. American Phytopathological Society Press: St Paul, MN.

Talbot, N. J., Kershaw, M., Wakley, G. E., de Vries, O. M. H., Wessels, J. G. H. and Hamer, J. E. (1996) MPG1 encodes a fungal hydrophobin involved in surface interactions during infection-related development of the rice blast fungus *Magnaporte grisea*. *Plant Cell*, 8, 985–999.

Torkkeli, M., Serimaa, R., Ikkala, O. and Linder, M. (2002) Aggregation and self-assembly of hydrophobins from *Trichoderma reesei*: Low-resolution structural models. *Biophysics Journal*, 83, 2240–2247.

Turner, J. E., Jennings, P. and Nicholson, P. (1999) Investigation of Fusarium infection and mycotoxin levels in harvested wheat grain. Home Grown Cereals Authority Project Report 207. HGCA, London, 15 pp.

Vaag, P. (1991) Immunological detection of *Fusarium* in barley and malt. *European Brewery Convention*, 23, 553–560.

Vaag, P. and Pederson, S. (1992) Practical experiences with immunological techniques for the detection of *Fusarium* in barley and malt. *European Brewery Convention* Biochemistry and Microbiology Groups Bulletin. EBC, Zoeterwoude, Netherlands.

Vaag, P., Preben, R., Knudson, A.-D., Pederson, S. and Meiling, E. A. (1993) A simple and rapid test for gushing tendency in brewing materials. *Proceedings of the European Brewery Convention Congress*, 24, 155–162.

Vogel, E. H. (1949) Some aspects of gushing beer. *Communications. MBAA*, 10, 6–7.

Waalwijk, C., van der Heide, R., de Vries, I., van der Lee, T., Schoen, C., Costrel de Corainville, G., Hauser-Hahn, I., Kastelein, P., Kohl, J., Lonnet, P., Demarquet, T. and Kema, G. H. J. (2004) Quantitative detection of *Fusarium* species in wheat using TaqMan. *European Journal of Plant Pathology*, 110 (5/6), 481–494.

Weideneder, A. (1992) Untersuchungen zum malzverursachten Wildwerden (Gushing) des Bieres. Doctoral thesis. Technical University Munich.

Wershofen, T. (2004) Gushing Ein überschäumend spritziges Erlebnis. *Brauwelt*, 35, 1061–1063.

Wessels, J. G. H. (1994) Developmental regulation of fungal cell wall formation. *Annual Review of Phytopathology*, 32, 413–437.

Wessels, J. G. H. (1996) Fungal hydrophobins, protein that function at an interface. *Trends Plant Science*, 1, 9–15.

Wessels, J. G. H. (1997) Hydrophobins: Proteins that change the nature of the fungal surface. *Advances in Microbial Physiology*, 38, 1–45.

Wessels, J. G. H., de Vries, O. M., Asgeirsdottir, S. A. and Schuren, F. H. J. (1991) Hydrophobin genes involved in formation of aerial hyphae and fruit bodies in *Schizophyllum*. *Plant Cell*, 3 (8), 793–799.

Wilt, P. (1986) Nucleation rates and bubble stability in water-carbon dioxide solutions. *Journal of Colloid and Interface Science*, 112, 530–538.

Winkelmann, L. (2004) Das Gushing-Puzzle eine Erfolgsgeschichte. *Brauwelt*, 25, 749–750.

Woesten, H. A. B. and de Vocht, M. L. (2000) Hydrophobins, the fungal coat unravelled *Biochim. Biophysics Acta*, 1469, 79–86.

Woesten, H. A. B. and Wessels, J. G. H. (1997) Hydrophobins, from molecular structure to multiple functions in fungal development. *Myoscience*, 38, 363–374.

Woesten, H. A. B., de Vries, O. M. H. and Wessels, J. G. H. (1993) Interfacial self-assembly of a fungal hydrophobin into a hydrophobic rodlet layer. *Plant Cell*, 5, 1567–1574.

Woesten, H. A. B., Schuren, F. H. J. and Wessels, J. G. H. (1994) Interfacial self-assembly of a hydrophobin into an amphipathic protein membrane mediates fungal attachment to hydrophobic surfaces. *EMBO Journal*, 13, 5848–5854.

Woesten, H. A. B., Ruardy, T. G., van der Mei, H. C., Busscher, H. J. and Wessels, J. G. H. (1995) Interfacial self-assembly of a Schizophyllum commune hydrophobin into an insoluble amphiphatic membrane depends on surface hydrophobicity *Colloids Surf. B: Biointerfaces*, 5, 189–195.

Woesten, H. A. B., van Wetter, M-A., Lugones, L. G., van der Mei, H. C., Busscher, H. J. and Wessels, J. G. H. (1999) How a fungus escapes the water to grow into the air. *Current Biology*, 9, 85–86.

Xu, X. M., Parry, D. W., Nicholson, P., Thomsett, M. A., Simpson, D., Edwards, S. G., Cooke, B. M., Doohan, F., van Maanen, A., Moretti, A., Tocco, G., Mule, G., Hornok, L., Giczey, G., Tatnell, J. and Ritieni, A. (2003) Is the amount of mycotoxins in cereal grains related to the quantity of Fusarium DNA?. *Aspects Applied Biology* (68), 101–108.

Yli-Mattila, T., Paavanen-Huhtala, S., Parikka, P., Hietaniemi, V., Jestoi, M. and Rizzo, A. (2004) Real-time PCR detection and quantification of *Fusarium poae* as compared to mycotoxin production in grains in Finland. *Proceedings of 2nd International Symposium on Fusarium Head Blight*, 2, 422–425.

Yount, D. E., Gillary, E. W. and Hoffmann, D. C. (1984) A microscopic investigation of bubble formation nuclei. *Journal of Acoustical Society of America*, 76, 1511–1521.

Zepf, M. (1998) Gushing. Ursachenfindung anhand von Modellversuchen. Doctoral thesis. Technical University Munich.

Zepf, M. and Geiger, E. (1999) Gushing problems with calcium oxalate I. *Brauwelt*, 139 (48), 2302–2304.

Zepf, M. and Geiger, E. (2000) Gushing problems with calcium oxalate II. *Brauwelt*, 140 (6/7), 222–223.

Zapf, M. W., Theisen, S., Vogel, R. F. and Niesen, L. (2006) Cloning of wheat LTP1500 and two *Fusarium culmorum* hydrophobins in *Saccharomyces cerevisiae* and assessment of their gushing inducing potential in experimental work fermentation. *Journal of the Institute of Brewing*, 112 (3), 237–245.

Zarnkow, M. and Back, W. (2001) Neue Erkenntnisse über gushingauslösende Substanzen. *Brauwelt*, 141 (9/10). 363–367, 370

7
Beer color
Thomas H. Shellhammer

Color perception

Color is a human visual perception utilizing a narrow portion of the electromagnetic spectrum (380–780 nm). Light itself has no color and color does not exist by itself; it only exists in the mind of the viewer (Delgado-Vargas and Lopez, 2003). Color perception exists in two stages. The first is a purely physical phenomenon that requires three elements: a source of light, an object and a detector (an eye, a diode, etc.) while the second stage is a complicated and incompletely known process whereby the human eye transmits information that the brain will interpret as color (Hari et al., 1994).

In order to evaluate the color of an object it must be illuminated and upon interacting with the object the incident light can be transmitted, reflected, refracted, absorbed and/or scattered. With liquid samples that are not opaque, as is generally the case with beer, the primary interest lies in how light is transmitted through and absorbed by the product, as opposed to reflected light. The wavelength of light absorbed by a liquid medium is typically complimentary to that perceived. For instance, an object that is perceived as green will absorb light that is red and visa versa. However, this simplistic model works well only when the material displays an absorbance maximum at a particular wavelength. If all wavelengths of light are absorbed the product will appear black and if none of the light is absorbed it will appear colorless (if no scattering take place, as with filtered water) or white (if substantial scattering takes place, as with milk). The Lambert-Beer law (Equation (7.1)) predicts the quantity of absorbed light in a liquid medium.

$$A = \alpha \cdot l \cdot c \qquad (7.1)$$

where

$$\alpha = \frac{4\pi k}{\lambda}$$

A = absorbance, α = molar absorptivity, k = extinction coefficient, λ = wavelength of light, l = material thickness and c = concentration of the absorbing material. The Lambert-Beer law is valid at lower concentrations, in the absence of significant scattering, and only if individual wavelengths are used. An implication of this law is that equal amounts of absorption result when light passes through equal thicknesses of material, thus reducing path length by half reduces absorbance by half as well. When measuring materials with very high absorbance it is preferable to reduce path length rather than diluting to reduce concentration since the diluent may affect the chemistry of the system (buffer strength, for instance) and alter its color (Smythe and Bamforth, 2000).

The human components involved in color perception include the eye, the nervous system and the brain. Within the eye, the light sensitive region on the retina is composed of rods and cones. The rods allow us to see in dim light conditions (maximum sensitivity at ~500nm) but do not confer color vision. The cones, on the other hand show less sensitivity to light but great sensitivity to color. In the human eye there are many more rods (~100 million) than cones (~3 million). Furthermore, there exist three types of cone receptors designated as red (R), green (G) and blue (B) which exist in unequal quantities – 64% R, 32% G and 4% B (Delgado-Vargas and Lopez, 2003). Genetic differences among humans result in variation in color sensitivity as well as color blindness. Because genes for red and green color receptors are located on the X chromosome (of which men have only one and women have two), red-green color blindness is predominant in men (7% of the population) and much lower in women (<1%) (Montgomery, 1995). Altered color perception at such a high frequency in the human population is one obvious reason why instrumental color measurement is necessary.

Measuring color

The principal attributes of any object's color are hue, lightness and saturation (Wyszecki and Stiles, 1967).

1. *Hue* is the quality we normally identify as an object's color, such as red, green, yellow and so forth. Hues form what we describe as the color wheel. The human eye can identify more than a million different hues (Loughrey, 2000).
2. *Lightness* is a term related to the concept of light and dark and is used to classify colors by separating those that are bright, midtone or dark.
3. *Saturation* or chroma is completely separate from hue and lightness. It can be defined as purity of color; that is as a color moves away from a central neutral gray its saturation increases as it becomes more vivid and less dull.

Often these three color attributes are described using a three-dimensional model (Figure 7.1).

As mentioned earlier, the physical requirements for measuring color are a source of light, the object and an observer. The most obvious source of variation with instrumental measurements is the source of light, for instance unaltered light for use with visual eye examinations, three colored lights used in colorimeters, or monochromatic light as with a spectrophotometer. Clearly the human observer varies as well. The CIE (Commission Internationale de

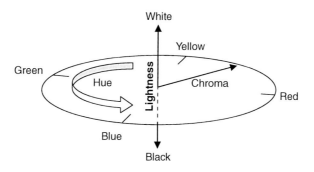

Figure 7.1
Three-dimensional color model.

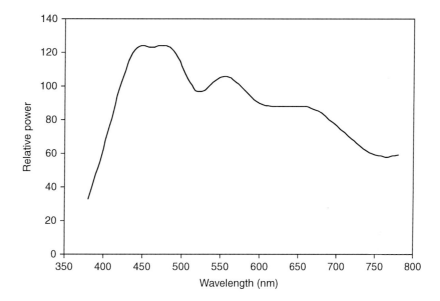

Figure 7.2
Spectral power distribution, $S_{(\lambda)}$, of CIE illuminant C (ASTM, 2001).

l'Eclairage) system developed by the International Commission on Illumination standardized the illuminant source as well as the observer. Three standard light sources, CIE A, CIE B and CIE C, were initially proposed of which CIE C is used most often for food color measurements. CIE C has a spectral distribution, $S_{(\lambda)}$, (Figure 7.2) that approximates that of overcast skylight and correlates with a temperature of approximately 6740 K (Delgado-Vargas and Lopez, 2003).

Standard observers were also proposed by CIE based on experiments with a small number of people possessing normal color vision in order to assess the spectral sensitivity of the human eye (Figure 7.3). In 1931 the CIE first defined the 2° Standard Color Observer and in 1964 defined a second color observer

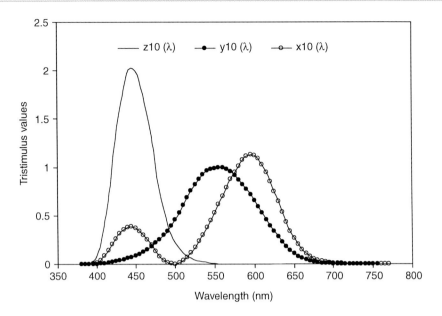

Figure 7.3
CIE color matching functions $x_{10(\lambda)}, y_{10(\lambda)}, z_{10(\lambda)}$ for the 10° Standard Observer (ASTM, 2001).

known as the 10° Supplementary Color Observer. Most color applications use the second observer (Loughrey, 2000).

The CIE XYZ numbers, known as the tristimulus values, can be calculated for a non-opaque liquid's color as it appears to a defined observer using a defined illuminant. The three physical elements, the light, object and observer, can be combined to create a set of XYZ tristimulus values using the following generic form.

$$\begin{matrix} \text{Illuminant} \\ \text{spectral} \\ \text{energy} \end{matrix} \times \begin{matrix} \text{Object's} \\ \text{transmittance} \end{matrix} \times \begin{matrix} \text{Observer} \\ \text{response} \end{matrix} = \begin{matrix} X \\ Y \\ Z \end{matrix} \quad (7.2)$$

In the case of beer color measurements, a spectrophotometer measures the spectral transmission, $T_{(\lambda)}$, of the sample (Figure 7.4) at distinct wavelengths. These data can be combined with the spectral power distribution, $S_{(\lambda)}$, and the standard observer spectral sensitivities, $x_{10(\lambda)}, y_{10(\lambda)}, z_{10(\lambda)}$, at the same wavelengths using Equation (7.2) and integrated over the entire visible spectrum to yield the CIE tristimulus values, X, Y and Z.

$$X = k \sum_{\lambda=380}^{780} T_{(\lambda)} S_{(\lambda)} x_{10(\lambda)} \quad (7.3)$$

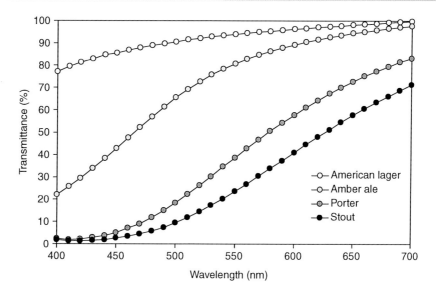

Figure 7.4
Transmission spectra for four different colored commercial beer styles (Shellhammer and Bamforth, 2008).

$$Y = k \sum_{\lambda=380}^{780} T_{(\lambda)} S_{(\lambda)} y_{10(\lambda)} \qquad (7.4)$$

$$Z = k \sum_{\lambda=380}^{780} T_{(\lambda)} S_{(\lambda)} z_{10(\lambda)} \qquad (7.5)$$

Where k is a normalizing factor calculated as

$$k = \frac{100}{\sum_{\lambda=380}^{780} S_{(\lambda)} y_{10(\lambda)}} \qquad (7.6)$$

One minor drawback to the CIE XYZ system is that these tristimulus values do not relate easily to intuitive color values or color differences. Alternative scales, such as the CIELAB and L*a*b*, overcome this drawback. These systems rely on opponent color theory in which signals from the human eye en route to the brain become coded as one brightness signal and two hue signals. Brightness is represented by luminosity (L) or light versus (opposing) dark, with a value of 0 indicating black and 100 indicating white. The chromatic signals are separately red versus green, called a, and yellow versus blue, called b. Using this nomenclature, positive a values indicate redness while negative a values indicate greenness. The same positive and negative relationship for the b value relates to yellowness and blueness, respectively. The L^*, a^* and b^* values can be calculated by transforming the CIE XYZ tristimulus values as follows:

$$L^* = 116\left(\frac{Y}{Y_n}\right)^{1/3} - 16 \qquad (7.7)$$

$$a^* = 500\left[\left(\frac{X}{X_n}\right)^{1/3} - \left(\frac{Y}{Y_n}\right)^{1/3}\right] \qquad (7.8)$$

$$b^* = 200\left[\left(\frac{Y}{Y_n}\right)^{1/3} - \left(\frac{Z}{Z_n}\right)^{1/3}\right] \qquad (7.9)$$

In Equations (7.7)–(7.9), white point values are defined as $X_n = 97.285$, $Y_n = 100.00$, $Z_n = 116.145$ (American Society for Testing and Materials, (ASTM), 2001). The process of determining tristimulus color using spectrophotometric transmission measurements is an approved standard method by the American Society of Brewing Chemists (ASBC) (2004a). Color expressed as $L^*a^*b^*$ is somewhat more intuitive than CIE XYZ and the magnitude of color differences can easily be determined as:

$$\Delta E = \sqrt{\left(\Delta L^*\right)^2 + \left(\Delta a^*\right)^2 + \left(\Delta b^*\right)^2} \qquad (7.10)$$

Color differences where ΔE is greater than one are distinguishable by the human eye (Hughes and Baxter, 2001).

Standard methods for measuring beer color

Historically, beer color was determined by visual comparisons against a set of color standards. Joseph Lovibond developed a set of standards and a tintometer in 1893 while solutions of potassium chromate have been used as reference standards in the early 20th century. The use of standardized Lovibond colored glasses made it possible to compare measurements from different Lovibond Tintometers. After a series of modifications, these standardized comparator discs were accepted by the European Brewing Convention in 1951 (Hughes and Baxter, 2001). So ingrained was this approach that many brewers and maltsters today still refer to color values in degrees Lovibond. The tintometer approach, while satisfactory at the time, had inherent flaws due to variation in color perception by the human observer. As previously mentioned, red–green color blindness is present at roughly 7% of the male population. Furthermore, instability of the glass (Strien and Drost, 1979) or liquid color standards over time, for instance orange dichromate slowly reducing to green chromium (III) (Hughes and Baxter, 2001), would lead to false color estimates. The solution to both of these issues was single wavelength measurements using a precision spectrophotometer.

The range of beer color is broad, from pale yellow lagers to black stouts (Table 7.1). Some styles utilizing caramel malts result in reddish and amber hues. Regardless of style, the transmission and absorption spectra do not show intensity maxima (Figures 7.4 and 7.5). Consequently, selection of a single wavelength was initially directed by the prevalent beer style being measured.

Table 7.1
Beer color across a range of beer styles

Style	Color	Color units	
		SRM	EBC
American/European light lager	Yellow/straw	2–4	4–8
American/European 100% malt Pilsner (Jurado, 2002a)	Golden	3–9	6–18
British pale ale	Amber	10–15	20–30
Munich-style helles	Golden/amber	3–15	6–30
Red beers (Jurado, 2002b)	Brown/amber/reddish	9–54	18–108
American porter	Dark brown	20–30	40–60
Irish stout	Black	35–70	70–140

Presently, absorbance at 430 nm for single-wavelength beer color measurements is the standard method of the American Society of Brewing Chemists (ASBC) and the European Brewery Convention (EBC) (ASBC, 2004b; EBC, 2004).

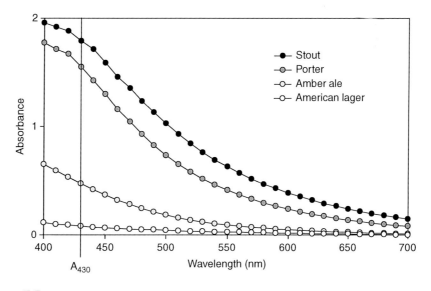

Figure 7.5
Absorbance spectra for four different colored commercial beer styles. Standard methods of beer analysis utilize absorbance at 430 nm (Shellhammer and Bamforth, 2008).

The European Brewery Convention (EBC) settled on measuring the absorbance at 430 nm since this wavelength (violet) is complementary to the transmission color of paler beers (yellow) that dominated continental Europe. The UK initially focused on absorbance at 530 nm (green) as this wavelength showed most variation to the amber ales prevalent in this region (Hughes and Baxter, 2001) and corresponded to the color of acidified potassium dichromate solution which was used for calibration (Smedley, 1995a,b). The UK industry later changed and adopted the A_{430} measurement.

The ASBC beer color standard method was developed in the 1950s using a half-inch cell. Absorbance readings were manipulated such that they were close to °Lovibond values, the standard at the time (Equation (7.11)).

$$\text{ASBC color} = A_{430}\left(\frac{1}{2}\text{inch cell}\right) \times 10 \quad (7.11)$$

Since most modern cuvettes have a path length of 10 mm, color measurements using a 10 mm cuvette are multiplied by 1.27 to accommodate for the shorter path length.

$$\text{ASBC color} = A_{430}(10\,\text{mm cell}) \times 10 \times 1.27 \quad (7.12)$$

The ASBC method is often referred to as being the (US) Standard Reference Method (SRM), and color values may be expressed as °SRM to denote the method. Color expressed as °SRM agrees closely with color in °Lovibond and the two can be used interchangeably.

One assumption that must be satisfied with beer color measurements is that the sample is free of turbidity that would scatter incident light. To test this assumption, the ASBC color analysis is also performed at 700 nm. Should the $A_{700} \leq 0.039 \times A_{430}$ the beer is considered "free of turbidity" and the color of beer can be determined from its absorbance at 430 nm. If the sample is not "free of turbidity" it requires further clarification by centrifugation or filtration followed by a repeat absorbance measurement at 430 nm. The EBC standard method for beer color also uses absorbance at 430 nm but with a slight modification (Equation (7.13)).

$$\text{EBC color} = A_{430}(10\,\text{mm cell}) \times 25 \quad (7.13)$$

Thus, for the same color, EBC units will be approximately twice (1.97 times) as large as SRM color units.

The brewing industry retains A_{430} readings as the conventional standard for online and offline color measurements because this technique affords rapid, straightforward, and easily transferable measurements to automated online color monitoring. While the use of a single wavelength color measurement is satisfactory in representing color for some beers, such as pale yellow lager beer, it becomes less satisfactory for darker and reddish beers. Furthermore, it is possible for two beers with identical EBC or ASBC colors to be visually different (Smythe and Bamforth, 2000; Smedley, 1995a,b) because of different transmission spectra outside of A_{430} (Figure 7.6). While the sample spectra in Figure 7.6 display

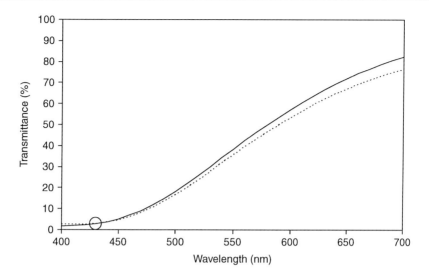

Figure 7.6
Transmission spectra for two beers that have different tristimulus color but have identical absorbance at 430 nm. Color = 29.7°SRM, 38.8°EBC. ΔE = 1.98 (Shellhammer and Bamforth, 2008).

identical A_{430} readings, they are distinctly different visually ($\Delta E > 1$). Tristimulus measurements of beer color overcome this flaw with single wavelength measurement techniques. However, this approach requires examining the entire visible spectrum which can be time intensive if rapid, online reading is required. Given that no distinct peaks appear in beer spectra, the sigmoidal nature of the transmission spectra (Figures 7.4 and 7.6) can be closely approximated by measuring as few as five individual wavelengths, interpolating the entire visible spectrum, and speeding up the data collection and processing time (Smedley, 1992).

Origins of beer color

Among the various ingredients used for beer production, beer color is determined primarily by the selection of the grains used in formulation, and more specifically the type of processing these grains have undergone. Barley contains very low concentrations of pigmented substances, and it is the malting process that results in color formation. The germination and kilning phases of the malting process set the stage for and determine the extent of color formation from Maillard browning reactions and in some cases caramelization and pyrolysis reactions. Secondary to these heat driven color-forming reactions, the oxidation of polyphenols derived from barley husk or hop vegetative matter can contribute to color formation during beer storage/ageing. Additionally, oxidation of polyphenols can lead to enhanced protein–polyphenol interaction and the formation of nonbiological haze. The scattering of light via this haze indirectly affects a consumer's perception of color as well as its physical measurement. Yeast does not contribute to beer color directly, other than color loss during production via adsorption of colored materials to their cell wall. Indirectly, yeast can affect beer color via turbidity in products where it remains present at packaging, such as with hefeweizen.

Maillard browning reactions are key to beer color and they originate in the barley endosperm during malting but can resume during wort boiling. The two key components of this reaction are reducing sugars (principally maltose) and free amino acids or amino groups of amino acids that comprise protein (Figure 7.7). While initial Maillard reaction steps are reasonably well defined, the later reactions leading to formation of colorful melanoidins are less so. Briefly, an aldose **1** or ketose reacts with molecules having amine functionality to form an imine (**2**, Schiff base) which isomerizes to a more stable amino ketone (**3**, Amadori compound). Further reactions of compounds **1** and **2** are strongly pH dependent (Rizzi, 1994). Under alkaline conditions (>pH 7), compounds **2** and **3** fragment and further react to form melanoidins. Under acidic conditions (<pH 5), the Amadori compound **3** will eliminate the original amine (R^1NH_2) to generate a 3-deoxyosone **4**. At moderate acidities (pH 5–7), deamination leads to an isomeric 1-deoxyosone **5**. The deoxyosones are highly reactive Maillard reaction intermediates and their further reactions lead to important flavor compounds, such as 5-hydroxymethylfurfural **6** in the case of glucose, or 3-furanones **7**. These low molecular weight compounds have characteristic caramel, toasted aromas, exhibit low detection thresholds, and are colorless. The reaction of furfural **6** and 3-furanone **7** (R = H) was observed to produce yellow dicyclic **8** and an orange tricyclic chromophors (Hoffman, 1998).

In addition to pH, the Maillard reaction is principally driven by time and temperature and modulated to a lesser extent by moisture. In the range of 60–100°C, an increase of 1°C can increase the rate of the Maillard reaction by more than 10% (Rizzi, 1994). Increasing heat exposure to higher temperatures promotes the production of more advanced Maillard reaction products because of their higher energy of activation thus resulting in greater color formation. The reaction also proceeds most rapidly at intermediate moisture contents (0.4–0.6 A_w).

While some color comes from low molecular weight Maillard reaction intermediates, clearly the bulk of the color formed from the Maillard reaction results from melanoidins which are high molecular weight polymeric structures. There are three main proposals for the structures of melanoidins. First is that melanoidins consists of repeating polymers of furans and/or pyrroles. Second is that low molecular weight colored substances can crosslink proteins via e-amino groups of lysine or arginine to produce high-molecular-weight colored melanoidins. And thirdly that the melanoidin skeleton is mainly built up of sugar degradation products, formed in the early stages of a Maillard reaction, polymerized through aldol-type condensation, and possibly linked by amino compounds (Cammerer et al., 2002). Colors produced by melanoidins are yellow, orange and red initially and turn to brown as the Maillard reaction is allowed to proceed (Nursten, 2005). Lightly kilned malt displays yellow colors characteristic of light lager beer while intensely kilned products display amber and brown hues characteristic of British ales or Vienna lagers (Table 7.2).

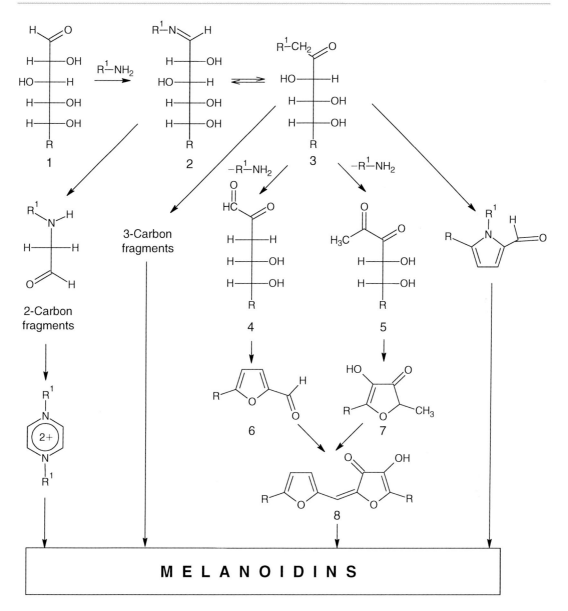

Figure 7.7
The basic Maillard reaction scheme leading to pigmented melanoidins (adapted from Rizzi, 1994).

Under conditions of high sugar concentration and high temperatures, caramelization and pyrolysis reactions can occur in addition to Maillard browning. At temperatures greater than 120°C, caramelization reactions proceed as thermal decomposition of sugars in the absence of amino nitrogen. Hydryoxymethyl furfural **6** (R = CH_2OH) is a key flavor component of caramelized maltose (Fadel and Farouk, 2002) but is colorless. Extensive heating

Table 7.2
Appearance and flavor of malted barley

Malt type	Color (SRM)	Appearance	Beer type	Flavor
Standard Malts				
Wheat	1	Pale straw	Weizen	Malty
Pale lager	2	Pale yellow	Light lagers	Cereal, DMS
Pale ale	3	Yellow, golden	Ales	Biscuit, toasted
Vienna	4	Amber	Dark lagers	Nutty toffee
Color/Caramel Malts				
Munich	10–20	Amber, brown	Amber beer	Intense malty
Cara pils	5–15	Pale	Lagers	Sweet, biscuit
Caramel/Crystal	20–120	Amber, brown, red	Ales and lagers	Toffee, nutty, burnt
Roasted Malts				
Chocolate	350	Brown, black	Porters and stouts	Coffee
Black	400–600	Black	Porters and stouts	Neutral
Roasted barley	300–800	Black	Irish stout	Bitter, burnt

leads to caramelization products that are reddish and/or brown in color. The rate of browning from caramelization is correlated with pH with the reaction proceeding slowly under acidic conditions and rapidly under basic conditions (Del Pilar Buera et al., 1987). Above 200°C, pyrolysis reactions predominate. These reactions amount to scorching or burning of sugars along with dramatic molecular breakdown resulting in strong, burnt aroma and flavor and the production of intensely black pigments.

All malts possess some degree of color, from very pale yellow (2°SRM) in lightly kilned Pilsner malts to extremely dark in roasted malts (800°SRM) (Table 7.2). From the standpoint of adding color to beer, dark, specialty malts can be categorized in three groups: color, caramel and roasted malts (Coghe et al., 2003). Color malts (high-dried malts) are the only malts produced in a kiln. The extent of color formation is due primarily to time and temperature during the curing stage, but is also influenced by the degree of modification at the conclusion of germination with greater modification leading to a large pool of Maillard browning reactants and therefore a greater potential for browning. These malts provide yellow to brown colors (Table 7.2).

In the production of caramel malt, the kilning operation is replaced with a combination of stewing and roasting in a roasting drum. Green malt, having just completed germination (45% moisture), is placed in a drum roaster and heated to a temperature which is optimal for enzymatic saccharification,

65–75°C (Gretenhart, 1997). During the stewing phase all of the starch is converted into fermentable, reducing sugars such as maltose, maltotriose and to minor degree glucose, along with nonfermentable dextrins. Following the stewing phase, the temperature is increased substantially to 80–145°C where caramelization and Maillard reactions take place rapidly because of the high concentrations of precursors and the intense heat. Because of the substantial contribution of caramelization reactions these malts tend to be amber with red hues, contribute significant color and are often used in the production of "red" beers.

Roasted malts receive even greater heat, in some cases within 20° of their combustion temperature (248°C), and this results in substantial color formation from pyrolysis reactions. These malts are used principally as coloring agents as they offer limited fermentable extract. Roasted malts offer colors which range from brown to black. The flavor that black malt provides is exceptionally strong, and, if the malt is overground, or elevated using less than perfect pneumatic equipment, it will break into fines. The resultant fines produce an exceptionally bitter, astringent note. High molecular weight extracts of this type can be used in small amounts to trim color of pale beer without impacting flavor just prior to packaging. Conversely, low molecular weight extracts can have the complimentary effect of adding flavor with minimal color.

The darker malts not only have less extract, they may also have lost their husk during rotary kilning, and this can have a negative effect on runoff. Furthermore, since the protein has been degraded, the use of these dark malts can result in poor foam formation unless the formulation is augmented to compensate. Grinding the dark malts more coarsely, to avoid fines, may require the brewer to add a second short saccharification rest during mashing after the normal conversion. The second rest would be at ~80°C for about 10 min, to ensure that all the starches have been converted to dextrins. Taking this step can have a positive effect on wort run off, depending on the mashing scheme being used.

In addition to melanoidins, malt also contributes polyphenols to wort along with the enzyme peroxidase which can modify the color of phenolic material during mashing and runoff. Further extraction of hop-derived polyphenols can occur during wort boiling if a brewer uses whole or pelletized hops. Oxidation of these polyphenols is a secondary contributor to beer color resulting in reddish-amber hues that can be apparent in lighter colored beers, particularly during storage. Mashing regimes that limit oxygen exposure and the extent of peroxidase action result in reduced color formation during mashing from enzymatic oxidation. Oxidation of barley and hop polyphenols during beer storage will result in increased reddish, brown color. Color changes as a result of polyphenol oxidation are most apparent in pale lager beers following extended storage post-packaging. In darker beers, oxidative browning is masked by the colors from the colored and roasted malts. Such phenolic oxidation will be promoted by high levels of dissolved oxygen in the packaged beer as well as the presence of soluble iron imparted by brewhouse equipment, brewing water or diatomaceous earth filtration media. Fining with polyphenol adsorbents such as polyvinylpolypyrolidone (PVPP), prior to packaging helps mitigate oxidative browning by reducing levels of potential browning polyphenolic precursors.

In very light colored beers, riboflavin (yellow or orange-yellow with peak absorbance at 445 nm) can contribute significantly to beer color (Briggs et al., 2004). Riboflavin is found at levels less than 1.5 ppm in beer (Baxter and Hughes, 2001).

Coloring agents such as malt extracts and caramel coloring can be added post-fermentation as a means of adjusting the beer's final color. These products are intensely colored with colors ranging from 250 to 3500°SRM for concentrated malt extracts and 5000 to 30,000°SRM for caramels (Smedley, 1995a,b). Caramel coloring is produced by cooking a carbohydrate solution at very high temperatures with or without ammonium or sulfite compounds (Comline, 2006). The type of caramel used for coloring beer, Caramel Color III or E150c (in Europe) utilizes ammonium compounds (such as ammonium hydroxide, carbonate, bicarbonate and phosphate) but not sulfites as reactants to promote caramelization (Kamuf et al., 2000). Using caramel coloring or high molecular weight malt extracts offers brewers a convenient means of ensuring consistent beer color without modifying beer flavor.

References

American Society of Brewing Chemists (ASBC), Beer-10A (2004a) Tristimulus analysis (colorimetric or spectrophotometric). *Methods of Analysis*, 9th edition. American Society of Brewing Chemists: St. Paul, MN.

American Society of Brewing Chemists (ASBC), Beer-10A (2004b) Spectrophotometric color method. *Methods of Analysis*, 9th edition. American Society of Brewing Chemists: St. Paul, MN.

American Society for Testing and Materials (ASTM), E-308-01 (2001) Standard practice for computing the colors of objects by using the CIE System, subcommittee E12.04 on color and appearance analysis.

Baxter, E. D. and Hughes, P. S. (2001) *Beer: Quality, safety and nutritional aspects*. The Royal Society of Chemistry: Cambridge, UK. p. 138.

Briggs, D. E., Boulton, C. A., Brookes, P. A. and Stevens, R. (2004) *Brewing Science and Practice*. Woodhead Publishing Limited: Cambridge, UK.

Cammerer, B., Jalyschko, W. and Kroh, W. L. (2002) Intact carbohydrate structures as part of the melanoidin skeleton. *Journal of Agricultural and Food Chemistry*, 50 (7), 2083–2087.

Coghe, S., Vanderhaegen, B., Pelgrims, B., Basteyns, A. and Delvaux, F. R. (2003) Characterization of dark specialty malts: New insights in color evaluation and pro- and antioxidative activity. *Journal of the American Society of Brewing Chemists*, 61 (3), 125–132.

Comline, P. (2006) Caramel for beer. *The Brewer and Distiller*, 2 (1), 17–19.

Delgado-Vargas, F. and Lopez, O. P. (2003) *Natural Colorants for Food and Nutraceutical Uses*. CRC Press: Boca Raton, FL. p. 325.

Del Pilar Buera, M., Chrife, J., Resnik, S. L. and Lozano, R. D. (1987) Nonenzymatic browning in liquid model systems of high water activity: Kinetics of color changes due to caramelization of various single sugars. *Journal of Food Science*, 52 (4). 1059–1062, 1073.

European Brewery Convention (EBC), Section 9 Beer Method, 9.6 Colour of Beer (2004) Spectrophotometric method: Instrumental method. *Analytica – EBC*. Verlag Hans Carl Getranke Fachverlag: Nurnburg, Germany.

Fadel, H. H. M. and Farouk, A. (2002) Caramelization of maltose solution in presence of alanine. *Amino Acids*, 22 (2), 199–213.

Gretenhart, K. E. (1997) Specialty malts. *MBAA Technical Quarterly*, 34 (2), 102–106.

Hari, R. K., Patel, T. R. and Marting, A. M. (1994) An overview of pigment production in biological systems: Functions, biosynthesis, and applications in food industry. *Food Reviews International*, 10, 49–70.

Hoffman, T. (1998) Characterization of the most intense coloured compounds from Maillard reactions of pentoses by application of colour dilution analysis. *Carbohydrate Research*, 33, 203–213.

Hughes, P. S. and Baxter, E. D. (2001) *Beer: Quality, safety and nutritional aspects*. Royal Society of Chemistry: Cambridge, UK.

Jurado, J. (2002a) Prevailing pilsner. *The Brewer International*, 2 (2), 20–26.

Jurado, J. (2002b) The renaissance of red. *The Brewer International*, 2 (4), 30–37.

Kamuf, W., Nixon, A. and Parker, O. (2000) Caramel color. In: *Natural Food Colorants* (G. J. Lauro and F. J. Francis eds), pp. 187–218. Marcel Dekker, Inc.: New York, NY.

Loughrey, K. (2000) The measurement of color. In: *Natural Food Colorants* (G. J. Lauro and F. J. Francis eds), pp. 187–218. Marcel Dekker, Inc.: New York, NY.

Montgomery, G. (1995) *Breaking the Code of Color*. Howard Hughes Medical Institute: Chevy Chase, MD. p. 62.

Nursten, H. (2005) *The Maillard Reaction. Chemistry, Biochemistry and Implications*. The Royal Society of Chemistry: Cambridge, UK. p. 214

Rizzi, G. P. (1994) The Maillard reaction in foods. In: *Maillard Reactions in Chemistry, Food, and Health* (T. P. Labuza, G. A. Reineccius, V. Monnier, J. O'Brien and J. Baynes eds), pp. 11–19. The Royal Society of Chemistry: Cambridge, UK.

Shellhammer, T. H. and Bamforth, C. W. (2008) Assessing color quality of beer. In: *Color Quality of Fresh and Processed Foods* (C. Culver and R. Wrolstad eds). ACS Symposium Series 983, Amercian Chemical Society: Washington, DC. pp 192–202.

Smedley, S. M. (1992) Colour determination of beer using tristimulus values. *Journal of the Institute of Brewing*, 98, 497–504.

Smedley, S. (1995a) Towards closer colour control in the brewing industry. Part 2: A better approach to colour measurement. *Brewers' Guardian*, 124 (11), 44–47.

Smedley, S. (1995b) Towards closer colour control in the brewing industry. *Brewers' Guardian*, 124 (10), 42–45.

Smythe, J. E. and Bamforth, C. W. (2000) Shortcomings of standard instrumental methods for assessing beer color. *Journal of the American Society of Brewing Chemists*, 58 (4), 165–166.

Strien, J. V. and Drost, B. W. (1979) Photometric determination of beer and wort color. *Journal of the American Society of Brewing Chemists*, 37 (2), 84–88.

Wyszecki, G. and Stiles, W. S. (1967) *Color Science. Concepts and Methods, Quantitative Data and Formulas*. John Wiley & Sons: New York, NY.

8
Beer and health

Charles W. Bamforth

Many people consider it foolish to even address the topic of beer in the context of nutrition and health, believing that beer, from start to finish, is simply a vehicle for relaxation and enjoyment. However, in an era when all components of dietary intake are more scrutinized than ever before and also when purveyors of wine are championing their product on a health-boosting platform, it is very much necessary to assess what is known about the composition and impact of beer as a foodstuff.

Wright et al. (2008a,b) addressed the perception that consumers have of the relative nutritional value and healthfulness of various beverages, including beer and wine. Generally wine was perceived as being more nutritionally robust than beer and more in keeping with a healthful diet, although purchase decisions for both types of beverage were primarily made according to factors such as taste, what the person is doing activity-wise, location, time of day and cost, with consideration of healthfulness and even calories being relegated to being of much less significance.

Since the CBS television program *60 Minutes* ran a story on the health-beneficial properties of red wine in 1991, the belief has caught on that this beverage in particular might represent more than simply a pleasurable aspect of one's lifestyle. Suddenly the wine industry, notably that in California, seized the moment and celebrated the opportunity afforded by their products to reduce the risk of atherosclerosis. In reality, the health benefits of alcoholic beverages had been implied, professed or unquestioningly accepted for thousands of years but here for the first time, the notion was thrust into the public consciousness.

Detailed studies around the world have now confirmed that the active ingredient in wine is actually ethanol and, as such, beer and other alcoholic drinks at an equivalent dose level are just as beneficial in countering atherosclerosis. Claims that polyphenols such as resveratrol have an especial role to play are now questioned, for the quantities required to have an impact exceed by orders of magnitude the amounts found in any wine.

The reality is that, comparing beer with wine, the former is actually more substantial in nutritive terms (Table 8.1). Thus beer is a significant source of certain B vitamins, minerals (especially silica, which is one of the potential

Table 8.1
The relative nutrient composition of beer and wine

Nutrient	Units	White table wine (one 5 fluid ounce serving)	Red table wine (one 5 fluid ounce serving)	Beer, regular (one 12 fluid ounce serving	Reference daily intake (males 19–50)
Proximates					
Water	g	127.7	127.1	327.4	3.7 liters per day
Energy	Kcal	122	125	153	1800
Protein	g	0.1	0.1	1.64	56
Fat	g	0	0	0	
Ash	g	0.29	0.41	0.57	
Carbohydrate	g	3.82	3.84	12.64	130
Fiber	g	0	0	0*	38
Sugars	g	1.41	0.91	0	
Minerals					
Calcium	mg	13	12	14	1000
Iron	mg	0.4	0.68	0.07	6
Magnesium	mg	15	18	21	350
Phosphorus	mg	26	34	50	580
Potassium	mg	104	187	96	4700
Sodium	mg	7	6	14	1500
Zinc	mg	0.18	0.21	0.04	9.4
Copper	mg	0.006	0.016	0.018	0.7
Manganese	mg	0.172	0.194	0.028	2.3
Selenium	µg	0.1	0.3	2.1	45
Vitamins					
Vitamin C	mg	0	0	0	90
Thiamine	mg	0.007	0.007	0.018	1.2
Riboflavin	mg	0.022	0.046	0.086	1.3
Niacin	mg	0.159	0.329	1.826	16
Pantothenic acid	mg	0.066	0.044	0.146	5
Vitamin B_6	mg	0.074	0.084	0.164	1.3

(Continued)

Table 8.1
(Continued)

Nutrient	Units	White table wine (one 5 fluid ounce serving)	Red table wine (one 5 fluid ounce serving)	Beer, regular (one 12 fluid ounce serving	Reference daily intake (males 19–50)
Folate	µg	1	1	21	400
Vitamin B_{12}	µg	0	0	0.07	2.4
Other					
Ethanol	g	15.1	15.6	13.9	

From Bamforth (2008). The dietary guidelines of the United States Department of Agriculture (http://www.health.gov/DIETARYGUIDELINES/dga2005/report/HTML/D8_Ethanol.htm) recommend a maximum of one drink per day for women and two for men, a drink being a 12 ounce serving of a regular beer or 5 ounces of wine (12% alcohol by volume). And it is stated that at this level there is no association of alcohol consumption with deficiencies of either macronutrients or micronutrients and, furthermore, there is no apparent association between consuming one or two alcoholic beverages daily and obesity. *it is now recognized that beer does contain soluble fiber.

mechanisms by which moderate beer consumption counters osteoporosis (Powell et al., 2005). Beer contains antioxidants, including polyphenols and phenolic acids such as ferulic acid, which has been shown to be more effectively taken into the body from beer than it is from tomatoes (Bourne et al., 2000). Beer contains some soluble fiber as well as polysaccharide degradation products that are likely to function as prebiotics (Bamforth and Gambill, 2007).

Several studies have highlighted the merits of moderate alcohol consumption. Guallar-Castillon et al. (2001) found a lower incidence of sub-optimal health in daily moderate drinkers of beer (and wine). Hospitalization is less for daily moderate drinkers (Longnecker and MacMahon, 1988). Richman and Warren (1985) showed that beer drinkers have significantly reduced rates of sickness – one drink per day gives 15% less disability than was encountered in the general population. There is, however, a view that the benefits of moderate consumption of alcoholic drinks is not seen until middle age (Klatsky and Udaltsova, 2007).

The beneficial impact of beer on the body might be through two mechanisms, firstly a direct dose-response effect on different bodily functions and secondly an indirect effect through the impact of components of beer (especially alcohol, but speculatively also some of the hop components, Cooper, 1994) in boosting the morale and perceived well-being of the drinker (Baum-Baicker, 1985b; Marmot et al., 1993; Pohorecky, 1990; Vasse et al., 1998). Such a mellowing influence counters stress-related illnesses (Morrell, 2000). Cleophas (1999) concludes that there is a significant psychological component in the beneficial relationship between moderate alcohol consumption and mortality.

Studies into the relationship between consumption of an individual foodstuff (in this case, beer) and bodily response are not trivial, it being essential that the researcher endeavor to remove as many interfering factors as possible. As Butterworth (1993) says

> Additional methodological problems are presented by a number of "confounding factors" such as age, sex, body mass index, diet, physical activity, smoking, coffee consumption, educational attainment, type A/B behavior, socio-economic status, and medical history, that may be factors in particular health problems in persons who have been the subjects of the reported studies. For example, a generally poor nutritional condition could possibly play a significant role in various health problems associated with heavy drinkers.

Studies based on self-reporting of dietary intake are less controlled than those in which feeding trials are conducted with defined diets in laboratories. However few and far between are the latter types of study, and where they do occur they are conducted with experimental animals and not the human, so are less simple to extrapolate.

A further criticism of survey-based studies is the reliability of individuals' reporting of alcohol intake (Dawson, 1998; McCann et al., 1999). As Klatsky (2001) says

> In data based upon surveys, systematic "underestimation" (lying) probably tends to lower the apparent threshold for harmful alcohol effects.

Rimm et al. (1996) highlight the greater reliability of case-control studies, in which regulated observations are made relating consumption of a specific alcoholic beverage to specific ailments of the body. From such investigations it is evident that wine, beer and spirits all confer a reduction in coronary heart disease. Rimm and co-workers (1996) write

> We conclude that if any type of drink does provide extra cardiovascular benefit apart from its alcohol content, the benefit is likely to be modest at best or possibly restricted to certain sub-populations.

Atherosclerosis

The mechanism by which alcohol counters the risk of atherosclerosis is complex. Alcohol lowers LDL cholesterol in the blood plasma and stimulates an increased level of HDL cholesterol (HDL_2 and HDL_3) and apolipoproteins A-I and A-II (Clevidence et al., 1995; Goldberg et al., 1995; Jansen et al., 1995; Parker et al., 1996). Alcohol also reduces the risk of blood clotting by lowering the level of fibrinogen in plasma (Stefanick et al., 1995) as well as lessening the aggregation tendencies of blood platelets (Renaud et al., 1992).

Diverse laboratories have reported U-shaped curves (e.g. Doll et al., 1994) or J-shaped curves (e.g. Tsugane et al., 1999) to illustrate the impact that different extents of alcohol intake have on coronary heart disease, but also on all causes of mortality. It appears that the J-shape depicts the relationship between alcohol intake and total mortality, with the U-shape better describing that between alcohol consumption and coronary heart disease. In other words the benefits of alcohol in countering atherosclerosis extend to substantial daily consumption, whereas if all health considerations are considered, then the J-shaped curve speaks to a daily consumption of between 1 and 3 units if the risk of mortality is to reach its nadir.

Mukamal et al. (2003) emphasized that frequency of alcohol consumption is of highest significance. In a study of 38,077 male health professionals over a 12 year period it was shown that men who consumed alcohol 3–4 or 5–7 days per week had a decreased risk of myocardial infarction when compared to those who drank less than once per week. The risk was similar for men taking 10 g alcohol per day or 30 g or more per day, and whether the beverage was beer, red wine, white wine or liquor was irrelevant.

Cleophas (1999) and Rimm and colleagues (1996) summarized the literature comparing the relative impact of beer, wine and spirits on coronary heart disease. These authors and others have emphasized the challenges that exist in this type of study, and stress that confounding influences may often be at play. For example, Renaud et al. (1992) observed that wine drinking tends to be associated with habits that are of themselves responsible for the additional benefits claimed to be associated with the drinking of wine. Wine drinkers often are wealthier and have a healthier lifestyle and superior health care (Galobardes et al., 2001; Klatsky et al., 1997). Wine drinkers probably have a superior overall diet as compared to many beer drinkers (Tjonneland et al., 1999). They may also exercise more (Woodward and Tunstall-Pedoe, 1995). Australian men who preferred beer to wine also drank larger quantities, smoked more and had a generally less healthy diet (Burke et al., 1995). Watten (1999) reported a positive correlation between smoking and beer consumption. Barefoot et al. (2002) highlight the significance of confounding factors as establishing the apparent difference in health impact of wine and beer. Wine drinkers had healthier diets than did those drinking beer and they were less likely to smoke. Wine drinkers took more servings of fruit and vegetables and fewer servings of red or fried meats. The diets of wine drinkers contained less cholesterol and saturated fat, but more fiber. Johansen et al. (2006) analyzed several million check-out receipts at supermarkets in Denmark to show that people buying wine bought more fruit and vegetables, poultry, low-fat cheese and milk. Purchasers of beer bought more sausages, chips, sugar, butter and soft drinks.

It is quite apparent that behavioral forces must be considered most carefully if a meaningful interpretation is going to be made of the impact of beer on the body. For example, hazardous drinking (defined as those occasions when five or more drinks are consumed daily) is associated more with beer than with other types of alcoholic beverage (Rogers and Greenfield, 1999). Predictably enough, this correlates with younger, unmarried males. Mortensen et al. (2001) found that those buying wine in preference to beer tend to have higher IQ's, higher parental educational attainment and higher socio-economic status. Clearly there are financial factors at play here (many wines are substantially more expensive than any beer when quantified on a cents per unit of alcohol basis), and there is no question that wine has "stolen the moral high ground" (Bamforth, 2008). Mortensen and colleagues found that beer drinkers did less well in assessment of psychiatric and health-related behavior than did wine drinkers and concluded that the apparently superior health benefits of wine over beer were related to better social and psychological performance.

Thus Gaziano et al. (1999) concluded that beer, wine and liquor are equally advantageous (in moderation) for countering atherosclerosis, and similar conclusions were drawn by Renaud et al. (1993), Klatsky (1994), Klatsky et al. (1997), Hein et al. (1996), and Rimm et al. (1996).

It is remarkable that the pronounced benefits of one type of beverage as opposed to another seem to largely reflect the type of beverage preferred in the country concerned. Thus a study in Honolulu where the majority of the population consumes beer showed a significant inverse correlation between coronary heart disease and beer drinking (Yano et al., 1977). Similar results were obtained in other beer-drinking countries such as Germany (Keil et al., 1997), the Czech Republic (Bobak et al., 2000) and Wales (Fehily et al., 1993). Hoffmeister et al. (1999) suggested that if Europeans stopped drinking beer there would be a decrease in life expectancy of 2 years – and much unhappiness.

Klatsky et al. (1997) felt that there might be modest extra benefits linked to drinking both beer *and* wine. For men the inverse relationship between drinking and coronary heart disease was significant for beer, for women it was significant for wine. However when analysis was controlled for total intake of alcohol, a significant relationship only remained for beer use by men. As Klatsky (2001) says

> it seems likely that ethyl alcohol is the major factor with respect to CHD risk. There seem to be no compelling health-related data that preclude personal preference as the best guide to choice of beverage.

Despite the strong evidence for ethanol being the key ingredient of alcoholic beverages in relationship to countering atherosclerosis, there remains considerable conviction on the part of many that other components are significant, too, notably certain polyphenolics and phenolics. However, it is essential to stress that such molecules will only be of benefit if they enter the body and target the tissues to be protected. Few are the number of studies that unequivocally demonstrate a direct uptake of antioxidants into the body.

Bourne et al. (2000) demonstrated that ferulic acid in alcohol-free beer (employed to preserve the sobriety of test subjects) could be detected in the urine. A series of polyphenolics also entered the urine (Walker and Baxter, 2000). To reach the kidney a material must be absorbed via the digestive system. Whilst this is not an absolute guarantee that the antioxidant fully accesses all key tissues, it does appear that the ferulic acid is capable of absorption by the body. Indeed, Bourne et al. found that this particular antioxidant is less well absorbed when it is presented to the body in the form of the tomato. The presence of a nutraceutical in high quantities in a food does not necessarily guarantee that such a food is a better source of a material than another food in which it is present in lower quantities: the matrix in which a molecule is delivered is important. Gorinstein et al. (2007) measured an increase in blood antioxidant levels and a favorable change in lipid and anticoagulant levels after consumption of 330 ml of beer per day for a month.

Alcohol *per se* stimulates the uptake of antioxidants into the body (Ghiselli et al., 2000).

The main antioxidants in alcoholic beverages such as wine and beer are the polyphenols (Frankel et al., 1993). A major class amongst the polyphenols, the flavanoids, is said to reduce risk of diseases of the cardiovascular system in humans (Hertog et al., 1993).

An inverse link between cardiovascular disease and the level in the diet of another type of molecule, namely the B-vitamin folate, has been reported (Riddell et al., 2000). Increased folate leads to decreased homocysteine (Eikelboom et al., 1999; McDowell and Lang, 2000), this latter molecule deriving from the metabolism of methionine being associated with increased cardiovascular disease, perhaps because homocysteine is toxic to endothelial cells (Bellamy et al., 1999; Langman and Cole, 1999).

Ubbink et al. (1998) showed a relationship between the lowering of total homocysteine concentration in blood serum and the folic acid content of beer.

The hop constituent xanthohumol has a strong inhibitory effect on the enzymes in liver microsomes that convert diglycerides to triglycerides and so it may reduce the extent of atherosclerosis (Tabata et al., 1997). Xanthohumol is also active in the oxidation of low-density lipoprotein (Miranda et al., 2000a). However, it is questionable whether most beers are hopped sufficiently for this effect to be significant. Nevertheless, the vast majority of beers do contain the bitter iso-α-acids, sometimes in relatively large quantities, and these have been shown to benefit blood lipid status in a mouse model system (Miura et al., 2005; Shimura et al., 2005).

Hypertension and stroke

Increased blood pressure (hypertension) is twice as common in heavy drinkers as in light drinkers (Kannel and Ellison, 1996). Beer drinking in particular has been said to be associated with higher blood pressure (Nevill et al., 1997). Four pints daily over a 3–4 days period had a demonstrable effect (Potter and Beevers, 1984). However, Moline et al. (2000) say that there is an inverse relationship between flavonoid levels in the blood and the incidence of hypertension. Keil et al. (1993) report that two-fifths of studies demonstrate that the blood pressure of non-drinkers is higher than is the case in those consuming 10–20 g alcohol per day. Thadhani et al. (2002) state that low or moderate alcohol intake leads to a lesser incidence of hypertension in women, irrespective of the alcoholic beverage type. Truelsen et al. (1998) insist that wine is better than beer in this context.

Hypertension is the biggest risk factor for strokes. Whilst it has been reported that moderate drinking (<60 g alcohol per day) presents a slightly increased risk of stroke as compared to non-drinkers (van Gijn et al., 1993), others claim that there is a reduced risk of stroke for light to moderate drinkers (Berger et al., 1999; Gill et al., 1988; Gill et al., 1991; Palomaeki et al., 1993; Rodgers et al., 1993; Sacco et al., 1999; Stampfer et al., 1988). It is heavy drinking (>6 drinks per day) or binge drinking that presents an increased risk of stroke (Hillbom et al., 1999; Juvela et al., 1995; Wannamethee and Shaper, 1996).

The digestive system

Alcohol in moderation stimulates the appetite (Hetherington et al., 2001). However alcohol suppresses saliva flow (Enberg et al., 2001), leading to a meal being perceived as being somewhat drier than if it is accompanied by a non-alcoholic drink. Beer, with its much higher water content, would be expected to be better than wine for hydrating the bolus.

Beer, but less so than wine, stimulates the production of the hormone gastrin that promotes the flow of gastric juices in the stomach (Chari et al., 1993).

It is now understood that the bacterium *Helicobacter pylori* induces ulceration of the stomach and duodenum, as well as causing stomach cancer. Alcohol protects against infection by *H. pylori* (Brenner et al., 1997, 1999, 2001; Ogihara et al., 2000). The hop β-acid lupulone inhibits growth of *H. pylori* (Ohsugi et al., 1997).

There is an increased risk of pancreatitis in heavy drinkers (Dreiling et al., 1952; Haber et al., 1995; Sakorafas and Tsiotou, 2000), Schmidt (1991) showed that the drinking of distilled spirits, but not wine or beer, is a risk factor for pancreatitis.

Alcohol speeds up the rate of emptying and filling of the gall bladder and so people with moderate alcohol intake develop fewer gallstones (Leitzmann et al., 1999). Again it appears that frequency of intake is an important factor, with daily intake of alcohol having the most benefit. The type of alcoholic beverage was irrelevant.

There seems to be a link between cirrhosis of the liver and excessive consumption of liquor, but not of beer or wine (Gruenewald and Ponicki, 1995; Kerr et al., 2000). However excessive intake of any alcoholic beverage can cause fatty infiltrations and a swelling in the liver (Kishi et al., 1996). Binge and prolonged drinking can cause alcoholic hepatitis (Maher, 2002).

Water-soluble melanoidins from roasted malt promote the activity of detoxifying enzymes (NADPH-cytochrome c reductase and glutathione S transferase) in intestinal cells (Faist et al., 2002).

Hop polyphenols can inhibit the growth of *Streptococci*, thereby delaying the onset of dental caries (Tagishara et al., 1997). Nakajima et al. (1998) showed that dark beers particularly inhibit the synthesis of a polysaccharide that anchors such bacteria to the teeth. By contrast, beers that contain significant levels of residual sugar and especially low pH (<4.0) have potentially harmful effects on teeth (Nogueira et al., 2000).

Alcohol attenuates the increase in blood glucose concentration in subjects given a glucose load by increasing the sensitivity of susceptible cells to insulin (Facchini et al., 1994). This in turn reduces demand on the pancreas. Moderate drinkers displayed a reduced risk of developing non-insulin dependent diabetes (Perry et al., 1995). Stampfer et al. (1988) found a lower incidence of non-insulin dependent diabetes in moderate drinkers. Rimm et al. (1995) reported that moderate alcohol consumption by healthy people could be associated with increased insulin sensitivity and a reduced risk of diabetes.

Components of beer, including quercetin and the iso-α-acids, inhibit aldose reductase (Shindo et al., 2002). This enzyme is important for removing sorbitol produced from glucose accumulating in diabetics and which causes damage to eyes and kidneys.

The fiber content of beer may contribute to flatus. Bolin and Stanton (1998) report a correlation between men's beer drinking and the aroma of the resultant flatus.

The reproductive system

Juhl et al. (2001) refute any link between alcohol consumption and fecundity. Beer contains a range of health-promoting isoflavanoids (phytoestrogens) (Gavaler, 1998; Lapcik et al., 1998; Walker, 2000). The xanthohumol found in hops does not have estrogenic activity but another hop component, 8-prenylnaringenin, does (De Keukeleire et al., 1997). The latter substance is the most potent estrogen yet identified. Milligan et al. (2000) investigated the relative estrogenic, androgenic and pro-estrogenic activities of 8-prenylnaringenin in comparison to 6-prenylnaringenin, 6, 8-diprenylnaringenin and 8-geranylnaringenin. While the latter three exhibited some estrogenicity, their potency was less than 1% of that of 8-prenylnaringenin. Promberger et al. (2001) have concluded that the risk from these types of material for the hormonal status of men is negligible, for the simple reason that these substances are actually found at extremely low levels in beer. The level of the principle isoflavanoid, isoxanthohumol, found in beer (1.5 mg/l or less) is some 20-fold less than the effective human dose for anti-cancer treatments (Forster and Koberlein, 1998). By contrast, Milligan et al. (1999) suggest that beer may account for around 10% of the daily intake of phytoestrogens. Such phytoestrogens are understood to counter breast and prostate cancer, as well as cardiovascular disease (Knight and Eden, 1996).

Hops (as opposed to beer *per se*) are more effective than other widely used plant preparations in alleviating post-menopausal symptoms (Dixon-Shanies and Shaikh, 1999). Hop extracts suppress menopausal hot flushes (Goetz, 1990). Hops are a constituent of some herbal breast enhancement preparations for women.

Koletzko and Lehner (2000) describe the received wisdom that moderate beer consumption may help in the initiation and success of breast-feeding. It may be that an as yet unidentified barley polysaccharide promotes prolactin secretion. Perhaps too the relaxing effects of alcohol and hop components have a beneficial impact on lactogenesis. Mennella (2001) maintains that alcohol consumption slightly reduces milk production.

Brain and cognitive function

Alcohol is metabolized as it is being drunk and thus there are likely to be greater risks of accumulating undesirably high levels of alcohol when it is taken in the form of short doses of high concentration as opposed to lower alcohol products, that is the majority of beers. Tremoliere et al. (1975) compared the rate of alcohol utilization in the body after challenging with beer, wine or spirits. Fourteen subjects with empty stomachs were given 0.5 g alcohol per kg body weight as beer, wine or whiskey over periods of 15 days to 1 month. Ethanol from beer was oxidized twice as fast as that in wine and seven times faster than that in whisky. Absorbed alcohol passes into the blood in 15–30 minutes if the stomach is empty, but in 1–3 hours if the stomach is full. There is a constant rate of metabolism of approximately 100 mg/h/kg body weight. Thus a dose of 750 ml of beer on an empty stomach by an adult male weighing 70 kg would not allow the French drinking limit of 0.8 g/l of blood to be reached in that person.

There is no consensus agreement about the prime causative agents of hangovers (Pradalier and Ollat, 1991). In part they are likely to be due to a build up of acetaldehyde produced by the oxidation of alcohol (Wall et al., 2000). The aldehyde interacts with components of brain cells to exert its effect. Congeners such as traces of methanol are also oxidized by alcohol dehydrogenase and the resultant formaldehyde is even more unpleasant in its effects than acetaldehyde.

Pain in the brain may also be induced by biogenic amines found in relatively small quantities in beer. Cerutti et al. (1989) measured 15–200 μg/l histamine, 0.7–35.5 mg/l tyramine, 0.5–07 mg/l cadaverine, 3.1–5.6 mg/l putrescine and <0.1–0.8 mg/l β-phenylethylamine in beer. Levels in beer are found to be lower than those in other foodstuffs, including wine and cheese (Gloria, 2003; Table 8.2). Migraine attacks were more frequently associated with the consumption of sparkling and red wines and spirits than with beer (Nicolodi and Sicuteri, 1999).

A study of multiple sets of twins born between 1917 and 1927 revealed a J-shaped relationship between alcohol consumption and cognitive function

Table 8.2
Amine content of beer, wine and cheese

	Spermine	Spermidine	Putrescine	Cadaverine	Histamine	Tyramine	Tryptamine
Lager	n.d.–1.41	n.d.–6	0.85–9.8	0.15–2.6	n.d.–0.9	0.3–3.1	n.d.–0.8
Stout	n.d.–2.05	0.31–1.38	1.99–5.84	0.3–1.37	n.d.–0.85	0.48–36.8	n.d.–10.1
Ice	n.d.–0.3	0.6–0.8	3.9–4.5	0.1–0.2	n.d.	0.7–1.4	n.d.
Bock	n.d.–1.73	0.25–2.1	1.55–6.3	0.15–1.72	n.d.–1.46	0.81–5.05	n.d.–3.5
Non-alcoholic	n.d.–1.2	1.35–2.3	2.3–4.95	n.d.–0.5	n.d.–0.62	0.6–3.3	n.d.–1.41
Pinot Noir	n.d.–2.38	n.d.–2.35	2.43–203	n.d.–2.07	n.d.–23.98	n.d.–8.31	n.d.–5.51
Cabernet	n.d.–1.17	n.d.–4.03	3.15–23.6	n.d.–1.51	n.d.–10.1	n.d.–7.53	n.d.
Blue	–	–	9.6–23.7	42.3–227	n.d.–409	2.2–166	n.d.–110
Cheddar	–	–	n.d.–99.6	n.d.–40.8	n.d.–154	n.d.–153	n.d.–30
Gorgonzola	–	–	1.2–124	5.8–428	1.7–191	8.9–255	2.4–43
Gouda	n.d.–1.13	n.d.–1.35	n.d.–107	n.d.–99.5	n.d.–30.5	n.d.–67	n.d.–88
Mozzarella	n.d.–1.31	n.d.–1.06	n.d.–1.37	n.d.–2.34	n.d.–11.3	n.d.–41	n.d.–10
Parmesan	0.07–0.09	n.d.–0.15	n.d.–4.3	n.d.–9.8	n.d.–27.2	n.d.–29	n.d.–1.7
Provolone	0.07–0.97	n.d.–2.38	n.d.–8.7	n.d.–111	n.d.–8.2	n.d.–10.9	n.d.–1.08
Swiss	–	–	–	–	n.d.–250	n.d.–180	n.d.–1.6

Quantities are mg/l for beer and wine; mg/100 g for cheese. n.d. = not detectable; – – = not determined.
Derived from Gloria (2003).

with moderate drinkers performed significantly better than abstainers or heavy drinkers (Christian et al., 1995). There was no indication that this was linked to any specific type of alcoholic beverage.

Moderate drinkers are said to be more outgoing and enthusiastic about life; less stressed; perform some tasks better after a drink; enjoy fewer incidences of depression; and fare better when elderly, including better cognitive function (Baum-Baicker, 1985).

Ruitenberg et al. (2002) found that light to moderate drinking (1–3 alcoholic drinks of any type per day) is significantly associated with a lower risk of dementia in those aged 55 years and above.

Kidney and urinary tract

Alcohol dehydrates the whole body (except the brain, which swells) through a diuretic impact on the kidney (Olson, 1979). This speaks to the advisability of drinking plenty of water before sleeping after drinking.

Beer is rather more diuretic than is water, with its content of organic acids and other yeast fermentation products and polyphenols believed to be significant (Buday and Denis, 1974; Piendl and Wagner, 1985).

Some beers may be a particular problem for sufferers of gout when compared to other alcoholic drinks because they can contain significant quantities of purines (Eastmond et al., 1995).

On the other hand, beer is superior to water alone in "flushing out" the kidneys, thereby protecting the kidney against stones (Curhan et al., 1996, 1998; Krieger et al., 1996; Shuster et al., 1985). Hirvonen et al. (1999) report that every bottle of beer consumed per day reduces risk of kidney stones by 40%.

Nagao et al. (1999) report that stale beer is less prone than is fresh beer to promote the rate of urination.

Age

Moderate alcohol consumption may be associated with better cognitive function in the elderly (Cervilla et al., 2000; Dufouil et al., 1997). Moderate consumption of wine and beer reduces the incidence of macular degeneration (Obisesan et al., 1998). Moderate consumption of beer stimulates appetite and promotes bowel function in the elderly (Dufour et al., 1992).

Regular consumption of alcohol lowered the risk of incurring Alzheimer's disease (Cupples et al., 2000).Although there is now some doubt about the significance of aluminum as an agent promoting Alzheimer's disease (Roberts et al., 1998), silicon, present in abundance in beer, is believed to be a key agent in eliminating aluminum from the body (Bellia et al., 1996).

Beer (half a pint daily) was employed alongside other "treats" to enhance the environment in a ward nursing elderly psycho geriatric men (Volpe and Kastenbaum, 1967). The outcome was increased sociability, less need for medication, return to continence and less need for physical restraint. Beer was interpreted by the patients as a symbol of trust.

Jugdaohsingh et al. (2002) stress how beer, bananas and string beans are rich sources of silicon in the diet. Rico et al. (2000) suggest that silicon is readily absorbed from beer, thereby protecting against osteoporosis. The hop component 8-prenylnaringenin is claimed to counter osteoporosis (Miyamoto et al., 1998).

Mukherjee and Sorrell (2000) show that moderate alcohol consumption has positive effects on bone mineral density in elderly women, and say that this is probably mediated by a decrease in bone remodeling.

Xanthohumol and humulone inhibit bone resorption (Tobe et al., 1997; Honma et al., 1998).

Cancer

Evidence suggests that alcohol consumption needs to be really rather substantial for it to be a causative factor in cancer. A 1989 report of the Committee on Diet and Health of the National Academy of Sciences warns against *excessive consumption* of alcohol, rather than avoidance. The literature is contradictory on the relationship between alcohol consumption and cancer, some papers even suggesting that the consumption of alcoholic beverages (including beer) may actually lessen the incidence of certain cancers.

A significant factor in bodily tissue damage is exerted through the action of free radicals. At higher concentrations ethanol is metabolized by a cytochrome P450 system that generates significant radical formation (Lieber, 1994; Nordmann, 1994). However Gasbarrini et al. (1998) suggest that a beer-containing diet (as opposed to an ethanol-containing one) reduced the prevalence of markers of oxidation risk in rats.

Most widely publicized of the potential carcinogens are the nitrosamines, but a remarkably rapid response by maltsters and brewers to the first report of significant levels of nitrosamines in beer (Spiegelhalder et al., 1979) means that levels of nitrosamines are now extremely low and at a fraction of the level present in the seventies (Sen et al., 1996). For moderate beer drinkers these residual levels of nitrosamines should not present a hazard (Tricker and Preussmann, 1991).

The carcinogenic chloropropanols are found in the more intensely heated grist materials used for brewing, however they do not survive into beer (Long, 1999).

Heavy drinkers may have an increased risk of developing oropharyngeal and lower esophageal cancer (Day et al., 1993; Kabat et al., 1993; Kune et al., 1993). However the risks are greater with "stronger" alcoholic beverages and those who smoke heavily (Ishii et al., 2001).

Non-drinkers may be at increased risk of lung cancer (Woodson et al., 1999). However, in a study claimed by the authors to be inconclusive, Bandera et al. (2001) find that, of eleven studies relating beer drinking to lung cancer, five suggested a positive association, two indicated possibly weak support, but four found no association.

Freudenheim et al. (1995) and Zhang et al. (1999) report no relationship between alcohol consumption and breast cancer, but Ferraroni et al. (1998) say that there is an increased risk, especially for wine as opposed to beer. Hebert et al. (1998) report an increased risk of breast cancer recurrence in beer drinkers.

Excessive beer drinking is claimed to be associated with an increased risk of colon cancer (Hsing et al., 1998). By contrast, Riboli et al. (1991) say that, while there is a link between beer drinking and rectal cancer, there is no relationship for colon cancer.

An increased risk of carcinoma in the upper digestive tract has been reported in beer and spirit drinkers (Gronbaek et al., 1998), however Albertsen and Gronbaek (2002) and Tavani et al. (1994) say that there is no link between alcohol consumption (of any type) and prostate cancer. Harris and colleagues (2003) report that PSA (an antigen that is diagnostic of prostate problems) was proportionately decreased by an increase in a man's beer intake.

Certain substances found in beer might counter cancer. Several unique polyphenol compounds have been found in hops and beer and which impart chemo-preventative properties in part by inhibiting cytochrome P450 (Henderson et al., 2000). The compounds include xanthohumol, 8-prenylnaringenin and isoxanthohumol.

Miranda et al. (1999) found that prenylated flavonoids, notably xanthohumol, inhibited the *in vitro* proliferation of a range of cancer cells. Hop extracts at very low concentration efficiently inhibit breast cancer cells (Zava et al., 1998).

8-Prenylnaringenin inhibits enzymes involved in the development of prostate cancer and also blocks the development of new blood vessels needed for tumor growth (De Keukeleire et al., 2001).

Humulone inhibits the growth of skin tumors in mice (Yasukawa, 1995) and also leukemia cells (Honma et al., 1998). Xanthohumol and isoxanthohumol, but also beer itself, inhibits the mutagenic impact of heterocyclic amines (Arimoto-Kobayashi et al., 1999; Miranda, 2000b). Beers with the highest content of polyphenol are the ones most capable of blocking nitrosation events in the rat (Pignatelli et al., 1983). Yoshikawa et al. (2002) found pseudouridine in a diversity of beers and demonstrated its anti-mutagenic activity. Actually pseudouridine seemed to account for only 3% of the antimutagenic impact of beer.

Allergy

Tyramine in the range 3.6–7.4 mg/l and histamine in the range 3–3.2 mg/l have been found in beers (Gorinstein et al., 1999). Tyramine can cause a rise in blood pressure by constricting the vascular system and increasing the heart rate (Gloria, 2003). Such amines can induce migraines and hypertensive crises (Crook, 1981; Zee et al., 1981).

Ehler et al. (2002) report a direct allergic reaction of individuals to ethanol itself.

Sulfite is added to some beers to enhance shelf-life and this can trigger reactions in sensitive people. However it is more commonly used as a preservative in sodas and other non-alcoholic drinks, and also of course in wine. The need to label it in the US if present at levels greater than 10 mg/l means that it is customarily avoided in this country.

As most beers are derived from barley or wheat, then it is generally recommended that they be avoided by sufferers of celiac disease who react to prolamin proteins (such as hordein and gluten) and peptides derived from gluten (Campbell, 1992). However there has been much debate about the extent to which beer actually presents a problem in this context (Kasarda, 2003). There is much less protein in beer than in the grist materials from which it is derived owing to the malting and brewing processes. Indeed many beers have as part of their grists non-prolamin containing adjuncts such as rice, corn and sugar.

Kasarda writes "beer made from wheat, barley or rye … might have some toxic peptides in it. I suspect that beer has low toxicity, perhaps even none, but further scientific studies would need to be carried out to prove this."

Cizkova et al. (2005) measured higher levels of prolamins in malt than in barley, presumably because of the release of reactive peptides during the hydrolysis of hordein. They found much less prolamin in beer than in malt (<0.1%), with the most reported being approximately 20 mg/l in a wheat beer. Most lagers contained less than 1.5 mg/l reactive protein. Ellis et al. (1990) measured 1.5 mg of the sensitive proteins per pint of beer. It is recommended that celiac patients limit daily intake of the relevant protein material to between 10 and 100 mg (Hischenhuber et al., 2006). On this basis it would seem that a pint of most beers would deliver approximately 15% of this lower limit, but only 1.5% of the upper limit.

Other proteins in beer may also have an allergic impact, for example a lipid transfer protein (LTP; Asero et al., 2001). Saliently, not all beers caused a reaction even thought most if not all contain LTP.

Kortekangassavolainen et al. (1993) report on allergic reactions to brewing yeast.

References

Albertsen, K. and Gronbaek, M. (2002) Does amount or type of alcohol influence the risk of prostate cancer? *Prostate*, 52, 297–304.

Arimoto-Kobayashi, S., Sugiyama, C., Harada, N., Takeuchi, M., Takemura, M. and Hayatsu, H. (1999) Inhibitory effects of beer and other alcoholic beverages on mutagenesis and DNA adduct formation induced by several carcinogens. *Journal of Agriculture and Food Chemistry*, 47, 221–230.

Asero, R., Mistrello, G., Roncarolo, D., Amato, S. and van Ree, R. (2001) A case of allergy to beer showing cross-reactivity between lipid transfer protein. *Annals of Allergy and Asthma Immunology*, 87, 65–67.

Bamforth, C. W. and Gambill, S. C. (2007) Fiber and putative prebiotics in beer. *Journal of the American Society of Brewing Chemistry*, 65, 67–69.

Bamforth, C. W. (2008) *Grape vs Grain*. Cambridge University Press.

Bandera, E. V., Freudenheim, J. L. and Vena, J. E. (2001) Alcohol consumption and lung cancer: A review of the epidemiological evidence. *Cancer Epidemiology Biomarkes Prevent*, 10, 813–821.

Barefoot, J. C., Gronbaek, M., Feaganes, J. R., McPherson, R. S., Williams, R. B. and Siegler, I. C. (2002) Alcoholic beverage preference, diet, and health habits in the UNC Alumni Heart Study. *American Journal of Clinical Nutrition*, 76, 466–472.

Baum-Baicker, C. (1985) The psychological benefits of moderate alcohol consumption: A review of the literature. *Drug and Alcohol Dependence*, 15, 305–322.

Bellamy, M. F., McDowell, I. F. W., Ramsey, M. W., Brownlee, M., Newcombe, R. G. and Lewis, M. J. (1999) Oral folate enhances endothelial function in hyperhomocysteinaemic subjects. *European Journal of Clinical Investigation*, 29, 659–662.

Bellia, J. P., Birchall, J. D. and Roberts, N. B. (1996) The role of silicic acid in the renal excretion of aluminium. *Annals of Clinical and Laboratory Science*, 26, 227–233.

Berger, K., Ajani, U. A., Kase, C. S., Gaziano, J. M., Buring, J. E., Glynn, R. J. and Hennekens, C. H. (1999) Light-to-moderate alcohol consumption and the risk of stroke among US male physicians. *The New England Journal of Medicine*, 341, 1557–1564.

Bobak, M., Skodova, Z. and Marmot, M. (2000) Effect of beer drinking on risk of myocardial infarction: Population based case-control study. *British Medical Journal*, 320, 1378–1379.

Bolin, T. D. and Stanton, R. A. (1998) Flatus emission patterns and fibre intake. *European Journal of Surgery*, 164, 115–118.

Bourne, L., Paganga, G., Baxter, D., Hughes, P. and Rice-Evans, C. (2000) Absorption of ferulic acid from low alcohol beer. *Free Radical Research*, 32, 273–280.

Brenner, H., Rothenbacher, D., Bode, G. and Adler, G. (1997) Relation of smoking and alcohol and coffee consumption to active *Helicobacter pylori* infection: Cross sectional study. *British Medical Journal*, 315, 1489–1492.

Brenner, H., Rothenbacher, D., Bode, G. and Adler, G. (1999) Inverse graded relation between alcohol consumption and active infection with *Helicobacter pylori*. *American Journal of Epidemiology*, 149, 571–576.

Brenner, H., Bode, G., Adler, G., Hoffmeister, A., Koenig, W. and Rothenbacher, D. (2001) Alcohol as a gastric disinfectant? The complex relationship between alcohol consumption and current Helicobacter pylori infection. *Epidemiology*, 12, 209–214.

Buday, A. Z. and Denis, G. (1974) The diuretic effect of beer. *Brewers Digest*, 49 (6), 56–58.

Burke, V., Puddey, I. B. and Beilin, L. J. (1995) Mortality associated with wines, beers and spirits. *British Medical Journal*, 311, 1166a.

Butterworth, K. R. (1993) Overview of the biomedical project on alcohol and health in alcohol and overweight. In: *Health Issues Related to Alcohol Consumption* (P. M. Verschuren ed.), pp. 1–16. ILSI Europe: Brussels.

Campbell, J. A. (1992) Dietary management of Celiac Disease: Variations in the gluten-free diet. *J Can Dietet Ass*, 53, 15–18.

Cerutti, G., Finoli, C. and Vecchio, A. (1989) Non-volatile amine biogenesis from wort to beer. *Monatsschrift fur Brauwissenschaft*, 42, 246–248.

Cervilla, J. A., Prince, M., Lovestone, S. and Mann, A. (2000) Long-term predictors of cognitive outcome in a cohort of older people with hypertension. *British Journal of Psychiatry*, 177, 66–71.

Chari, S., Teyssen, S. and Singer, M. V. (1993) Alcohol and gastric acid secretion in humans. *Gut*, 34, 843–847.

Christian, J. C., Reed, T., Carmelli, D., Page, W. F., Norton, J. A. and Breitner, J. C. S. (1995) Self-reported Alcohol intake and cognition in aging twins. *Journal of Studies on Alcohol*, 56, 414–416.

Cizkova, H., Dostalek, P., Hochel, I., Gabrovska, D. and Rysova, J. (2005) Beer: A nutritional support for coeliacs? *Proceedings of the European Brewing Convention Congress*, 941–946.

Cleophas, T. J. (1999) Wine, beer and spirits and the risk of myocardial infarction: A systematic review. *Biomed Pharmaceutical*, 53, 417–423.

Clevidence, B. A., Reichman, M. E., Judd, J. T., Muesing, R. A., Schatzkin, A., Schaeffer, E. J., Li, Z. L., Jenner, J., Brown, C. C., Sunkin, M., Campbell, W. S. and

Taylor, P. R. (1995) Effects of alcohol consumption on lipoproteins of premenopausal women: A controlled diet study. *Arteriosclerosis, Thrombosis, and Vascular Biology*, 15, 179–184.

Cooper, T. J. (1994) Medical considerations of moderate alcohol consumption. *Proceedings of the 23rd Institute of Brewing Convention (Australia and NZ Section)*, 32–37.

Crook, M. (1981) Migraine: A biochemical headache? *Biochemical Reviews*, 9, 351–357.

Curhan, G. C., Willett, W. C., Rimm, E. B., Spiegelman, D. and Stampfer, M. J. (1996) Prospective study of beverage use and the risk of kidney stones. *American Journal of Epidemiology*, 143, 240–247.

Curhan, G. C., Willett, W. C., Speizer, F. E. and Stampfer, M. J. (1998) Beverage use and risk for kidney stones in women. *Annals of Internal Medicine*, 128, 534.

Dawson, D. A. (1998) Volume of ethanol consumption: Effects of different approaches to measurement. *Journal of Studies on Alcohol*, 59, 191–197.

Day, G. L., Blot, W. J., Austin, D. F., Bernstein, L., Greenberg, R. S., Prestonmartin, S., Schoenberg, J. B., Winn, D. M., McLaughlin, J. K. and Fraumeni, G. F. (1993) Racial differences in risk of oral and pharyngeal cancer: Alcohol, tobacco and other determinants. *Journal of the National Cancer Institute*, 85, 465–473.

De Keukeleire, D., Milligan, S. R., Du Cooman, L. and Heyerick, A. (1997) Hop-derived phytoestrogens in beer? *Proceedings of the European Brewing Convention Congress, Maastricht*, 239–246.

De Keukeleire, D., Milligan, S. R., Kalita, J. G., Pocock, V., De Cooman, L., Heyerick, A., Rong, H. and Roelens, F. (2001) Prenylated hop flavonoids are key agents in relation to health beneficial properties of beer. *Proceedings of the European Brewing Convention Congress, Budapest*, 82–91.

Dixon-Shanies, D. and Shaikh, N. (1999) Growth inhibition of human breast cancer cells by herbs and phytoestrogens. *Onc Rep*, 6, 1383–1387.

Doll, R., Peto, R., Hall, E., Wheatley, K. and Gray, R. (1994) Mortality in relation to consumption of alcohol: 13 years' observations on male British doctors. *British Medical Journal*, 309, 911–918.

Dreiling, D. A., Richman, A. and Fradkin, N. F. (1952) The role of alcohol in the etiology of pancreatitis: A study of the effect of intravenous ethyl alcohol on the external secretion of the pancreas. *Gastroenterology*, 20, 636–646.

Dufouil, C., Ducimetiere, P. and Alperovitch, A. (1997) Sex differences in the association between alcohol consumption and cognitive performance. EVA study group. *Epidemiology of vascular ageing. American Journal of Epidemiology*, 146, 405–412.

Dufour, M. C., Archer, L. and Gordis, E. (1992) Alcohol and the elderly. *Clin Ger Med*, 8, 127–141.

Eastmond, C. J., Garton, M., Robins, S. and Riddoch, S. (1995) The effects of alcoholic beverages on urate metabolism in gout sufferers. *British Journal of Rheumatology*, 34, 756–759.

Ehler, I., Hipler, U.-C., Zuberbier, T. and Worm, M. (2002) Ethanol as a cause of hypersensitivity reactions to alcoholic beverages. *Clinical and Experimental Allergy*, 32, 1231–1235.

Eikelboom, J. W., Lonn, E., Genest, J., Hankey, G. and Yusuf, S. (1999) Homocyst(e)ine and cardiovascular disease: A critical review of the epidemiologic evidence. *Annals of Internal Medicine*, 131, 363–375.

Ellis, H. J., Freedman, A. R. and Ciclitira, P. J. (1990) Detection and estimation of the barley prolamin content of beer and malt to assess their suitability for patients with celiac disease. *Clinica Chimica Acta*, 189, 123–130.

Enberg, N., Alho, H., Loimaranta, V. and Lenander-Lumikari, M. (2001) Saliva flow rate, amylase activity, and protein and electrolyte concentrations in saliva after acute alcohol consumption. *Oral Surgery Oral Medicine Oral Pathology Oral Radiology Endodontics*, 92, 292–298.

Facchini, F., Chen, I. Y.-D. and Reaven, G. M. (1994) Light to moderate alcohol intake is associated with enhanced insulin sensitivity. *Diabetes Care*, 17, 115–119.

Faist, V., Lindenmeier, M., Geisler, C., Erbersdobler, H. F. and Hofmann, T. (2002) Influence of molecular weight fractions isolated from roasted malt on the enzyme activities of NADPH-cytochrome c-reductase and glutathione-S-transferase in caco-2 cells. *Journal of Agriculture and Food Chemistry*, 50, 602–606.

Fehily, A. M., Yarnell, J. W. G., Sweetnam, P. M. and Elwood, P. C. (1993) Diet and incident ischemic heart disease – the Caerphilly Study. *British Journal of Nutrition* 69, 303–314.

Ferraroni, M., Decarli, A., Franceschi, S. and La Vecchia, C. (1998) Alcohol consumption and risk of breast cancer: A multicentre Italian case-control study. *European Journal of Cancer*, 34, 1403–1409.

Forster, A. and Koberlein, A. (1998) The location of xanthohumol from hops during beer production. *Brauwelt*, 138, 1677–1679.

Frankel, E. N., Kanner, J., German, J. B., Parks, E. and Kinsella, J. E. (1993) Inhibition of oxidation of human low density lipoprotein by phenolic substances in red wine. *Lancet*, 341, 454–457.

Freudenheim, J. L., Marshall, J. R., Graham, S., Laughlin, R., Vena, J. E., Swanson, M., Ambrosone, C. and Nemoto, T. (1995) Lifetime alcohol consumption and risk of breast cancer. *The Nutrition and Cancer Journal*, 23, 1–11.

Galobardes, B., Morabia, A. and Bernstein, M. S. (2001) Diet and socioeconomic position: Does the use of different indicators matter? *International Journal of Epidemiology*, 30, 334–340.

Gasbarrini, A., Addolorato, G., Simoncini, M., Gasbarrini, G., Fantozzi, P., Mancini, F., Montanari, L., Nardini, M., Ghiselli, A. and Scaccini, C. (1998) Beer affects oxidative stress due to ethanol in rats. *Digestive Diseases and Sciences*, 43, 1332–1338.

Gavaler, J. S. (1998) Alcoholic beverages as a source of estrogens. *Alcohol Health & Research World*, 22, 220–227.

Gaziano, J. M., Hennekens, C. H., Godfried, S. L., Sesso, H. D., Glynn, R. J., Breslow, J. L. and Buring, J. E. (1999) Type of alcoholic beverage and risk of myocardial infarction. *The American Journal of Cardiology*, 83, 52–57.

Ghiselli, A., Natella, F., Guidi, A., Montanari, L., Fantozzi, P. and Scaccini, C. (2000) Beer increases plasma antioxidant capacity in humans. *The Journal of Nutritional Biochemistry*, 11, 76–80.

Gill, J. S., Shipley, M. J., Hornby, R. H., Gill, S. K. and Beevers, D. G. (1988) A community case-control study of alcohol consumption in stroke. *International Journal of Epidemiology*, 17, 542–547.

Gloria, M. B. A. (2003) Amines. In: *Encylopedia of Food Sciences and Nutrition Second edition* (B. Caballero, L. C. Trugo and P. M. Finglas eds), volume 1, pp. 173–181. Academic Press: London.

Goetz, P. (1990) Traitement des bouffées de chaleur par insuffisance ovarienne par l'extrait de houblon. *Revue de Phytothérapie Pratique* (4), 13–15.

Goldberg, D. M., Hahn, S. E. and Parkes, J. G. (1995) Beyond alcohol: Beverage consumption and cardiovascular mortality. *Clinica Chimica Acta*, 237, 155–187.

Gorinstein, S., Zemser, M., Vargas-Albores, F., Ochoa, J. L., Paredes-Lopez, O., Scheler, C., Salnikow, J., Martin-Belloso, O. and Trakhtenberg, S. (1999) Proteins and amino acids in beers, their contents and relationships with other analytical data. *Food Chemistry*, 67, 71–78.

Gorinstein, S., Caspi, A., Libman, I., Leontowicz, H., Leontowicz, M., Tashma, Z., Katrich, E., Jastrzebski, Z. and Trakhtenberg, S. (2007) Bioactivity of beer and its influence on human metabolism. *International Journal of Food Science and Nutrition*, 58, 94–107.

Gronbaek, M., Becker, U., Johansen, D., Tonnesen, H., Jensen, G. and Sorensen, T. I. A. (1998) Population based cohort study of the association between alcohol intake and cancer of the upper digestive tract. *British Medical Journal*, 317, 844–848.

Gruenewald, P. J. and Ponicki, W. R. (1995) The relationship of alcohol sales to cirrhosis mortality. *Journal of Studies on Alcohol*, 56, 635–641.

Guallar-Castillon, P., Rodriguez-Artalejo, F., Ganan, L. D., Banegas, J. R. B., Urdinguio, P. L. and Cabrera, R. H. (2001) Consumption of alcoholic beverages and subjective health in Spain. *Journal of Epidemiology and Community Health*, 55, 648–652.

Haber, P., Wilson, J., Apte, M., Korsten, M. and Pirola, R. (1995) Individual susceptibility to alcohol pancreatitis: Still and enigma. *Journal of Laboratory and Clinical Medicine*, 125, 305–312.

Harris, A., Gray, M., Slaney, D., Turley, M., Fowles, J. and Weinstein, P. (2003) Ethnic differences in diet and associations with surrogate markers of prostate disease in New Zealand. *Epidemiology*, 14 (5). Supplement: S27.

Hebert, J. R., Hurley, T. G. and Ma, Y. S. (1998) The effect of dietary exposures on recurrence and mortality in early stage breast cancer. *Breast Cancer Research and Treatment*, 51, 17–28.

Hein, H. O., Suadicani, P. and Gyntelberg, F. (1996) Alcohol consumption, serum low density lipoprotein cholesterol concentration, and risk of ischaemic heart disease: Six year follow up in the Copenhagen male study. *British Medical Journal*, 312, 736–741.

Henderson, M. C., Miranda, C. L., Stevens, J. F., Deinzer, M. L. and Buhler, D. R. (2000) In vitro inhibition of human P450 enzymes by prenylated flavonoids from hops, Humulus lupulus. *Xenobiotica*, 30, 235–251.

Hertog, M. G. L., Feskens, E. J. M., Hollman, P. C. H., Katan, M. B. and Kromhout, D. (1993) Dietary antioxidant flavonoids and risk of coronary heart disease – the Zutphen elederly study. *Lancet*, 342, 1007–1011.

Hetherington, M. M., Cameron, F., Wallis, D. J. and Pirie, L. M. (2001) Stimulation of appetite by alcohol. *Physiology and Behavior*, 74, 283–289.

Hillbom, M., Numminen, H. and Juvela, S. (1999) Recent heavy drinking of alcohol and embolic stroke. *Stroke*, 30, 2307–2312.

Hirvonen, T., Pietinen, P., Virtanen, M., Albanes, D. and Virtamo, J. (1999) Nutrient intake and use of beverages and the risk of kidney stones among male smokers. *American Journal of Epidemiology*, 150, 187–194.

Hischenhuber, C., Crevel, R., Jarry, B., Mäki, M., Moneret-Vautrin, D. A., Romano, A., Troncone, R. and Ward, R. (2006) Safe amounts of gluten for patients with wheat allergy or coeliac disease. *Alimentary Pharmacology and Therapeutics*, 1, 559–575.

Hoffmeister, H., Schelp, F. P., Mensink, G. B. M., Dietz, E. and Bohning, D. (1999) The relationship between alcohol consumption, health indicators and mortality in the German population. *International Journal of Epidemiology*, 28, 1066–1072.

Honma, Y., Tobe, H., Makishima, M., Yokoyama, A. and Okabe-Kado, J. (1998) Induction of differentiation of myelogenous leukemia cells by humulone, a bitter in the hop. *Leukemia Research*, 22, 605–610.

Hsing, A. W., McLaughlin, J. K., Chow, W. H., Schuman, L. M., Chien, H. T. C., Gridley, G., Bielke, E., Wacholder, S. and Blot, W. J. (1998) Risk factors for colorectal cancer in a prospective study among US white men. *International Journal of Cancer*, 77, 549–553.

Ishii, H., Kato, S., Yokoyama, A. and Maruyama, K. (2001) Alcohol and cancer of the aerodigestive tract. In: *Alcohol in Health and Disease* (D. P. Agarwal and H. K. Seitz eds), pp. 501–515. Marcel Dekker: New York.

Jansen, D. F., Nedeljkovic, S., Feskens, E. J. M., Oistojic, M. C., Grujic, M. Z., Bloemberg, B. P. M. and Kromhout, D. (1995) Consumption, alcohol use and cigarette smoking as determinants of serum total and HDL cholesterol in two Serbian cohorts of the seven countries study. *Arteriosclerosis, Thrombosis, and Vascular Biology*, 15, 1793.

Johansen, D., Friis, K., Skovenborg, E. and Gronbaek, M. (2006) Food buying habits of people who buy wine or beer: Cross sectional study. *British Medical Journal*, 332, 519–521.

Jugdaohsingh, R., Anderson, S. H. C., Tucker, K. L., Elliott, H., Kiel, D. P., Thompson, R. P. H. and Powell, J. J. (2002) Dietary silicon intake and absorption. *American Journal of Clinical Nutrition*, 75, 887–893.

Juhl, M., Andersen, A. M. N., Gronbaek, M. and Olsen, J. (2001) Moderate alcohol consumption and waiting time to pregnancy. *Human Reproduction*, 16, 2705–2709.

Juvela, S., Hillbom, M. and Palomaki, H. (1995) Risk factors for spontaneous intracerebral hemorrhage. *Stroke*, 26, 1558–1564.

Kabat, G. C., Ng, S. K. C. and Wynder, E. L. (1993) Tobacco, alcohol intake and diet in relation to adenocarcinoma of the esophagus and gastric cardia. *Cancer Causes and Control*, 4, 123–132.

Kannel, W. B. and Ellison, R. C. (1996) Alcohol and coronary heart disease: The evidence for a protective effect. *Clinica Chimica Acta*, 246, 59–76.

Kasarda, D.D. (2003) Celiac disease and safe grains. http://wheat.pw.usda.gov/ggpages/topics/Celiac.vs.grains.html (accessed August 19, 2007)

Keil, U., Swales, J. D. and Grobbee, D. E. (1993) Alcohol Intake and its relation to hypertension. In: *Health Issues Related to Alcohol Consumption* (P. M. Verschuren ed.), pp. 17–42. ILSI Press: Brussels.

Kerr, W. C., Fillmore, K. M. and Marvy, P. (2000) Beverage-specific alcohol consumption and cirrhosis mortality in a group of English-speaking beer-drinking countries. *Addiction*, 95, 339–346.

Kishi, M., Maeyama, S., Koike, J., Aida, Y., Yoshida, H. and Uchikoshi, T. (1996) Correlation between intrasinusoidal neutrophilic infiltration and ceroid-lipofuscinosis in alcoholic liver fibrosis with or without fatty change: Clinicopathological comparison with nutritional fatty liver. *Alcoholism: Clinical and Experimental Research*, 20, A366–A370.

Klatsky, A. L. (1994) *Epidemiology of coronary heart disease: Influence of alcohol Alcoholism: Clinical and Experimental Research*, 18, 88–96.

Klatsky, A. L. (2001) Alcohol and cardiovascular diseases. In: *Alcohol in Health and Disease* (D. P. Agarwal and H. K. Seitz eds), pp. 517–546. Marcel Dekker: New York.

Klatsky, A. L. and Udaltsova, N. (2007) Alcohol drinking and total mortality risk. *Annals of Epidemiology*, 17, S63–S67.

Klatsky, A. L., Armstrong, M. A. and Friedman, G. D. (1997) Red wine, white wine, liquor, beer and risk for coronary artery disease hospitalisation. *The American Journal of Cardiology*, 80, 416–420.

Knight, D. C. and Eden, J. A. (1995) Phytoestrogens: A short review. *Maturitas*, 22 (3), 167–175.

Koletzko, B. and Lehner, F. (2000) Beer and breastfeeding. *Short and long term effects of breast feeding on child health. Advances in Experimental Medicine and Biology*, 478, 23–28.

Kortekangassavolainen, O., Lammintausta, K. and Kalimo, K. (1993) Skin prick test reactions to brewers yeast (*Saccharomyces cerevisiae*) in adult atopic dermatitis patients. *Allergy*, 48, 147–150.

Krieger, J. N., Kronmal, R. A., Coxon, V., Wortley, P., Thompson, L. and Sherrard, D. J. (1996) Dietary and behavioural risk factors for urolithiasis: Potential implications for prevention. *American Journal of Kidney Diseases*, 28, 195–201.

Kune, G. A., Kune, S., Field, B., Watson, L. F., Cleland, H., Merenstein, D. and Vitetta, L. (1993) Oral and pharyngeal cancer, diet, smoking, alcohol and serum vitamin

A and beta-carotene levels: A case control study in men. *The Nutrition and Cancer Journal*, 20, 61–70.

Langman, L. J. and Cole, D. E. C. (1999) Homocysteine. *Critical Reviews in Clinical and Laboratory Sciences*, 36, 365–406.

Lapcik, O., Hill, M., Hampl, R., Wahala, K. and Adlercreutz, H. (1998) Identification of isoflavanoids in beer. *Steroids*, 63, 14–20.

Leitzmann, M. F., Giovannucci, E. L., Stampfer, M. J., Spiegelman, D., Colditz, G. A., Willett, W. C. and Rimm, E. B. (1999) Prospective study of alcohol consumption patterns in relation to symptomatic gallstone disease in men. *Alcoholism: Clinical and Experimental Research*, 23, 835–841.

Lieber, C. S. (1994) Mechanisms of ethanol-drug-nutrition interactions. *Journal of Toxicology – Clinical Toxicology*, 32, 631–681.

Long, D. E. (1999) From cobalt to chloropropanol: De tribulationibus aptis cervisiis imbibendis. *Journal of the Institute of Brewing*, 105, 79–84.

Longnecker, M. and MacMahon, B. (1988) Associations between alcoholic beverage consumption and hospitalisation, 1983 National Health Interview Survey. *American Journal of Public health*, 78, 115–153.

Maher, J. J. (2002) Treatment of alcoholic hepatitis. *J Gast Hepat*, 17, 448–455.

Marmot, M. G., Rose, G., Shipley, M. J. and Thomas, B. J. (1981) Alcohol and mortality: A U-shaped curve. *Lancet*, 1, 580–583.

McCann, S. E., Marshall, J. R., Trevisan, M., Russell, M., Muti, P., Markovic, N., Chan, A. W. K. and Freudenheim, J. L. (1999) Recent alcohol intake as estimated by the health habits and history questionnaire, the Harvard semiquantitative food frequency questionnaire and a more detailed alcohol intake questionnaire. *American Journal of Epidemiology*, 150, 334–340.

McDowell, I. F. W. and Lang, D. (2000) Homocysteine and endothelial dysfunction: A link with cardiovascular disease. *Journal of Nutrition*, 130, 369S–372S.

Mennella, J. (2001) Alcohol's effect on lactation. *Alcohol Research & Health*, 25, 230–234.

Milligan, S. R., Kalita, J. C., Heyerick, A., Rong, H., De Cooman, L. and De Keukeleire, D. (1999) Identification of a potent phytoestrogen in hops (*Humulus lupulus* L.) and beer. *Journal of Clinical Endocrinology and Metabolism*, 84, 2249–2252.

Milligan, S. R., Kalita, J. C., Pocock, V., Van de Kauter, V., Stevens, J. F., Deinzer, M. L., Rong, H. and De Keukeleire, D. (2000) The endocrine activities of 8-prenylnaringenin and related hop (Humulus lupulus L.) flavonoids. *J Clin Endocrin Metab*, 85, 4912–4915.

Miranda, C. L., Stevens, J. F., Helmrich, A., Henderson, M. C., Rodriguez, R. J., Yang, Y. H., Deinzer, M. L., Barnes, D. W. and Buhler, D. R. (1999) Antiproliferative and cytotoxic effects of prenylated flavonoids from hops (*Humulus lupulus*) in human cancer cell lines. *Food and Chemical Toxicology*, 37, 271–285.

Miranda, C. L., Stevens, J. F., Ivanov, V., McCall, M., Frei, B., Deinzer, M. L. and Buhler, D. R. (2000a) Antioxidant and prooxidant actions of prenylated and non-prenylated chalcones and flavanones in vitro. *Journal of Agriculture and Food Chemistry*, 48, 3876–3884.

Miranda, C. L., Aponso, G. L. M., Stevens, J. F., Deinzer, M. L. and Buhler, D. R. (2000b) Prenylated chalcones and flavanones as inducers of quinone reductase in mouse Hepa 1c1c7 cells. *Cancer Letters*, 49, 21–29.

Miura, Y., Hosono, M., Oyamada, C., Odai, H., Oikawa, S. and Kondo, K. (2005) Dietary isohumulones, the bitter components of beer, raise plasma HDL-cholesterol levels and reduce liver cholesterol and triacylglycerol contents similar to PPAR alpha activations in C57BL/6 mice. *British Journal of Nutrition*, 93, 559–567.

Miyamoto, M., Matsushita, Y., Kiyokawa, A., Fukuda, C., Iijima, Y., Sugano, M. and Akiyama, (1998) Prenylflavonoids: A new class of non-steroidal phytoestrogen (part 2). Estrogenic effects of 8-isopentenylnaringenin on bone metabolism. *Planta Medica*, 64, 516–519.

Moline, J., Bukharovich, I. F., Wolff, M. S. and Phillips, R. (2000) Dietary flavonoids and hypertension: Is there a link? *Medical Hypotheses*, 55, 306–309.

Morrell, P. (2000) Re: Does copper in beer protect the heart? *British Medical Journal*, 320, 1378–1379.

Mortensen, E. L., Jensen, H. H., Sanders, S. A. and Reinisch, J. M. (2001) *Better psychological functioning and higher social status may largely explain the apparent health benefits of wine: A study of wine and beer drinking in young Danish adults Archives of Internal Medicine*, 161, 1844–1848.

Mukamal, K. J., Conigrave, K. M., Mittleman, M. A., Camargo, C. A., Stampfer, M. J., Willett, W. C. and Rimm, E. B. (2003) Roles of drinking pattern and type of alcohol consumed in coronary heart disease in men. *The New England Journal of Medicine*, 348, 109–118.

Mukherjee, S. and Sorrell, M. F. (2000) Effects of alcohol consumption on bone metabolism in elderly women. *American Journal of Clinical Nutrition*, 72, 1073.

Nagao, Y., Kodama, H., Yamaguchi, T., Yonezawa, T., Taguchi, A., Fujino, S., Morimoto, K. and Fushiki, T. (1999) Reduced urination rate while drinking beer with an unpleasant taste and off flavor. *Bioscience Biotechnology and Biochemistry*, 63, 468–473.

Nakajima, Y., Murata, M., Watanabe, F., Niki, K. and Homma, S. (1998) Formation of an inhibitor of carcinogenic glucan synthesis in dark beer. *Lebensm.Wiss. u.-Technol*, 31, 503–508.

Nevill, A. M., Holder, R. L., Fentem, P. H., Rayson, M., Marshall, T., Cooke, C. and Tuxworth, W. (1997) Modelling the associations of BMI, physical activity and diet with arterial blood pressure: Some results from the Allied Dunbar national fitness survey. *Annals of Human Biology*, 24, 229–247.

Nicolodi, M. and Sicuteri, F. (1999) Wine and migraine: Compatibility or incompatibility. *Drugs under Experimental and Clinical Research*, 25, 147–153.

Nogueira, F. N., Souza, D. N. and Nicolau, J. (2000) In vitro approach to evaluate potential harmful effects of beer on health. *Journal of Dental Research*, 28, 271–276.

Nordmann, R. (1994) Alcohol and antioxidant systems. *Alcohol Alcoholism*, 29, 513–522.

Obisesan, T. O., Hirsch, R., Kosoko, O., Carlson, L. and Parrott, M. (1998) Moderate wine consumption is associated with decreased odds of developing age-related macular degeneration in NHANES-1. *Journal of the American Geriatrics Society*, 46, 1–7.

Ogihara, A., Kikuchi, S., Hasegawa, A., Kurosawa, M., Miki, K., Kaneko, E. and Mizukoshi, H. (2000) Relationship between *Helicobacter pylori* infection and smoking and drinking habits. *Journal of Gastroenterology and Hepatology*, 15, 271–276.

Ohsugi, M., Basnet, P., Kadota, S., Isbii, E., Tamora, T., Okumura, Y. and Namba, T. (1997) Antibacterial activity of traditional medicines and an active constituent lupulone from *Humulus lupulus* against *Helicobacter pylori*. *Journal of Traditional Medicine*, 14, 186–191.

Olson, R. E. (1979) Absorption, metabolism and excretion of ethanol. In: *Fermented Food Beverages in Nutrition* (C. F. Gastineau, W. J. Darby and T. B. Turner eds), pp. 197–211. Academic Press: New York.

Palomaeki, H. and Kaste, M. (1993) Regular light-to-moderate intake of alcohol and the risk of ischemic stroke: Is there a beneficial effect? *Stroke*, 24, 1828–1832.

Parker, D. R., McPhillips, J. B., Derby, C. A., Gans, K. M., Lasater, T. M. and Carleton, R. A. (1996) High density lipoprotein cholesterol and types of alcoholic beverages consumed among men and women. *American Journal of Public health*, 86, 1022–1027.

Perry, I. J., Wannamethee, S. G., Walker, M. K., Thomson, A. G., Whincup, P. H. and Shaper, A. G. (1995) Prospective study of risk factors for development of non-insulin dependent diabetes in middle aged British men. *British Medical Journal*, 310, 560–564.

Piendl, A. and Wagner, I. (1985) Biergenuss und Diurese. *Baruindustrie*, 70, 1082–1087.

Pignatelli, B., Scriban, R., Descotes, G. and Bartsch, H. (1983) Inhibition of endogenous nitrosation of proline in rats by lyophilized beer constituents. *Carcinogenesis*, 4, 491–494.

Pohorecky, L. A. (1990) Interaction of alcohol and stress at the cardiovascular level. *Alcohol*, 7, 537–541.

Potter, J. F. and Beevers, D. G. (1984) Pressor effect of alcohol in hypertension. *Lancet*, 1, 119–122.

Powell, J. J., McNaughton, S. A., Jugdaohsingh, R., Anderson, S. H. C., Dear, J., Khot, F., Mowatt, L., Gleason, K. L., Sykes, M., Thompson, R. P. H., Bolton-Smith, C. and Hodson, M. J. (2005) A provisional database for the silicon content of foods in the United Kingdom. *British Journal of Nutrition*, 94, 804–812.

Pradalier, A. and Ollat, H. (1991) Migraine and alcohol. *Headache Quarterly – Current Treatment and Research*, 2, 177–186.

Promberger, A., Dornstauder, E., Fruhwirth, C., Schmid, E. R. and Jungbauer, A. (2001) Determination of estrogenic activity in beer by biological and chemical means. *Journal of Agriculture and Food Chemistry*, 49, 633–640.

Renaud, S. C., Beswick, A. D., Fehily, A. M., Sharp, D. S. and Elwood, P. C. (1992) Alcohol and platelet aggregation: The Caerphilly prospective heart disease study. *American Journal of Clinical Nutrition*, 55, 1012–1017.

Renaud, S., Criqui, M. H., Farchi, G. and Veenstra, J. (1993) Alcohol drinking and coronary heart disease. In: *Health Issues Related to Alcohol Consumption* (P. M. Verschuren ed.), pp. 81–124. ILSI Press: Washington, DC.

Riboli, E., Cornee, J., MacQuartmoulin, G., Kaaks, R., Casagrande, C. and Guyader, M. (1991) Cancer and polyps of the colorectum and lifetime consumption of beer and other alcoholic beverages. *American Journal of Epidemiology*, 134, 157–166.

Richman, A. and Warren, R. A. (1985) Alcohol consumption and morbidity in the Canada Health Survey. *Drug and Alcohol Dependence*, 15, 255–282.

Rico, H., Gallego-Lago, J. L., Hernández, E. R., Villa, L. F., Sanchez-Atrio, A., Seco, C. and Gérvas, J. J. (2000) Effect of silicon supplement on osteopenia induced by ovariectomy in rats. *Calcified Tissue International*, 66, 53–55.

Riddell, L. J., Chisholm, A., Williams, S. and Mann, J. I. (2000) Dietary strategies for lowering homocysteine concentrations. *American Journal of Clinical Nutrition*, 71, 1448–1454.

Rimm, E. B., Chan, J., Stampfer, M. J., Colditz, G. A. and Willett, W. C. (1995) Prospective study of cigarette smoking, alcohol use and the risk of diabetes in men. *British Medical Journal*, 310, 555–559.

Rimm, E. B., Klatsky, A., Grobbee, D. and Stampfer, M. J. (1996) Review of moderate alcohol consumption and reduced risk of coronary heart disease: Is the effect due to beer, wine or spirits? *British Medical Journal*, 312, 731–736.

Roberts, N. B., Clough, A., Bellia, J. P. and Kim, J. Y. (1998) Increased absorption of aluminium from a normal dietary intake in dementia. *Journal of Inorganic Biochemistry*, 69, 171–176.

Rodgers, H., Aitken, P. D., French, J. M., Curless, R. H., Bates, D. and James, O. F. W. (1993) Alcohol and stroke: A case control study of drinking habits past and present. *Stroke*, 24, 1473–1477.

Rogers, J. D. and Greenfield, T. K. (1999) Beer drinking accounts for most of the hazardous alcohol consumption reported in the United States. *Journal of Studies on Alcohol*, 60, 732–739.

Ruitenberg, A., van Swieten, J. C., Witteman, J. C. M., Mehta, K. M., van Duijn, C. M., Hofman, A. and Breteler, M. M. B. (2002) Alcohol consumption and risk of dementia: The Rotterdam Study. *Lancet*, 359, 281–286.

Sacco, R. L., Elkind, M., Boden-Albala, B., Lin, I. F., Kargman, D. E., Hauser, W. A., Shea, S. and Paik, M. C. (1999) The protective effect of moderate alcohol consumption on ischemic stroke. *The Journal of the American Medical Association*, 281, 53–60.

Sakorafas, G. H. and Tsiotou, A. G. (2000) Etiology and pathogenesis of acute pancreatitis: Current concepts. *Journal of Clinical Gastroenterology*, 30, 343–356.

Schmidt, D. N. (1991) Apparent risk factors for chronic and acute pancreatitis in Stockholm County: Spirits but not wine and beer. *International Journal of Pancreatology*, 8, 45–50.

Sen, N. P., Seaman, S. W., Bergeron, C. and Brousseau, R. (1996) Trends in the levels of N-nitrosodimethylamine in Canadian and imported beers. *Journal of Agriculture and Food Chemistry*, 44, 1498–1501.

Shimura, M., Hasumi, A., Minato, T., Hosono, M., Miura, Y., Mizutani, S., Kondo, K., Oikawa, S. and Yoshida, A. (2005) Isohumulones modulate blood lipid status through the activation of PPAR alpha. *Biochim Biphys Acta Molecular Cell Biology Lipids*, 1736, 51–60.

Shindo, S., Tomatsu, M., Nakda, T., Shibamoto, N., Tachibana, T. and Mori, K. (2002) Inhibition of aldose reductase activity by extracts from hops. *Journal of Institute of Brewing*, 108, 344–347.

Shuster, J., Finlayson, B., Scheaffer, R. L., Sierakowski, R., Zoltek, J. and Dzegede, S. (1985) Primary liquid intake and urinary stone disease. *The Journal of Chronic Diseases*, 38, 907–914.

Spiegelhalder, B., Eisenbrand, G. and Preussmann, R. (1979) Contamination of beer with trace quantities of N-nitrosodimethylamine. *Food and Cosmetics Toxicology*, 17, 29–31.

Stampfer, M. J., Colditz, G. A., Willett, W. C., Manson, J. E., Arky, R. A., Hennekens, C. H. and Speizer, F. E. (1988) A prospective study of moderate alcohol drinking and risk of diabetes in women. *American Journal of Epidemiology*, 128, 549–558.

Stefanick, M., Legault, C., Tracy, R. P., Howard, G., Kessler, C. M., Lucas, D. L. and Bush, T. L. (1995) Distribution and correlates of plasma fibrinogen in middle aged women – initial findings of the postmenopausal estrogen-progestin interventions (PEPI) study. *Arteriosclerosis, Thrombosis, and Vascular Biology*, 15, 2085–2093.

Tabata, N., Ito, M., Tomoda, H. and Omura, S. (1997) Xanthohumols, diacylglycerol acyltransferase inhibitors, from *Humulus lupulus*. *Phytochemistry*, 46, 683–687.

Tagashira, M., Uchiyama, K., Yoshimura, T., Shirota, M. and Uemitsu, N. (1997) Inhibition by hop bract polyphenols of cellular adherence and water-insoluble glucan synthesis of Mutans Streptococci. *Bioscience Biotechnology and Biochemistry*, 61, 332–335.

Tavani, A., Negri, E., Franceschi, S., Talamini, R. and Lavecchia, C. (1994) Alcohol consumption and risk of prostate cancer. *Nut Canc*, 21, 25–31.

Thadhani, R., Camargo, C. A., Stampfer, M. J., Curhan, G. C., Willett, W. C. and Rimm, E. B. (2002) Prospective study of moderate alcohol consumption and risk of hypertension in young women. *Archives of Internal Medicine*, 162, 569–574.

Tjonneland, A., Gronbaek, M., Stripp, C. and Overvad, K. (1999) Wine intake and diet in a random sample of 48763 Danish men and women. *American Journal of Clinical Nutrition*, 69, 49–54.

Tobe, H., Muraki, Y., Kitamura, K., Komiyama, O., Sato, Y., Sugioka, T., Maruyama, H. B., Matsuda, E. and Nagai, M. (1997) Bone resorption inhibitors from hop extract. *Bioscience Biotechnology and Biochemistry*, 61, 158–159.

Tremoliere, J., Caridroit, M., Scheggia, E., Sautier, C., Geissler, C., Carre, L. and Fontan, M. (1975) Metabolisation de l'ethanol chex l'homme normal ingerant biere, vin ou whisky. Cahiers de Nutrition et de Dietique, supplement au fascicule 4, 10, 73–86.

Tricker, A. R. and Preussmann, R. (1991) Volatile and non-volatile nitrosamines in beer. *Journal of Cancer Research and Clinical Oncology*, 117, 130–132.

Truelsen, T., Gronbaek, M., Schnohr, P. and Boysen, G. (1998) Intake of beer, wine, and spirits and risk of stroke: The Copenhagen City Heart Study. *Stroke*, 29, 2467–2472.

Tsugane, S., Fahey, M. T., Sasaki, S. and Baba, S. (1999) Alcohol consumption and all-cause and cancer mortality among middle-aged Japanese men: Seven-year follow-up of the JPHC study cohort I. *American Journal of Epidemiology*, 150, 1201–1207.

Ubbink, J. B., Fehily, A. M., Pickering, J., Elwood, P. C. and Vermaak, W. J. H. (1998) Homocysteine and ischaemic heart disease in the Caerphilly cohort. *Atherosclerosis*, 140, 349–356.

Van Gijn, J., Stampfer, M. J., Wolfe, C. and Algra, A. (1993) The association between alcohol and stroke. In: *Health Issues Related to Alcohol consumption* (P. M. Verschuren ed.), pp. 43–79. ILSI Press: Brussels.

Vasse, R. M., Nijhuis, F. J. and Kok, G. (1998) Associations between work stress, alcohol consumption and sickness absence. *Addiction*, 93, 231–241.

Volpe, A. and Kastenbaum, R. (1967) Beer and TLC. *American Journal of Nursing*, 67, 100–103.

Walker, C. J. and Baxter, E. D. (2000) Health-promoting ingredients in beer. *The Master Brewers Association of the Americas Technical Quarterly*, 37, 301–305.

Walker, C.J. (200) Phytoestrogens in beer – good news or bad news? Brau Union International, 18, 38–39.

Wall, T. L., Horn, S. M., Johnson, M. L., Smith, T. L. and Carr, L. G. (2000) Hangover symptoms in Asian Americans with variations in the aldehyde dehydrogenase (ALDH2) gene. *Journal of Studies on Alcohol*, 61, 13–17.

Wannamethee, S. G. and Shaper, A. G. (1996) Patterns of alcohol intake and risk of stroke in middle-aged British men. *Stroke*, 27, 1033–1039.

Watten, R. G. (1999) Smokers and non-smokers: Differences in alcohol consumption and intake of other health-related substances in Norway – a general population study. *European Journal of Public Health*, 9, 306–308.

Woodson, K., Albanes, D., Tangrea, J. A., Rautalahti, M., Virtamo, J. and Taylor, P. R. (1999) Association between alcohol and lung cancer in the alpha-tocopherol, beta-carotene cancer prevention study in Finland. *Cancer Causes and Control*, 10, 219–226.

Woodward, M. and Tunstall-Pedoe, H. (1995) Alcohol consumption, diet, coronary risk factors and prevalent coronary heart disease in men and women in the Scottish heart health study. *Journal of Epidemiology and Community Health*, 49, 354.

Wright, C. A., Bruhn, C. M., Heymann, H. and Bamforth, C. W. (2008a). Beer consumers' perceptions of the health aspects of alcoholic beverages. *Journal of Food Science*, in press.

Wright, C. A., Bruhn, C. M., Heymann, H. and Bamforth, C. W. (2008b) Beer and wine consumers' perceptions of the nutritional value of alcoholic and non-alcoholic beverages. Journal of Food Science, in press.

Yano, K., Rhoads, G. G. and Kagan, A. (1977) Coffee, alcohol and risk of coronary heart disease among Japanese men living in Hawaii. *The New England Journal of Medicine*, 297, 405–409.

Yasukawa, K., Yamaguchi, A., Arita, J., Sakurai, S., Ikeda, A. and Takido, M. (1993) *Phytotherapy Research*, 7, 185–189.

Yoshikawa, T., Kimura, S., Hatano, T., Okamoto, K., Hayatsu, H. and Arimoto-Kobayashi, S. (2002) Pseudouridine, an antimutagenic substance in beer towards N-methyl-N'-Nitro-N-Nitrosoguanidine (MNNG). *Food and Chemical Toxicology*, 40, 1165–1170.

Zava, D. T., Dollbaum, C. M. and Blen, M. (1998) Estrogen and progestin bioactivity of foods, herbs, and spices. *Proceedings of the Society for Experimental Biology and Medicine*, 217, 369–378.

Zee, J. A., Simard, R. E. and Desmarais, M. (1981) Biogenic amines in Canadian, American and Eurpean beers. *Canadian Institute of Food Science and Technology Journal*, 14, 119–122.

Zhang, Y. Q., Kreger, B. E., Dorgan, J. F., Splansky, G. L., Cupples, L. A. and Ellison, R. C. (1999) Alcohol consumption and risk of breast cancer: The Framlingham study revisited. *American Journal of Epidemiology*, 149, 93–101.

Appendix
Practicalities of achieving quality

Charles W. Bamforth

This book has focused on quality and has delivered an in-depth account of the myriad of factors from raw materials, through processing to product ready for – and through – consumption, that influence perception and acceptability of beer. For the most part here, early in the 21st century, beers have never been more consistent. The achievement of this speaks not only to an awareness of the vast complexity of factors that impact quality but also an understanding of how to control them. The entire success or failure of a company stands on its ability to consistently deliver product to the satisfaction of the consumer.

Definitions of quality

There are many definitions of *quality*:

"Quality is the achievement of consistency and the elimination of unwanted surprises."

"The supply of goods that do not come back to customers who do."

"The extent of correspondence between expectation and realization."

"The match between what you want and what you get."

Quality is very much a personal issue. There are innumerable types of beer worldwide and no drinker is expected to like all of them equally. However, the problem is more subtle than that. For example, a beer leaving a brewery destined for export is expected to be inherently "clean and bright," whereas by the time it has traversed an ocean or two, encountered extremes of temperature and only emerged on supermarket shelves after prolonged storage in a warehouse, then it will certainly display aged aroma and may even be somewhat turbid. Whereas the brewer, trained to champion fresh beer that is in most (but not all) cases brilliantly clear, will be horrified to sample that product at retail, the customer may actually conclude that this is what the beer should taste and look like and, indeed, any effort by the brewer to lesson the deterioration may be met with complaint.

Responsibility for quality

Quality is not the exclusive preserve of a Quality Department. To achieve quality depends on a commitment from all employees – in other words a holistic quality environment. It is fashionable to speak of *Total Quality Management* (TQM), where everybody in a company has a commitment to excellence, manifesting itself in superiority in the organization. The provision of a comfortable, attractive and pleasing working environment, for example, will lead to dividends in quality performance through the enhanced motivation and pride of the workforce.

Quality must be to the forefront in every aspect of a company's operations: purchasing, manufacture, marketing, etc. It can only succeed if driven from the highest level of management.

Quality systems

TQM is, in part, achieved via the adoption of a quality system. Nowadays formalized approaches are available, embodied in international standards such as those of the ISO 9000 series. These are exercises in focusing, achieving compliance and ensuring that systems are documented. They represent good discipline. However it must be realized that they don't necessarily ensure that a product is "good" or "bad." They don't guarantee that process stages are necessarily the correct ones. All they seek to do is ensure that standard procedures are followed. It is up to the company to ensure that best practices prevail. To have such accreditation is really a stamp of approval that a company is paying heed to the need for a quality system.

There are companies that actively seek out suppliers who possess this type of accreditation. It is rather more far-sighted for companies to look for suppliers that have genuine quality systems in place, which in effect means the presence of, and unwavering adherence to, a Quality Manual. This should document everything that pertains to the product: specifications for the beer, for the raw materials from which it will be derived and for key control points in the process; standard operating procedures (SOP's) for everything from materials purchase, storage, handling and use, all the way through the brewery to beer shipment.

The most common quality management standard employed by breweries is *ISO9001:2000*. It specifies the quality system that proves the capability to manufacture and supply product to an established specification. It defines the activities that must be controlled, but not how they are to be controlled.

Appendix Practicalities of achieving quality

The key elements are

(a) A quality policy, usually signed by the senior executive, which speaks of the company's commitment to quality and to meet the needs of customers.
(b) A Quality Manual that outlines and maps the scope and elements of the quality management system and how these may be cross-referenced to relevant procedures and work instructions.
(c) Written procedures for all key operations that affect product quality.
(d) Systems for ensuring uniform use of updated documentation, including specifications, control diagrams and schemes, codes of practice, operating manuals, HACCP (see later).
(e) Secure records to demonstrate that the system is working effectively and meets legal and contractual requirements.
(f) A schedule of meetings to review the quality system.
(g) Training of all staff whose jobs impact product quality.
(h) Control of purchasing of all materials and services that can affect product quality; for example lists of approved suppliers, auditing regimes, etc.
(i) Calibration programs for equipment according to recognized standards.
(j) Internal audits to ensure that the system is effectively implemented and maintained.
(k) Systems for handling non-conforming product.
(l) Systems for handling complaints.
(m) Systems for adjusting standard procedures in order that problem recurrence is avoided.

Quality assurance versus quality control

Quality control (QC) is a reactive approach, one which is invariably associated with waste in that it seeks to respond to measurements that are made and effect corrections if the values are outwith specification. Much more effective is the quality assurance (QA) approach, in which systems are introduced that strive to *ensure* that the product at every stage in its production is within specification. The emphasis is one of *prevention* rather than *detection*. Brewers speak of *Right First Time*.

Analysis and quantification of the process and product are, of course, necessary. However, it is preferable that these measurements should be pro-active rather than re-active. For example, measuring and controlling wort gravity and oxygenation and viable yeast pitching rate are vastly preferable in the pursuit of controlled fermentation than is the monitoring of specific gravity in fermenter and response to deviations arising because the appropriate control strategies had not been put in place. Wherever possible measurements are either made automatically by an in-line or on-line device linked to a feedback loop (e.g. in-line measurement of turbidity to feedback-control a filter aid dosing pump) or manually by the person operating that part of the process, who can take ownership of correcting the issue, rather than waiting for analysis to be performed by a laboratory technician with the attendant delays and lack of sensitivity of response.

Specifications

Achievement of consistency demands that meaningful specifications are brought to beer. The need is for specifications that pertain to the finished product that allow a quantified or at least qualified gauge of how close any batch of product is to perfection. Specifications are needed to allow selection and handling of raw materials. And specifications are required at all points from raw materials to end product that guarantee that the stream at all stages is fit for use.

Specifications should be realistic and meaningful. They need to relate to what is expected in the product but not set to such a rigor that they are unachievable. For example if we demand that our small-pack beer display minimum aged character in 6 months of optimum storage, then the specification for oxygen needs to be as low as possible but nonetheless not so low that can't be achieved consistently in the equipment at our disposal. The same considerations apply to the specifications that we set on our raw materials.

The cost of quality

We can consider quality-related costs under the headings of internal failure, external failure, appraisal and prevention.

Internal failure costs are the ones associated with product being out of specification at any stage in the production process and before the beer has left the brewery for retail. They can be subdivided into:

- *Rework*: the cost of correcting matters to return a product to "normal,", for example rectifying carbon dioxide levels in bright beer tank out of specification.
- *Re-inspection*: the cost of checking re-worked material.
- *Scrap*: product that is beyond repair, for example beer that has become contaminated.
- *Analysis of failure*: the cost of investigating the cause of the internal failure.

External failure costs pertain to the implication of a product actually getting out into trade with a quality defect.

- *Recall*: the cost of investigating problems, recovering product and, probably above all else, the cost of lost reputation and market.
- *Warranty*: the cost of replacing product.
- *Complaints*: the cost of handling customer's objections.
- *Liability*: the implications of litigation.

Appraisal costs are the expense of analyses throughout the process, from raw materials to end product. They can be sub-divided into:

- *Inspection and test*: analytical methodology, whether in laboratory or on-line.
- *Internal auditing*
- *Auditing of suppliers*.

Appendix Practicalities of achieving quality

Table A.1
Elements of a quality assurance program

1. A defined quality system: what is worth measuring, when, where, how often and by whom
2. Established quality standards and procedures for raw materials with spot-checking and auditing protocols
3. Audits and surveys of processes and procedures
4. Projects – for example installation of in-line analysis and feedback control systems
5. Training and auditing of operators on QC checks
6. Collaborative exercises – for example inter-laboratory method checking
7. Specialized analysis and procedures within a QA laboratory
8. Complaints handling procedures
9. Quality awareness campaigns
10. Instrument checking and calibration

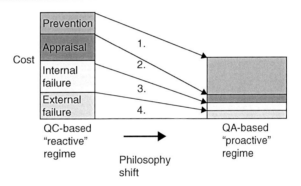

Figure A.1
The cost impact of moving to a QA strategy (derived from Bamforth, C.W. (2002) *Standards of Brewing*, Brewers Publications, Boulder, CO).

Prevention costs are the ones devoted to operating a well-run QA system (see Table A.1).

Figure A.1 illustrates that increased investment in prevention enables a reduction in appraisal costs but much more importantly allows for greatly reduced losses accruing from internal and external failure.

Statistical process control

Robust analysis in any production operation depends on the generation of reliable data and an appreciation of its relevance.

It is important to have tools that "locate" an individual measurement and its status in relation to other values. It is also necessary to quantify how "disperse" the data is – how broad is the spread?

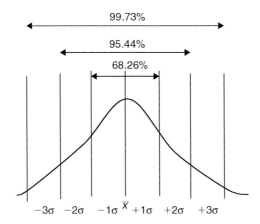

Figure A.2
Standard distribution (derived from Bamforth, C.W. (2002) *Standards of Brewing*, Brewers Publications, Boulder, CO).

Most values in a substantial-enough population are distributed as illustrated in Figure A.2. The *mean* (M) is the average of all the values (x_i), obtained by adding them up (Σ) and dividing by the number of measurements (n).

$$M = \frac{1}{n}\Sigma x_i$$

The extent to which the values are dispersed (i.e. the subtraction of the lowest value from the highest value) tells us how widely spread the data set is, but does not signify where the data is congregated. The deviation of any individual measurement from the mean is given by the expression $x_i - M$. The *standard deviation* (σ) is given by

$$\sigma = \sqrt{\frac{1}{n-1}\Sigma\,[x_i - M]^2}$$

σ^2 is called the *variance*.

The coefficient of variation (CV) is a simple way to illustrate variation

$$CV = \frac{100\sigma}{M}$$

It allows expression on a percentage basis of the error inherent in a method. The lower is CV, the more reliable is an analysis.

The normal distribution, as depicted in Figure A.2, is an illustrative way of describing probability, that is the likelihood of a value being a certain distance from the mean. Thus there is a 68.26% chance of the value being within one standard deviation of the mean, 95.44% probability of it being within 2 standard deviations and a 99.73% chance of it being within 3 standard deviations.

Process capability

No system is immune from variation and some degree of inconsistency or "*noise.*" There are two types of noise:

- Internal noise, examples of which include fluctuations within a batch of raw materials (malt, hops, yeast, water, etc.), sampling inconsistencies and wear and tear in machinery.
- External noise, such as human factors (different operators) and differences between sources of raw materials (e.g. harvest year).

If a process generates a product that has a stable mean value then it is said to be "in control." There may be much variation about the mean, but only within clearly understood limits. When a process deviates to a greater extent than this, leading to a shift in the mean or an increase in the variability, it is then "out of control."

Process capability is given by

$$\frac{\text{Upper limit of measurement} - \text{lower limit of measurement}}{6\sigma}$$

The denominator 6σ speaks to the breadth of a normal distribution (see Figure A.2) and has in recent years been used as the name for a well-known approach to quality.

Control charts

Charting data make it easier to spot when a process is moving out of control. Properly done, these charts allow the operator to determine whether any changes are within or outwith normal variation. If a change is observed in a parameter, but nonetheless it was entirely within normal fluctuation, it would be foolhardy to make adjustments. It is only those parameters whose value will lead to a go/no go decision that are measured on every batch.

The simplest type of control diagram to use is to plot data within a framework of action (reject) and warning lines (Figure A.3). Convention has it that these lines should be positioned at three standard errors ($3\sigma/\sqrt{n}$) above and below. This explains 99.73% of the variation in data for a normal distribution. It is often the case that two other "warning lines" within the action lines are used. These are set at two standard errors ($2\sigma/\sqrt{n}$). The action and warning lines are sometimes referred to as the reject quality limits (RQL) and acceptable quality limits (AQL) respectively.

Another type of plot frequently used is the CUSUM (cumulative sum) plot.

$$\text{CUSUM} = \Sigma_i (x_i - T)$$

where x_i is the ith measurement of a quantity and T is the target value (what the measurement ideally should be, or *specification*). So the CUSUM is the summation of all the deviations from normality.

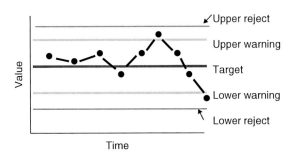

Figure A.3
Trend plotting (derived from Bamforth, C.W. (2002) *Standards of Brewing*, Brewers Publications, Boulder, CO).

CUSUM plots are valuable for highlighting where substantial changes are occurring in a parameter. When the slope of these plots trends upwards then this indicates that the average is somewhat above the target value. When the slope trends downwards this indicates that the average is somewhat below the target value. The more pronounced the slope, the more does the value deviate from target.

Standard methods of analysis

Worldwide, there are several sets of methods, with many overlaps but some significant differences, with origins lying in brewing practices in separate nations. The standard methods of the Institute of Brewing (now called the Institute of Brewing and Distilling, IBD) were originally developed in England for the analysis of ale-type beers and so, for example, use small-scale infusion mashes at constant temperature. The methods of the European Brewery Convention (EBC) are based on the production of lager-style products: their mashes (Congress mashes) have a rising temperature regime. There have been major efforts to merge the two different compendia. The Methods of Analysis of the American Society of Brewing Chemists (ASBC) owe more to the EBC than the IBD but they form the analytical reference point for all North American brewers, whether producing lagers or ales. Box A.1 offers a crib of the contents of the ASBC methods.

Methods are written in a standard format that offers the best opportunity for whosoever is pursuing it to be consistent in their approach. The method is then distributed to different laboratories, together with a set of samples that have been produced in a single location. Following analysis, the values obtained are returned to a central data coordinator.

There may be several sources of variance (defined earlier) in analytical data. Although not applicable in the establishment of standard procedures where carefully regulated samples are distributed from a central location, for routine samples one of the main risks is a associated with unrepresentative sampling.

Appendix Practicalities of achieving quality

Box A.1
A summary of the ASBC methods of analysis

1. Barley methods

1. Barley-1 Importance of representative sampling. Triers for static beds (e.g. boxcars) or "spout samplers" or "belt samplers" for grain under transfer. Grading on basis of malting vs. feed quality and on other attributes, such as damage.
2. Barley-2 Physical tests.
   - Variety by physical characteristics: appearance.
   - Test weight per bushel (US) is "the weight in pounds of the volume of grain required to fill level full a Winchester bushel measure of 2150.42 cubic inches capacity."
   - Assortment by sieving
   - Thousand kernal weight – weight of a 1000 kernels
   - Texture of endosperm – mealy or steely?
   - Skinned and broken kernels
   - Weathering and kernel damage
3. Barley-3 Germination
   Germinative energy – assessment of dormancy
   Germinative capacity – viability
   GE/GC/Water sensitivity simultaneously
4. Barley-4 Milling for analysis: obtaining a representative powder
5. Barley-5 A. Moisture by drying. B. Alternate oven method C. NIR
6. Barley-6 Extract – trying to predict what you will get after malting by use of enzymes
7. Barley – 7 Protein A Kjeldahl & N x 6.25 B. NIR c. Combustion and N x 6.25 d. Protein in whole grain by NIR
8. Barley-8 Potential Diastatic Power
9. Barley-9 Kernel brightness by instrument
10. Barley-10 Pre-germination by fluorescein dibutyrate
11. Barley-11 deoxynivalenol by GC
12. Sprout damage

2. Malt methods

1. Standard sampling, directly comparable with methods for barley.
2. Physical tests, very similar to those for barley (test weight, assortment by sieving, Thousand Kernel Weight, Foreign seeds and skinned and broken kernels, mealiness). Additional method is acrospire length – longer acrospires mean more modification.
3. Moisture (c.f. barley)
4. Extract: specific gravity from a small-scale mash; rate of filtration through filter paper loose predictor of wort separation
5. Soluble protein (Kjeldahl) and free amino N (ninhydrin) levels in wort
6. Diastatic Power: titration (by thiosulfate) of iodide not binding to starch. Rapid method: measurement of reducing sugars by PAHBAH. Automated flow analysis.
7. Alpha-amylase. Use of excess beta-amylase, measurement of residual starch by iodine
8. Use of Kjeldahl or combustion to measure protein (N x 6.25) (c.f. barley)
9. Colored malts; mash with standard malt and measure Extract and color
10. Nitrosamines by gas chromatography
11. Sulfur dioxide by color reaction with para-rosaniline hydrochloride
12. Modification using the friabilimeter; second sieving for gross under-modification
13. Deoxynivalenol by GC (see barley)

14. Dimethyl sulfide precursor: caustic hydrolysis of S-methylmethionine and GC measurement of DMS
15. Grist analysis

3. Adjunct methods

Cereals vs. syrups
Cereals-1 Sampling (c.f. barley and malt)
Cereals-2 Physical characteristics: color, odor, husks germs and foreign seeds, mold, infestation, assortment of grits
Cereals-3 Moisture by oven drying
Cereals-4 Oil by extraction with petroleum ether
Cereals-5 Extract: use of malt (method A) or added enzymes (method B)
Cereals-6 Nitrogen – either Kjeldahl or by combustion
Cereals-7 Ash – by ignition at 550°C
Sugars and Syrups-1 Sampling of containers – square root of total number of containers
Sugars and Syrups-2 Color, odor, taste
Sugars and Syrups-3 Clarity of 10% solution
Sugars and Syrups-4 Color of 10% solution
Sugars and Syrups-5 Extract: direct measurement of specific gravity
Sugars and Syrups-6 Moisture = 100 – Extract
Sugars and Syrups-7 Fermentable extract by fermentation (A) or rapid fermentation (B): measure change in Extract
Sugars and Syrups-8 Iodine reaction for residual starch (blue color) or dextrins (reddish)
Sugars and Syrups-9 Acidity by either potentiometric titration or using phenolphthalein as an indicator
Sugars and Syrups-10 pH
Sugars and Syrups-11 Protein (Kjeldahl)
Sugars and Syrups-12 Ash by ignition
Sugars and Syrups-13 Diastatic Power for malt syrups – same method as for malt
Sugars and Syrups-14 Total reducing sugars – Fehlings
Sugars and Syrups-15 Sucrose – measure reducing sugars before and after inversion using either acid or invertase
Sugars and Syrups-16 Dextrose in presence of other reducing sugars
Sugars and Syrups-17 Fermentable saccharides by chromatography – GC or HPLC
Sugars and Syrups-18 Fermentable carbohydrates by cation exchange HPLC

4. Brewers grains methods

1. Sampling – wet grains and dry grains
2. Preparation of sample
3. Moisture – wet samples, preliminary dried samples, dry grains, wet samples (rapid)
4. Available extract – wet grains and dry grains
5. Soluble extract – wet grains and dry grains
6. Feed analysis
7. Protein by combustion

5. Hops methods

1. Sampling "Oregon sampler" to take cylindrical core; statistical sampling. Grinding of cones or pellets
2. Physical examination: leaves and stems, color, luster, size of cones, condition of cones, lupulin (amount, color, stickiness), aroma, seeds

3. Aphids
4. Moisture – by distillation, vacuum drying, oven drying
5. Resins
6. Alpha and beta acids: extraction in toluene and measurement of absorbance or lead acetate conductimetric method
7. Alpha acids in hops and hop products by ion-exchange chromatography
8. Non-isomerized hop extracts. Isopropyl ether to extract prior to spectrophotometric or conductimetric analysis
9. Isomerized hop extracts – ion-exchange and HPLC analysis of iso-alpha-acids
10. Hop bitter acids in non-isomerized extracts by stepwise ion-exchange chromatography
11. Gradient elution of hop constituents using ion-exchange chromatography
12. Hop storage index. Increase in absorbance at 275 nm relative to 325 nm.
13. Total essential oils by steam distillation
14. c.f. method 6, but by HPLC
15. Iso-α-acids in isomerized pellets by HPLC
16. Iso-alpha, alpha and beta acids in hop extracts and isomerized hop extracts by HPLC

6. Wort methods

1. Sampling. Risk of infection, insolubilization, color change, etc. Best to analyze fresh – otherwise define storage conditions and attempt to standardize between labs. Any filtration through paper, not kieselguhr (but see 9).
2. Specific gravity. A. By pycnometer – now archived. B. By digital density meter – wort should be bright.
3. Extract. By calculation from specific gravity.
4. Apparent extract by hydrometer.
5. Fermentable extract. Use of yeast – either A. Over 48 h (or till fermentation complete), or B. Rapidly (5 h).
6. Iodine reaction archived.
7. Total acidity by potentiometry or indicator method.
8. pH.
9. Color. Must be bright, but filter paper insufficient. Standardized kieselguhr filtration. Then A_{430}.
10. Protein by Kjeldahl or combustion.
11. Reducing sugars.
12. Free amino nitrogen – ninhydrin reaction.
13. Viscosity – dynamic viscosity (cP) which is resistance to shear flow within a liquid; kinematic viscosity (cS) is a measure of time taken for a liquid to flow through an orifice under gravity.
14. Fermentable saccharides by chromatography. A. GC B. HPLC.
15. Magnesium by atomic absorption spectrophotometry (AAS).
16. Zinc by AAS.
17. Protein in unhopped wort by spectrophotometry – measured at 215 and 225 nm.
18. β-Glucan in Congress wort by fluorescence – Calcofluor.
19. Fermentable carbohydrates by cation exchange HPLC.

7. Beer methods

1. Sampling. Preparation: attemperation to 15–20°C – decarbonation either by manual swirling in Erlenmeyer or on a rotary shaker; Procedures for aseptic sampling from tank, line and finished package.
2. Specific gravity – pycnometers: archived, digital density meter.
3. Apparent extract hydrometer.

4. Alcohol – distillation, refractometry, gas chromatography, SCABA (alcohol by catalytic combustion, density by densitometer), enzymic (low alcohol beers).
5. Real extract – density of residue from distillation, by refractometry.
6. Calculations – Original extract, real degree of fermentation, apparent degree of fermentation, carbohydrate content.
7. Iodine reaction – archived.
8. Total acidity – titration with sodium hydroxide – end-point by potentiometry (pH meter) or indicator (phenolphthalein).
9. pH.
10. Color – spectrophotometry at 430 nm, photometer.
11. Protein – Kjeldahl, combustion.
12. Reducing sugars – Fehling's.
13. Dissolved carbon dioxide – pressure methods, manometry/volumetry.
14. Ash – ignition.
15. Total phosphorus – archived.
16. End fermentation – yeast to ferment away any residual sugars.
17. Dextrins – archived.
18. Iron – colorimetry with bipyridine or o-phenanthroline; atomic absorption spec; color formation with Ferrozine.
19. Copper – colorimetry with ZDBT or Cuprethol; AAS.
20. Calcium – titrimetry, AAS.
21. Total SO_2 – Monier-Williams (reaction with p-rosaniline hydrochloride).
22. Foam collapse – Sigma Value or Foam Flashing (for bottles).
23. Bitterness – A_{275} of iso-octane extracts (inc. automated flow); HPLC.
24. Heptyl hydroxybenzoate – archived.
25. Diacetyl – reaction with α-naphthol or dimethylglyoxime; GC.
26. Formazin standards for haze.
27. Physical stability – haze after chilling – elevated temperature storage.
28. Chill proofing enzymes – archived.
29. Lower boiling volatiles – GC for esters and alcohols.
30. Triangular taste test.
31. FAN – ninhydrin.
32. Viscosity.
33. Calories by calculation – from alcohol, real extract and ash.
34. Dissolved oxygen – colorimetry using indigo carmine.
35. Total polyphenols – colorimetry with Fe (III) in alkali.
36. Sodium by AAS.
37. Potassium by AAS.
38. Magnesium and calcium – AAS, sequential titration with EDTA.
39. Chloride – conductometric or mercurimetric titration.
40. Nitrosamines distillation or adsorption prior to GC.
41. Total carbohydrate by spectrophotometry (phenol reaction) or HPLC.
42. Aluminum by graphite furnace AAS.
43. Chloride, phosphate and sulfate by ion chromatography.
44. Dimethyl sulfide by GC and chemiluminescence detection.

8. Microbiological methods

Yeast

1. Representative sampling important: avoidance of stratification – taking samples at different stages in pumping. Rousing of holding tank contents. Taking samples from

different places in compressed cake. Must hold cold 0–2°C and analyze as soon as possible (c.f. autolysis, contaminant growth). Sample in lab jar must be well-mixed.
2. Examination. Use of your senses – color, flocculence, odor, taste. Microscope – cell shape, condition, cleanliness, cell abnormalities, damaged cells, contaminants.
3. Viability – methylene blue, methylene violet. Latter felt to be more reliable. Spores – grow yeast in acetate as main C source and then count spores after staining with malachite green. Respiratory deficient yeasts using a triphenyltetrazolium chloride overlay technique – if respiratory deficient don't reduce dye.
4. Counting – hemocytometer and microscope.
5. Yeast solids. Dry weight will give contribution from other solids. Spin down can allow differentiation between yeast and trub.
6. Viability by slide culture: only living organisms give colony on microscope slide. Compare with hemocytometer.
7. Yeast sporulation – good for identifying wild yeasts because brewing yeast sporulates poorly if at all – use of media to induce aerobic growth on acetate.
8. Killer yeast identification Plate test in which an agar plate seeded with a sensitive strain and then inoculated with killer strain. Clear zone of inhibition. Can use to test whether a yeast produces killer toxin.
9. Giant colony morphology. Growth on wort gelatin medium: production of characteristic morphologies.
10. Differentiation of ale and lager yeast. Growth on 5-bromo-4-chloro-3-indolyl-α-D-galactoside (X-α-gal) – agar plates – melibiase(lager yeasts) produce α-galactosidase which can break this down. Release indole which oxidatively polymerizes to give an insoluble blue-green dye that does not diffuse into agar. Or, growth on Bacto YM agar at 37°C – ale yeasts can grow. Or, growth in a broth containing melibiose – monitor pH drop by a pH indicator.
11. Flocculation. Helm assay – observe yeast sedimentation in a calcium sulfate solution at pH 4.5. Or absorbance method – similar, but measure yeast in suspension using absorbance at 600 nm.

Microbiological

1. Aseptic sampling – beer in process, packaged beer, process water, brewery air, equipment and surfaces. Important to get a representative sample.
2. Detection agar. Incubate separately for aerobic/anaerobic. Organisms collected by membrane filtration for beer, rinse water, etc. Cellulose acetate or cellulose nitrate membranes 0.45 μm. Put filter directly onto plate.
3. Differential staining. Gram positive/negative.
4. General culture media Universal beer agar medium and brewers' tomato juice medium. For all types of organism including brewing yeast. Nystatin suppresses growth of brewing yeast
5. Differential culture media
Nystatin (formerly Cycloheximide) medium (for bacteria).
Lee's multi-differential agar (LMDA) – general bacteria.
Raka Ray – Lactic acid bacteria.
Lysine medium – wild yeasts can use lysine but brewers yeast can't.
Lin's wild yeast differential medium – fuchsin-sulfite and crystal violet.
Barney-Miller brewery medium: Lacto and Pedio.
DeMan Rogosa Sharpe (MRS) – detects Lacto and Pedio.

MYGP + Cu – detects wild yeast.
CLEN – some wild yeast. Selective medium for megasphera and pectinatus.

9. Filter aids

1. Sampling.
2. pH of a suspension of the filter aid in water.
3. Impact of filter aid on odor and taste.
4. Iron pick-up by beer.

10. Packages and packaging materials

Bottles

1. Dimensions (height, outside diameter, out-of perpendicular, identification marks, glass distribution, weight, locking ring diameter, Reinforcing ring diameter, width of locking ring, throat diameter, finish.
2. Defects.
3. Color (amber, redness ratio).
4. Capacity (overflow, fill point).
5. Surface protective coatings (lubricity, coating quality, rub tests).

Bottle closures

1. Defects glossary and classification.
2. Test pressure (for crowns, pilfer-proof closures).
3. Gas retention capacity of crowns.
4. Resistance to pasteurization.
5. Removal torque procedures.
6. Crimp determination for crowns.

Cans

1. Defects glossary and classification.
2. Rusting tendency.
3. Dimensions (metal gauge thickness, flange width, filled can countersink depth).
4. Ends (curl opening, seaming chuck-fit, ring-pull-end pop and pull tests).
5. Capacity (overflow, headspace).
6. Enamel rater for evaluating metal exposure.

Fills

1. Total contents of bottles and cans by calculation from measured net weight.
2. Total contents of cans of known tare weight.

11. Sensory analysis

1. Terms and definitions – some key ones: ascending method of limits test, bias, category scaling, descriptive analysis, detection threshold, difference threshold, directional difference test, duo-trio test, hedonic, paired comparison test, paired preference test, ranking test, rating test, recognition threshold, reference standard, synergism, terminal threshold, threshold, triangular test.
2. Test room, equipment and conduct of test.

3. How to choose the appropriate method.
4. Selecting and training assessors.
5. Reporting data.
6. Paired comparison test – presentation of two samples for discussion of attributes.
7. Triangular test – pick out the different beer from three samples (two of one, one of another beer).
8. Duo-trio test – presentation of a reference sample, followed by a pair of beers; subject asked to pick out the non-reference beer from the pair.
9. Threshold of added substances – ascending method of limits test; assessors presented with a series of triangles of beers, with progressively increasing amounts of a given flavor substance. Aim is to identify where the flavor threshold lies.
10. Descriptive analysis.
11. Ranking test – rank attributes of a range of beers.
12. Flavor terminology and reference standards.
13. Difference-from-control – score magnitude of difference of attributes in a beer from those in a presented control.

The other sources of variance (replication error and systematic error) have diverse origins: different batches of chemicals, unconscious deviations from the laid down method, contaminating species in glassware or water, atmospheric conditions, human imperfections, etc.. The extent to which each of these matters depends on the method: the more robust the procedure, the less the scatter.

The usual approach to testing for errors was developed by Youden and it is used not only in setting up new recommended methods but also routinely for screening and comparing performance on existing methods between laboratories, for example the different laboratories across locations within a major brewing company. In this technique the various labs are sent a pair of samples, representing two different levels of the analyte under examination. Each lab would be asked to make their measurements on this parameter in the two beers, the values for beer one being the A series and those for beer two being B values. The collated data is then plotted as shown in Figure A.4. The circle has a radius that is determined by multiplying the standard deviations of the replication error. Essentially there is a 95% probability of a result falling within the circle. The digits on the plot indicate the data generated by different laboratories, in this example revealing that lab 2 is reliable for measure A though not B and lab 3 is underestimating in both instances. Table A.2 illustrates how the data is handled.

Two other values are important. The first of these is *repeatability* (r), which is an index of how consistently data can be generated by a single operator in a single location using a standard method. *The difference between two single results found on identical test material by one operator using the same apparatus within the shortest feasible time interval will exceed r on average not more than once in 20 cases in the normal and correct operation of the method.* It is given by $2 \times \sqrt{2} \times \sqrt(\sigma_r)$, where σ_r is the standard deviation for the procedure when assessed within a single laboratory. The second parameter is *reproducibility* (R), this being the maximum permitted differences between values reported by different labs using a stand-

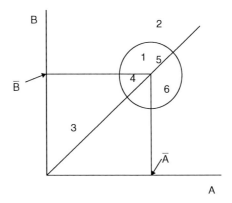

Figure A.4
Youden plot (derived from Bamforth, C.W. (2002) *Standards of Brewing*, Brewers Publications, Boulder, CO).

Table A.2
Procedure for calculating errors using the Youden method

	Laboratory 1	Laboratory 2	Laboratory n	Total	Average
Result A	A_1	A_2	A_n	ΣA	$\bar{A}$
Result B	B_1	B_2	B_n	ΣB	$\bar{B}$
A − B	X_1	X_2	X_n	ΣX	$\bar{X}$
A + B	Y_1	Y_2	Y_n	ΣY	$\bar{Y}$

$$\text{Standard deviation of the total error} = \frac{\Sigma(Y - \bar{Y})^2}{2(n-1)}$$

$$\text{Standard deviation of the replication error} = \frac{\Sigma(X - \bar{X})^2}{2(n-1)}$$

ard method. This is the value obtained in collaborative trials between labs and is given by $2 \times \sqrt{2} \times \sqrt{(\sigma_b^2 + \sigma_r^2)}$, where σ_b is the between lab standard deviation. In short, r is a measure of "within lab" error and R of "between lab error" and the lower r and R, the more reliable is the procedure.

Setting specifications and monitoring performance

The information gleaned in this type of inter-collaborative exercise helps the brewer achieve the goal of setting realistic and meaningful specifications for the product, raw materials and process samples.

When setting specifications there are two key requirements:

a. Knowledge of the r and R values, which indicate how reliable a method is.
b. Appreciating the true range over which a parameter can vary before a change is observed in quality.

Table A.3
Components of a malt specification

Parameter	Rationale
Moisture	Excess moisture reduces stability. Brewer does not wish to buy water
Hot water extract	Indicates potential extractable material obtainable in the brewhouse
Saccharification time	Indicates sufficiency of starch-degrading enzymes
Color	Correct color for finished beer style. Indication of kilning severity
Protein	Excess protein means less carbohydrate extract, plus increased haze risk
Kolbach Index (also known as Soluble Nitrogen ratio or Soluble/Total protein ratio)	Measure of protein modification – relevant re colloidal stability
Diastatic power	Total starch-degrading enzyme activity
Wort pH	Impacts extraction and stability
β-Glucan in wort	Predictor of wort separation and beer filtration problems
Friability	Assessment of extent of malt modification and predictor of wort separation problems
Partly unmodified grains ("homogeneity")	Predictor of wort separation and beer filtration problems
Dust and extraneous materials	Index of unusable and potential problematic material that would need to be screened out
Dimethyl sulfide precursor	Potential dimethyl sulfide levels in lager beers
Arsenic, lead, nitrosamines, ochratoxin A, deoxynivalenol	Food safety issues
Gushing	Predictor of over-foaming in end product

Note: Additionally the contract is likely to specify variety, minimum storage period, extent or otherwise of pesticide usage, freedom from genetically modified material.

What is measured?

Tables A.3–A.6 give typical analytical elements for malt, hops, in process and final product respectively. Incoming water will have a depth of analysis associated with it that far exceeds any of the above, but almost without exception it is the raw material whose analysis is left in its entirety to the supplier (Tables A.7 and A.8). The brewer should, however, constantly be checking its organoleptic quality and its pH, on arrival into the brewery and at all places where it is processed (carbon filtration, ion exchange, deaeration).

Hazard analysis critical control points (HACCP)

HACCP is a management system designed to assure the safety of food products. Several countries have legal requirements for HACCP.

Table A.4
Components of a hop specification (pellets)

Parameter	Rationale
Moisture	Stability
α-Acid	Potential bitterness
Pesticides	Food safety issue
Heavy metals	Food safety and product quality issue
Hop Storage Index	Extent of deterioration of hop

Table A.5
Key in-process checks

Check	Frequency	Rationale
(a) *Milled grist*		
Particle size distribution	Weekly	Confirmation of correct mill gaps
(b) *Mash*		
Iodine test	Every mash	Confirmation of starch conversion
(c) *Sweet wort*		
Specific gravity	Every batch	Monitoring extract recovery, cut-off and strength of wort to kettle
pH	Every batch	Cut-off point for wort collection. No cleaning agent contamination
Clarity	Varies	Some brewers believe it relates to product quality
(d) *Boiling and hopped wort*		
Specific gravity	Every batch	Strength of wort to next stage and per cent evaporation
pH	Every batch	No cleaning agent contamination
Bitterness	Varies	Hop utilization
Color	Varies	Predictor of color in final product
Clarity	Varies	Some brewers believe it relates to product quality
Oxygen in cooled wort	Every batch	Needed for yeast
(e) *Yeast*		
Viability	Every batch	To refine pitching rate and monitor deterioration in yeast performance
Cell count	Every batch	To define pitching rate
(f) *Fermentation and "green beer"*		
Specific gravity	Every batch	Progress of fermentation
pH	Every batch	Progress of fermentation
Total vicinal diketones (i.e. diacetyl and pentanedione and their precursors)	Every batch	Avoidance of flavor problems

Note: Monitoring of temperature at all stages is a given, as too is tasting and smelling. The efficiency of in-place cleaning systems should be evaluated by monitoring the strength of detergent and by screening swabs of tanks or post-rinse samples using traditional or rapid microbiological tests.

Table A.6
QC checks on beer

Parameter	Frequency
Clarity	Every batch
CO_2	Every batch
Foam	Pour – every batch
	Instrumental monthly
Haze breakdown	Small pack monthly
Color	Every batch
Ethanol	Every batch
Apparent extract (ergo original extract)	Monthly
Fermentable extract	Monthly
Bitterness	Every batch
Free amino nitrogen	Monthly
pH	Every batch
Volatile compounds	Monthly
Inorganics (notably iron, copper, sulfate, chloride, nitrate)	Monthly
Sulfur dioxide	Every batch if there are legal requirements to label if above a certain level (e.g. 10 ppm in US); otherwise monthly
Polyphenols	Monthly
O_2	Every batch
N_2 (if used)	Every batch
Microbiological status	Monthly
Taste clearance	Every batch
Flavor profile (or Trueness to Type)	Monthly

Source: From Bamforth, C.W. (2002) *Standards of Brewing*, Brewers Publications, Boulder, CO.

Pre-requisites for implementing HACCP are

(a) A broadly articulated policy declaring the company's commitment to making safe products.
(b) "Securing of the borders," meaning confidence that there is no threat from surrounding industries and from accidental or malicious contamination.
(c) Identification of those elements of the process that present the greatest risk, for example packaging locales. In the latter there needs to be rigorous policies regarding the presence of glass.

Table A.7
Extract from the National Primary Drinking Water Regulations

Component	Maximum contaminant level goal	Maximum contaminant level (mg/l unless stated)	Potential health effects	Sources of contaminant
Cryptosporidium or Giardia	Zero	99–99.9% removal/inactivation	Diarrhea, vomiting, cramps	Fecal waste
Legionella	Zero	Deemed to be controlled if Giardia is defeated	Legionnaire's disease	Multiplies in water heating systems
Coliforms (including Escherichia coli)	Zero	No more than 5% samples positive within a month	Indicator or presence of other potentially harmful bacteria	Coliforms naturally present in the environment; E. coli comes from fecal waste
Turbidity	n/a	<1 nephelometric turbidity unit	General indicator of contamination, including by microbes	Soil runoff
Bromate	Zero	0.01	Risk of cancer	Byproduct of disinfection
Chlorine	4	4	Eye/nose irritation; stomach discomfort	Additive to control microbes
Chlorine dioxide	0.8	0.8	Anemia; nervous system effects	Additive to control microbes
Haloacetic acids (e.g. trichloracetic)		0.06	Risk of cancer	Byproduct of disinfection
Trihalomethanes		0.08	Liver, kidney or central nervous system ills, risk of cancer	Byproduct of disinfection
Arsenic		0.05	Skin damage, circulation problems, risk of cancer	Erosion of natural deposits; runoff from glass and electronics production wastes
Asbestos	7 million fibers per liter	7 million fibers per liter	Benign intestinal polyps	Decay of asbestos cement in water mains; erosion of natural deposits

Appendix Practicalities of achieving quality

Copper	1.3	1.3	Gastrointestinal distress, liver or kidney damage	Corrosion of household plumbing systems; erosion of natural deposits
Fluoride	4	4	Bone disease	Additive to promote strong teeth; erosion of natural deposits
Lead	Zero	0.015	Kidney problems; high blood pressure	Corrosion of household plumbing systems; erosion of natural deposits
Nitrate	10	10	Blue Baby Syndrome	Runoff from fertilizer use, leaching from septic tanks, sewage, erosion of natural deposits
Nitrite	1	1	Blue Baby Syndrome	Runoff from fertilizer use, leaching from septic tanks, sewage, erosion of natural deposits
Selenium	0.05	0.05	Hair or fingernail loss, circulatory problems, numbness in fingers and toes	Discharge from petroleum refineries, erosion of natural deposits, discharge from mines
Benzene	Zero	0.005	Anemia; decrease in blood platelets; risk of cancer	Discharge from factories; leaching from gas storage tanks and landfills
Carbon tetrachloride	Zero	0.005	Liver problems; risk of cancer	Discharge from chemical plants and other industrial activities
Dinoseb	0.007	0.007	Reproductive difficulties	Runoff from herbicide use
Dioxin	Zero	0.00000003	Reproductive difficulties, risk of cancer	Emissions from waste incineration and other combustion; discharge from chemical factories
Alpha particles	Zero (as of 12/8/03)	15 picoCuries per liter	Risk of cancer	Erosion of natural deposits
Beta particles and photon emitters	Zero (as of 12/8/03)	4 millirems per year	Risk of cancer	Decay of natural and man-made deposits

Source: Reproduced from Lewis, M.J. and Bamforth, C.W. (2006) *Essays in Brewing Science*. Springer, New York. The full table can be found at http://www.epa.gov/safewater/mcl.html.

Table A.8
National Secondary Drinking Water Regulations

Contaminant	Secondary standard
Aluminum	0.050.2 mg/l
Chloride	250 mg/l
Color	15 color units
Copper	1 mg/l
Corrosivity	Non-corrosive
Fluoride	2 mg/l
Foaming agents	0.5 mg/l
Iron	0.3 mg/l
Manganese	0.05 mg/l
Odor	3 threshold odor number
pH	6.5–8.5
Silver	0.1 mg/l
Sulfate	250 mg/l
Total dissolved solids	500 mg/l
Zinc	5 mg/l

Note: These are non-enforceable guidelines regulating contaminants that may cause cosmetic effects (e.g. skin or tooth discoloration) or aesthetic effects (taste, odor, color). States may choose to adopt them as enforceable standards.
Source: Reproduced from Lewis, M. J. and Bamforth, C.W. (2006) *Essays in Brewing Science*. Springer, New York.

(d) Rigor in buildings and equipment, ensuring that they are properly cleaned and maintained. Consideration should be given *inter alia* to restroom facilities and pest control.
(e) Rigor in all suppliers of materials, with checking procedures at receipt.
(f) Precautions regarding the inviolability of vehicles employed for raw materials arriving at the brewery and beer and co-products leaving it.
(g) Proper training programs for all employees.
(h) Existence of a robust product recall system.

Customary steps in implementing HACCP are

(a) Expression and realization of the commitment to the policy and procedures of senior management.
(b) Definition of scope and extent, from raw material delivery to shipment of product from the brewery.
(c) Establishment of HACCP team(s).
(d) Generation of a process flow chart that includes all raw materials and process steps within the scope of HACCP.

(e) Identification by the HACCP team(s) of all potential hazards and determination of appropriate control measures. Hazards include chemical and microbiological contamination (e.g. deoxynivalenol in malt, caustic contamination from faulty in-place cleaning, glycol from leaking wort coolers) and physical risks (e.g. broken glass or filling tubes in packaging).
(f) Identification of the critical control points (CCP), viz the stages where controls must be introduced to minimize hazards.
(g) Establishment of what the critical limits are to be at each CCP – that is realistic specifications based on ready measurement capabilities.
(h) Establishment of response procedures for each CCP, that is corrective actions and records.
(i) Training of personnel.
(j) Updating whenever there are changes in raw materials, equipment or declared process procedures.

Index

2-phenylethanol 75, 81
3-methyl-2-butene-1-thiol (MBT) 77, 78
10 Supplementary Color Observer 216
α-glucan 151
β-acids 66
β-amylase 30, 34, 38, 118
β-damascenone 99
β-glucan 12, 21, 151
β-glucanase 34, 201
β-glucosidase 99
AAS (atomic absorbance spectroscopy) 154
Accelerated ageing 153
Acceptable quality limits (AQL) 261
Acetal formation 98
Acetaldehyde 75–6, 92, 98, 136, 238
Acetic acid bacteria 164, 167, 169, 179
Acetobacter 169
Acetomonas 169
Acetyl coenzyme A 75
Acid washing 167
Acidaminococcaceae 164
Adjunct selection 34
Adsorbents 201
Aerobacter 169
Aerobacter aerogenes 167
Aflatoxin B 1, 166
Alcohols 232, 235, 236, 237, 239, 240
  oxidation 97
Aldehydes 68, 79
  in beer 76
  formation 75
Aldol condensations 98
Aldose reductase 236
Aliphatic thiols 77
Alkaline phosphatase 176
Allergy 241–2

Alternaria 165, 189
Alternaria alternate 191
Alzheimer's disease 239
American Society for Testing and
  Materials (ASTM) 218
American Society of Brewing Chemists
  (ASBC) 218, 220, 262
  methods of analysis 263–9
Amino acid composition, of polypeptides:
  identification 125–30
  and protein Z 127
  related to foam 131
  related to haze 129
Amylase inhibitors 121
Amyloglucosidase 42, 170
Anthocyanogens 136, 138
Antigen 1, 117–18, 126, 132
Antigen 1a 122
Antigen 1b 122
Aphis blight *see Aphis humuli*
Aphis humuli 168
Appraisal costs 258
Aqua-Trace™ 175
Arabinoxylan 12, 21, 32, 33, 34, 123, 152
Arginine 20, 128, 130, 132, 222
Arrhenius' Law 95
Ascorbic acid 150
Asparagine 125
Aspartic acid 125, 126, 128, 130
Aspergillus 165, 189
Aspergillus niger 149
Atherosclerosis 232–5
ATP bioluminescence 175

Bacillus 174
Bacillus sp. 168
Bacteroides serpens 169

Barley 18, 29, 30, 31, 32, 34, 112, 122, 123, 135, 136, 188, 189, 190, 221, 224
   evaluation of processes, from barley to beer 100–5
   and malt 165–6
   treatment 199
Beading see Creaming
Beer foam physics 3
   bubble formation and size 3–4
   coalescence 6
   creaming 5–6
   disproportionation 6–8
   drainage 4–5
Beer production processes, impact of 72
   fermentation 75–7
   mashing and sweet wort separation 72–3
   maturation and finishing 77–8
   wort boiling and wort clarification 74–5
Belgian white beers 116
Bentonite 150
Bev-Trace™ 175
Bicinchoninic method (BCA) 20
Biological haze 111
Biological stability 112
Biuret method 115, 116
Black malts 225
Bradford assay 115, 116
Brain and cognitive function 237–9
Brettanomyces 167, 180
Brewery surfaces 168
Brewing process 26, 37–43
   manipulation 28–48
   outline 164–5
Brewing raw materials and beer flavor 62
   hop compounds 66–72
   malted barley 63–5
Bubble formation 3–4

Calcium oxalate 202
Cancer 240–1
Candida 167
Caramel malts 219, 224
Carbonyls binding:
   by amino groups in proteinaceous species 98
   by sulfur dioxide 98
Carica papaya 149
Carlsberg Research Laboratories 16

Carlsberg test 198
Cations 25
ChemScan™ 174
Chill haze 114, 128, 138
Chroma see Saturation
CIE (Commission Internationale de l'Eclairage) 214–15
Citrobacter 169
Cladosporium 165
Clarex 149
Coalescence 6, 7
Coefficient of variation (CV) 260
Coliform bacteria 167
Colloidal haze see Protein-polyphenol haze
Colloidal stability, of beer 111
   biological stability 112
   chill haze 114
   combined stabilization system 150
   effectiveness of beer stabilization, testing 152–4
   haze identification 154
   non-biological hazes 151–2
   polypeptides:
      amino acid composition, identification by 125–30
      functions 116–17
      hydrophobicity, identification by 130–2
      size, identification by 117–25
      summary 132–3
   polyphenols 134
      in beer 135–8
      detecting methods 135
      haze-forming reactions 138–40
   protein, in beer 114
      measurement 115–16
   stabilization treatments 140
      clarex 149
      copper or kettle finings 148
      isinglass finings 147–8
      lucilite TR 146–7
      papain 149
      Polyvinylpolypyrrolidine 144–6
      silica 140–4
      silica sol 144
      tannic acid 149
      treatments 150
   whole brewing process, for stability 112–14

Color, of beer:
   measurement 214–18
      standards methods for 218–21
   origins 221–6
   perception 213–14
Color malts 224
Combined stabilization system (CSS) 150
Committee on Diet and Health of the National Academy of Sciences 240
Concanavalin A 151
Control charts 261–2
Coomassie blue binding (CBB) 20
Copper finings 148
Cost of quality 258–9
Creaming 5–6
CUSUM (cumulative sum) plot 261, 262
Cycloheximide 172
Cysteine 100, 123, 125, 130, 192

Debaryomyces 164, 167
Dekkera 164, 167
Deoxynivalenol (DON) 166, 189, 196, 199
Detection, of microbiological instability of beer 171
   rapid methods 173
      biochemical methods 174–6
      molecular methods 176–8
      physical methods 173–4
   traditional methods 172–3
Digestive system 235–6
Dimethyl disulfide 77
Dimethyl sulfide (DMS) 61, 65, 74, 76, 81
Dimethyl trisulfide 77
Direct Epifluorescence Filter Technique (DEFT) 174–5
Dispense 171, 172
Disproportionation 6–8
Doubly modified Carlsberg test (M^2CT) 198
Drainage 4–5

E-2-nonenal 86, 88, 96, 98
EBC Haze Group 128
Electron spin resonance spectroscopy 92
ELISA (Enzyme Linked Immuno Sorbent Assay) 13, 19, 196, 197
End-point PCR 177
Enteric bacteria 167, 169
Enterobacter agglomerans see *Rahnella aqualitis*

Enterobacteriaceae 164
Enzymes 201
Epicoccum 165
Ergosterol 197
Escherichia 169, 174
Escherichia coli 171
Esters 25, 69, 75
Ethanol 27, 75, 112
Euchema cottonii 148
European Brewery Convention (EBC) 218, 220, 262
Eurotium 165
External failure costs 258
External noise 261
Extra cellular polysaccharides (EPS) 197

Fenton reaction 91
Fermentation 41, 75–7
   high gravity 169
   low calorie 170
   non-alcoholic 170
   traditional 169
FISH (fluorescent *in situ* hybridization) 177
Flavanoids 134, 135, 234
Flavor, of beer:
   beer production processes, impact of 72
      fermentation 75–7
      mashing and sweet wort separation 72–3
      maturation and finishing 77–8
      wort boiling and wort clarification 74–5
   brewing raw materials and beer flavor 62
      hop compounds 66–72
      malted barley 63–5
   flavor unit 61–2
   holistic flavor perception 81
   in-pack flavor changes 78–9
   taints and off-flavor 80
Flavor instability:
   processes evaluation from barley to beer in 100–5
   expectations, for flavor stability 104
   flavor change in beer, chemistry of 96
      acetal formation 98
      alcohols, oxidation of 97
      aldol condensations 98

Flavor instability (*continued*)
  amino groups in proteinaceous species, binding of carbonyls by 98
  carbonyl compounds reduction, by yeast 98–9
  enzymic oxidation, of unsaturated fatty acids , 96–7
  ester levels, changes in 100
  hydrogen peroxide removal, by peroxidases 99–100
  iso-α-acids 97
  non-enzymic oxidation, of unsaturated fatty acids 97
  polyphenols and melanoidins, oxygen radical scavenging by 99
  release of flavor active compounds, by enzymes from yeast 99
  strecker degradation, of amino acids 98
  sulfur compounds 100
  sulfur dioxide, binding of carbonyls by 98
  vicinal diketone release in beer 100
 process impacts on 101–3
 shelf life of beer, factors impacting 88
  fenton reaction 91
  hydrogen peroxide, sources of 91
  oxygen 88–9
  radicals producing radicals 92–5
  reactive oxygen species 89–90
  sulfur compounds 91–2
  transition metal ions 90–1
  temperature, impact of 95–6
Flavor unit (FU) 61–2
Flow cytometry 174
Foam, of beer:
 beer foam physics 3
  bubble formation and size 3–4
  coalescence 6, 7
  creaming 5–6
  disproportionation 6–8
  drainage 4–5
 cations 25
 foam measurement 8–12
 foam quality:
  beer components influencing 12–13
  features 2
 hop acids 22–4
 lipids 25–6
 negative/positive beer constituents 26–7
 non-starch polysaccharides 21–2
 optimizing foam quality, manipulating brewing process for 28
  brewing process 37–43
  palliative options, of gas composition, widgets and other devices 43–6
  raw material selection 28–37
  troubleshooting 47–8
 proteins:
  hordeins 16–17
  lipid binding proteins 19
  lipid transfer protein 13–16
  protein measurement 19–21
  protein Z 17–19
Foam analysis:
 procedures, selection of 10
 tests 9
Foam collapse time (FCT) 12
Foam measurement 8–12
Foam quality:
 beer components influencing 12–13
 features 2
 optimization 28
  adjunct selection 34
  brewing process 37–43
  hop selection 34–7
  malt lipoxygenase 33–4
  malt modification 32–3
  malt selection 28–32
  palliative options 43–6
  troubleshooting 47–8
Folin-Ciocalteu method 135
Forcing tests 152–3
Formaldehyde 150, 199
Free amino nitrogen (FAN) 72, 79, 115
Free radical 89
Fumosins B1 166
Fumosins B2 166
Fusarium 28, 112, 165, 166, 188, 189, 190, 195, 196, 199, 200, 201
Fusarium avenaceum 189, 191, 196, 197
Fusarium culmorum 189, 191, 194, 197
Fusarium damaged kernels (FDK) 199
Fusarium equiseti 189
Fusarium graminearum 166, 189, 191, 196, 197
Fusarium head blight (FHB) 188, 189
 etiology 189
Fusarium moniliforme 166
Fusarium poae 189, 191, 194, 197

Fusarium sambucinum 197
Fusarium sp. 189, 195, 200
Fusarium sporotrichiodes 189
Fusarium tricinctum 189, 197

Gas chromatography 176
Gas liquid chromatography (GLC) 135
Gelatin 150
Geotrichum candidum 200, 201
Geraniol 69, 72
Glucose oxidase 150
Glutamic acid 115, 118, 120, 125, 126, 128, 130, 132
Glutamine 125
Glutelin 15, 126
Gluten 117, 242
Glycine 126, 128, 130, 132, 147
Glycoproteins 38, 118, 119, 124
Glycosidases 99
Gram negative bacteria 164, 167
Gram positive bacteria 164, 168
Green malt 224–5
Gushing, in beer 185, 186
  physical background 187–8
  primary gushing 188
    causal agents 191–4
    control, of gushing 198–201
    gushing risk, determination of 195–8
    mycoflora 188–91
  risk, determination of 195
    Carlsberg test 198
    correlative factors 195–7
    hydrophobins 197
    laboratory gushing tests 197
    modified Carlsberg test 198
  secondary gushing 201
    bottle and bottle filling 203–4
    calcium oxalate 202
    isomerized hop extracts 203
    metal ions 202–3
    particles and ions derived from Kieselgur 204
    tensides 203
  terminology 186–7

Hafnia 164, 169, 174, 178
Hafnia proteus 167, 169, 178
Hanseniaspora 164
Hansenula 167

Hazard analysis critical control points (HACCP) 271–7
  pre-requisites for implementation 273
Haze 111
  identification 154
Haze-forming reactions:
  between polypeptides and polyphenols 138–40
Health and beer:
  age 239–40
  allergy 241–2
  atherosclerosis 232–5
  beer vs. wine 229, 230–1
  brain and cognitive function 237–9
  cancer 240–1
  digestive system 235–6
  hypertension and stroke 235
  kidney and urinary tract 239
  nutrient composition 230–1
  reproductive system 237
Helicobacter pylori 236
High gravity fermentation 169
Histamine 241
Histidine 192, 193
Holistic flavor perception 81
Homocysteine 235
Hop acids 22–4
Hop compounds 66–72
Hop oil:
  components, classification of 68
  terpene hydrocarbons in 68
Hop selection 34–7
Hop specification components 272
Hops 168
Horc 177
Hordein 12, 16–17, 30, 31, 118, 121, 123, 128, 242
Hordoindoline 19
Horseradish peroxidase 176
HPLC (High pressure liquid chromatography) 116, 135
Hue 214
Humulone 22, 240, 241
Hybridization 176
Hydrogen peroxide 150
  sources 91
Hydroperoxides 96
Hydrophobic particle mechanism 6
Hydrophobicity 13, 66, 121
  polypeptides identification by 130

Hydrophobins 166, 191–4, 197
  physiochemical properties 193
  secondary structure 194
Hydroxyethyl radical 92
Hypertension and stroke 235

Immunoanalysis 175–6
Impedance/conductance measurement 173
In-pack flavor changes 78–9
Institute of Brewing 262
Internal failure costs 258
Internal noise 261
Isinglass finings 147–8
Iso-α-acids 27, 36, 66, 67, 72, 74
  oxidation 97
Iso-humulones *see* Isomerized hop extracts
Isoleucine 76, 121
Isomerized hop extracts 203

Karyotyping 177
Kettle finings *see* Copper finings
Kettle hop aroma 71
Kidney and urinary tract 239
Kjeldahl method 115
Klebsiella 169
Kloeckera 167
Kluyveromyces 164
Kodak Scientific Imaging Systems 11
Kolbach Index 191

Laboratory gushing tests 197
Lacing 8–9
Lacing index method 11
Lactic acid bacteria (LAB) 167, 175
Lactobacillus 164, 169, 174, 176, 177, 178
Lactobacillus brevis 177
Lactobacillus casei 174
Lactobacillus lindneri 177
Lactobacillus plantarum 179, 200
Lager beers 70–1
Lambert-Beer law 213
LC-MS (liquid chromatography-mass spectroscopy) 116, 125
Leucine 119, 121, 126, 128
Leuconostoc 177
Light adsorbing particles (LAPs) 115
Lightness 214
Linalool 69, 70, 72
Linoleic acid 65, 72, 79, 94, 97
Lipid binding proteins 19

Lipid transfer protein (LTP) 13–16, 242
  LTP1 12, 13, 47, 121, 123
    average malt varietal content 31
  LTP2 13, 123
Lipids 25–6
Lipoxygenase (LOX) 26, 30, 34, 96
Lovibond, Joseph 218
Low alcohol fermentation *see* Non-alcoholic fermentation
Low calorie fermentation 170
Lowry assay 115
Lucilite TR 146–7
Lysine 119

Maillard browning reactions 222
Maillard reaction cascade 64, 65
Malt lipoxygenase 33–4
Malt modification 32–3
Malt selection 28–32
Malt specification component 271
Malted barley 63–5
  appearance and flavor 224
Malts 224, 225
  caramel 224
  color 224
  roasted 225
Mashing and sweet wort separation 72–3
Maturation, on beer flavor 77–8
Maturex® 77
Mean (M) 260
Megasphaera 164, 172, 177, 179
Megasphaera cerevisiae 171
Megasphaera sp. 169
Meilgaard 61
Melanoidins 122, 222, 225
  Maillard reaction scheme 223
  oxygen radical scavenging by 99
Metal ions 202–3
Methional 76, 77, 79
Methionine 76, 77, 192, 193
Microbial spoilage, overview of 163–4
Microbiological stability, of beer:
  brewing process, outline of 164–5
  detection 171
    rapid methods 173–8
    traditional methods 172–3
  dispense 171, 172
  fermentation:
    high gravity 169
    low calorie 170
    non-alcoholic 170
    traditional 169

improvement 178–80
microbial spoilage, overview of 163–4
packaging and packaged beer 170–1
quality control 180
raw materials 165
   barley and malt 165–6
   brewery surfaces 168
   hops 168
   pitching yeast 167
   sugars 168
   water 166–7
storage and finishing 170
wort 169
Microcalorimetry 173
Micrococcus 164
Microcolony method 174
Microdochium nivale 191
Microstar™ 175
Mishandled beer 186
Modified Carlsberg test (MCT) 198
Molecular probes 176–7
MRS (deMan, Rogosa and Sharpe) 172
MWCO (molecular weight cut off) 202
Mycoflora 188–91
Mycotoxins 166, 189–90, 191
   testing 196

National Primary Drinking Water
   Regulations 274–5
National Secondary Drinking Water
   Regulations 276
Negative/positive beer constituents 26–7
NIBEM 9, 11, 12
Nigrospora 166, 189
Nigrospora sp. 192
Nitrile rubber (NBR) 168
Nitrogen vs. CO_2 43–4
Nivalenol (NIV) 189
NMR (nuclear magnetic resonance) 121
Noise 261
Non-alcoholic fermentation 170
Non-biological hazes 111, 151–2
Non-specific lipid transfer proteins
   (ns-LTP) 194
Non-starch polysaccharides 12, 21–2
Non-thermal pasteurization 179
Nylon 150

Obesumbacterium proteus see *Hafnia proteus*
Ochratoxin A 166
Off-flavor and taints 80
Ostwald ripening see Disproportionation

Oxygen 88–9
   activation through addition of electrons
      90
   solubility of oxygen in worts of
      increasing strength 89

Palliative options 43–6
   nitrogen vs. CO_2 43–4
   PGA and chemical enhancement 44–7
   widgets and foam promoters 44
Papain 42, 114, 149, 186
PAPI (probable amylase/protease
   inhibitor) 121
Particle spreading mechanism 6
Pectinatus 164, 172, 176, 177, 179
Pectinatus cerevisiiphilus 171
Pectinatus cerevisiophilus 169
Pectinatus frisingensis 171
Pediococcus pentosaceus 200
Pediococcus 164, 169, 174, 176, 178
Pediococcus damnosus 167, 174, 177
Penicillium 165, 189
Pentachlorophenol 80
Perlon 150
Permanent haze see Protein-polyphenol
   haze
Peroxidases 225
   hydrogen peroxide removal by 99–100
Pichia 164, 167
Pitching yeast 167
Podosphaera castagnei 168
Polyacrylamide gel electrophoresis
   (PAGE) 175, 176
Polyamides 150
Polyclar® Brewbrite 145
Polymerase Chain Reaction (PCR) 177
Polypeptides:
   functions 116–17
   identification:
      by amino acid composition 125–30
      by hydrophobicity 130–2
      by size 117–25
   summary 132–3
Polyphenol oxidase 150
Polyphenols 12, 27, 134, 221
   in beer 135–8
   detecting methods 135
   haze-forming reactions 138–40
   oxygen radical scavenging by 99
Polyproline 139, 140, 145, 146
Polysaccharides 151, 122, 237
Polytetrafluoroethylene (PTFE) 168

Polyvinylpolypyrrolidine (PVPP) 114, 144–6, 147, 225
Precipitative tests 153
Prevention costs 259
Primary beer gushing 187, 188
   causal agents 191
      hydrophobins 191–4
   control, of gushing 198
      adsorbents 201
      beer, treatment of 201
      biological treatment 200–1
      chemical treatment 199–200
      enzymes 201
      physical treatment 199
   gushing risk, determination of 195
      Carlsberg test 198
      correlative factors 195–7
      hydrophobins 197
      laboratory gushing tests 197
      modified Carlsberg test 198
   mycoflora 188–91
Proanthocyanidins 136
Process capability 261
Process impacts, on flavor instability 101–3
Proline 118, 146
Proline endoproteinase 17, 18
Propylene glycol alginate (PGA) 114
   and chemical enhancement 44–7
Protease inhibitors 121, 122
Proteases 16, 33, 41, 42, 189, 191
Protein fingerprinting 175
Protein, in beer 114
   measurement 19–21, 115–16
"Protein rest" 112, 115, 118
Protein Z 12, 17–19, 47, 118
   importance 18–19
Protein Z4 18, 19, 20, 21, 29, 30, 31, 33, 39, 122
Protein Z7 18, 21, 29, 33, 122
Proteinase A 41, 43, 124
Protein–polyphenol haze 111
Proteins:
   hordeins 16–17
   lipid binding proteins 19
   lipid transfer protein 13–16
   protein measurement 19–21
   protein Z 17–19
Pulsed electric fields (PEF) 179
Puroindoline 19, 34
Pyrenophora teres 188
Pyrogallol redmolybdate (PRM) 20

Quality 255
   control charts 261–2
   cost of quality 258–9
   definitions 255–6
   HACCP 271–7
   process capability 261
   quality assurance vs. quality control 257
   quality systems 256–7
   responsibility for 256
   specification setting and monitoring performance 270–1
   specifications 258
   standard methods of analysis 262–70
   statistical process control 259–60
Quality control vs. quality assurance 257
Quality-related costs 258–9
Quality systems 256–7
Radical propagation 92
Radicals producing radicals 92–5
Rahnella 164, 169
Rahnella aqualitis 167, 169, 174
Raka-Ray 172
RAPD-PCR 178
Raw materials 165
   barley and malt 165–6
   brewery surfaces 168
   hops 168
   pitching yeast 167
   sugars 168
   water 166–7
Reactive oxygen species (ROS) 89–90
Real-time PCR (RT-PCR) 178
Red spider *see* Tetranychus telarius
Reject quality limits (RQL) 261
Reproductive system 237
Rhizopus 189
Rhodotorula 167
Rhodotorula sp. 196
Riboflavin 77, 226
Riboprinting *see* Ribotyping
Ribosomal RNA (rRNA) 171, 177
Ribotyping 177
Roasted malts 225
Ross and Clark procedure 9, 11
Rudin head retention 9, 11

S-methylmethionine (SMM) 74
   degradation 74
Saccharomyces 164, 167, 177, 180
Saccharomyces cerevisiae 177, 180

Saccharomyces diastaticus 174, 176, 177
Saccharomycodes 167
Salmonella/Shigella spp. 167
Saturated ammonium sulfate (SAS) 153
Saturated ammonium sulfate
    precipitation limit (SASPL) 153
Saturation 214
SDS-PAGE (sodium dodecyl
    sulfate-polyacrylamide gel
    electrophoresis) 116
Secondary beer gushing 187, 201
    bottle and bottle filling 203–4
    calcium oxalate 202
    isomerized hop extracts 203
    metal ions 202–3
    particles and ions derived from
        Kieselgur 204
    tensides 203
Selenomonas 164
Serine 126, 128, 130
Shelf life of beer, factors impacting 88
    fenton reaction 91
    hydrogen peroxide, sources of 91
    oxygen 88–9
    radicals producing radicals 92–5
    reactive oxygen species 89–90
    sulfur compounds 91–2
    transition metal ions 90–1
Sigma head value (SHV) 9, 11
Silica 140–4
Silica sol 144
Singlet oxygen 89
Sodium dodecyl sulfate (SDS) 116
Speciation 93
Spectrophotometer 216
Stabifix® 128
Stability, of beer:
    types 111
Stabilization treatment, of beer 140
    copper or kettle finings 148
    isinglass finings 147–8
    lucilite TR 146–7
    papain 149
        clarex 149
    Polyvinylpolypyrrolidine 144–6
    silica 140–4
    silica sol 144
    tannic acid 149
Standard deviation (σ) 260
Standard methods of analysis 262–70
Standard operating procedures (SOPs)
    256

Standard Reference Method (SRM) 220
Staphylococcus 164
Statistical process control 259–60
Stemphylium 189
Storage, of beer:
    compounds formed during 86–7
Streptococci 236
Streptomyces 201
Sugars 168
Sulfur compounds 91–2, 100
Summer-type gushing 186

Taints and off-flavor 80
Tannic acid 149
Tannin 134
Tannoids 135, 137
Temperature, impact of 95–6
Tensides 203
"Tetra hop" 22
Tetranychus telarius 168
Thermal pasteurization 179
Three-dimensional color model 215
Thyrosine 192, 193
Torulaspora 164, 167
Total Quality Management (TQM) 256
Traditional fermentation 169
Trans-2-nonenal *see* E-2-nonenal
Transition metal ions 90–1
Treatment, of beer 201
Trichlorophenol (TCP) 80
Trichoderma 166, 189
Trichoderma reesei 194
Troubleshooting, with beer foams 47–8
Tryptophan 125, 192, 193
Turbidometry 173–4
Tyramine 241

UBA (Universal Beer Agar) 172
Unsaturated fatty acids:
    enzymic oxidation 96–7
    non-enzymic oxidation 97
    oxidation 94

Valine 121
Variance 260
Vermicon identification technology
    (VIT) 177
Vibrio cholerae 167
Vicinal diketone 76
    releases in beer 100
Viton 168

Water 166–7
Whole brewing process 112–14
Widgets and foam promoters 44
Wild beer 186
Wild yeast 164, 167
Williopsis 164
Wine vs. beer 229, 230–1
Winter-type gushing 186
Wort 169
   boiling and clarification 74–5

Xanthohumol 235

Yeast 41, 98
Youden 269
Zearalone 166
Zygosaccharomyces 167
Zymomonas 164, 169
Zymomonas spp. 169
Zymophilus 164

Edwards Brothers Malloy
Ann Arbor MI. USA
March 20, 2013